# FIRST BOOK on PWM Dc Power Conversion

# CIRCUITS and OPERATION of PWM CONVERTERS

**Byungcho Choi and Syam Kumar Pidaparthy***

Kyungpook National University
Daegu, Korea

LS Automotive Technologies*
Anyang-si, Korea

가디언북

**GuardianBook Publishing**

Published by GuardianBook Publishing, Daegu, Korea.



ISBN 979-11-87021-11-7
krw: 28,000
Date of Publication: September 1st 2022
Published by GuardianBook © 2022
Phone: 070-4226-8351
E-mail: gdpub@naver.com

Printed in the Republic of Korea.

10 9 8 7 6 5 4 3 2 1

# PREFACE

Most industrial and consumer electronics are powered from a dc source whose voltage level is fixed in advance. For examples, telecommunication electronics usually receive their input power from a 48 V dc source and cellular phones operate with 3 V rechargeable batteries. The circuits inside such instruments, however, require several dc voltages that are different from the voltage level of the input source. Accordingly, each instrument contains specially-engineered circuitry that is capable of changing the input voltage to another dc level, as required by the various internal circuits.

The process of changing the voltage level of a dc power source into another value is referred to as the dc-to-dc power conversion. The dc-to-dc power conversion is performed in many different ways, each with a distinctive circuit technique. The most popular scheme is the dc-to-dc power conversion circuit that employs the pulsewidth modulation (PWM) technique. The dc-to-dc power conversion based on the PWM technique is called the PWM dc-to-dc power conversion. Functional units designed for the PWM dc-to-dc power conversion are called PWM dc-to-dc converters

Conventionally, PWM dc-to-dc converters were treated as a standalone unit powered from an ideal voltage source and connected to a resistive load. This ideal operational scenario provided many useful outcomes, which are generally adaptable to most standalone PWM converters even operating with a non-ideal voltage source and non-resistive load. The individual dc power conversion with a single PWM dc-to-dc converter has evolved into complex dc power distribution systems, where a number of PWM converters and intermediate filter stages are combined together, both physically and functionally, to offer efficient and reliable dc power conversions.

The first author of the current book has published a couple of textbooks on the PWM dc-to-dc power conversion: the first edition and second edition of the book - *Pulsewidth Modulated Dc-to-Dc Power Conversion.* The first edition was published in 2013, aiming to be a textbook for senior and graduate students, majoring in power electronics with focus on the PWM dc-to-dc power conversion. The second edition of the book was published in 2021, in which the scope of the book was stretched to include the multi-stage parallel-module dc power distribution systems. For the first time, the second edition of the book documented the dynamic analyses and design techniques of practical dc power distribution systems in a textbook format.

The first author of the current book believed that the previous books are suited for graduate students, professional researchers, and experienced engineers. At the same time, he saw that the presentations of the earlier books need to be revised into more appropriate formats for undergraduate students and inexperienced practitioners. The author decided to write an accompanying publication to offer an easy stepping stone to the previous professional textbooks. The new publication is set to be a three-volume series, successively over next several years.

The reference book for the series is the second edition of *Pulsewidth Modulated Dc-to-Dc Power Conversion.* Theoretical presentations of the second edition will be paraphrased in easier and reader-friendly formats, along with many informative illustrations, without losing any technical and pedagogical contents of the reference book. The refined contents will be grouped into three technological categories:

- Volume I: Circuits and operation of PWM converters
- Volume II: Dynamics and control of standalone PWM converters
- Volume III: Analysis and design of Dc power distribution systems

Each volume will present the theoretical and technical contents in more explanatory and stimulating manners with many design examples.

In this book series, computer simulations will be extensively used in order to illuminate theoretical discussions and to validate design results. Validity and performance of most design examples will be supported by computer simulations. Computer models will describe all engineering details so that each example can immediately be converted into working hardware. License-free copies of software will be employed for unrestricted uses of the simulation examples for both practicing engineers and motivated students.

BYUNGCHO CHOI AND SYAM KUMAR PIDAPARTHY

*Daegu, Korea*
*August 2022*

# CONTENTS IN BRIEF

# CONTENTS

# CHAPTER 1

# PWM DC-TO-DC POWER CONVERSION



This introductory chapter presents an overview of the pulsewidth modulated (PWM) dc-to-dc power conversion. We discuss basic principles and unique natures of dc-to-dc power conversion circuits, operating with the PWM technique. The current chapter also presents features and issues of dc power distribution systems for modern electronic instruments and systems. As one digression, we discusses features and utilities of commercial software as a simulation tool for power electronics circuits and systems. Finally, this chapter outlines the scopes and topics of the three books, which will be successively published under the title *Books on PWM Dc Power Conversion*, and highlights the contents of the current book.

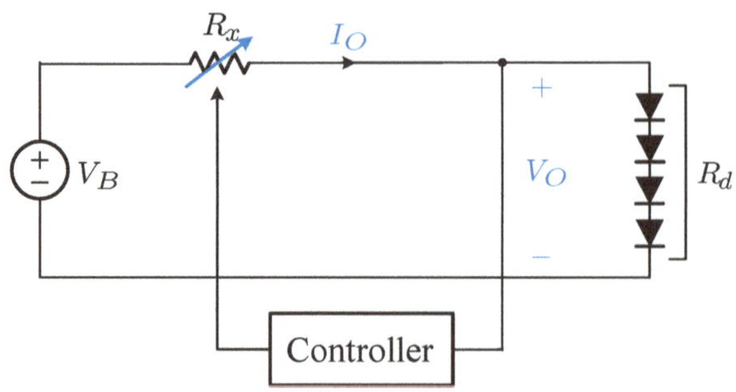


**Figure 1.1** Conventional approach to lighting LED string.

## 1.1 PWM DC-TO-DC POWER CONVERSION

The PWM dc-to-dc power conversion is generally described as the process of changing the voltage level of a dc source using PWM technique. However, a more definitive and precise description is necessary to comprehend natures and features of the PWM dc-to-dc power conversion circuit.

### 1.1.1 Dc-to-Dc Power Conversion

For an accurate description of the dc-to-dc power conversion, this section presents two different approaches to energizing an LED string using a dc voltage sourced from a battery. It is presumed that the LED string requires a strict 12V for operation, while the battery voltage is varied between 18 V and 30 V depending on the internal status and operational conditions. Figure 1.1 shows the first approach where a variable resistor and controller are employed between the battery and LED string. The controller is assumed to only draw a negligible current.

In Fig. 1.1, the controller adjusts the resistance of the variable resistor $R_x$ to meet the following relationship

$$V_O = \frac{R_d}{R_x + R_d} V_B = 12 \text{ V} \tag{1.1}$$

where $V_O$ is the voltage across the LED string, $R_d$ denotes the resistance of the LED string, and $V_B$ is the battery voltage that varies between $18\,\text{V} < V_B < 30\,\text{V}$. Figure 1.1 certainly fulfills the goal of providing a fixed dc voltage from a variable voltage source, however, it has one critical problem that renders this approach impractical.

The variable resistor is accompanied by an ohmic power loss

$$\boxed{P_{loss} = P_{in} - P_{out} = I_O V_B - I_O V_O = I_O(V_B - V_O)} \tag{1.2}$$

where $P_{in}$ is the input power drawn from the battery, $P_{out}$ is the output power delivered to the LED string, and $I_O$ is the current flowing from the battery to the

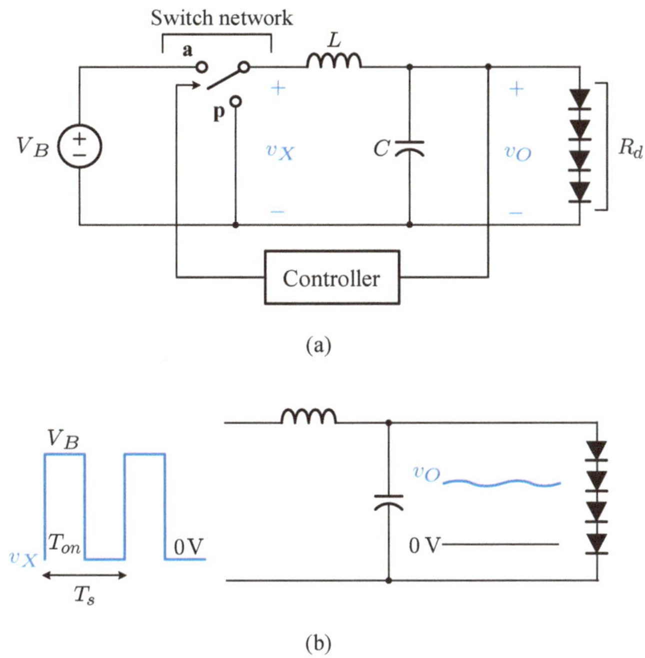


**Figure 1.2** Dc-to-dc power conversion applied to power LED string. (a) Circuit diagram. (b) Input and output waveforms of LC filter.

LED string. The power loss is given by the product of the load current and the difference between the battery voltage and LED string voltage. This power loss easily becomes significant. For example, when an LED string which consumes a 60 W power at 12 V voltage level is connected to a 30 V battery, the power loss is as large as $P_{loss} = (60/12)\,(30 - 12) = 90$ W. This loss is even larger than the power consumed in the LED string, $P_{out} = 60$ W.

The power loss is always transformed into heat and the resulting heat must be removed using an appropriate cooling system. The cooling system usually employs bulky heat sinks and noisy fans, consequently increasing the size, weight, and noises of the system. Accordingly, Fig. 1.1 cannot be used for applications where the dimension and weight should be limited, which is the case for most modern electronic instruments and systems.

Figure 1.2 shows an alternative approach where a *switch network* and LC filter are inserted between the battery and LED string. The switch network periodically changes its connection. Within each switching period $T_s$, the switch network maintains position **a** for $T_{on}$ and position **p** for the remaining switching period, $T_s - T_{on}$. This switch network is called the *single-pole double-throw* (*SPDT*) *switch* because it contains one pole which is always connected to one of the two contacts – the throw **a** and the throw **p**. With the switching action of the SPDT switch, the battery voltage is

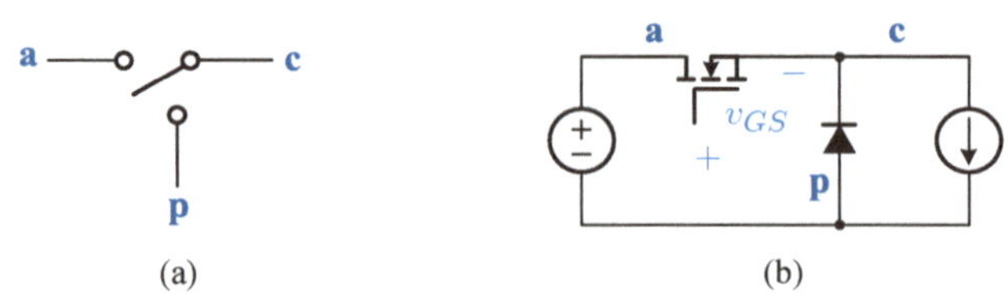


**Figure 1.3** Implementation of SPDT switch. (a) SPDT switch symbol. (b) MOSFET-diode pair as SPDT switch.

transformed into a rectangular waveform at the output of the SPDT switch, $v_X$ in Fig. 1.2. The rectangular waveform is then applied to the LC filter. The LC filter alters the rectangular waveform into a smoothly-filtered continuous voltage waveform, $v_O$ in Fig. 1.2.

The SPDT switch is usually implemented using an active semiconductor switch and passive semiconductor switch. Figure 1.3 shows an example of such implementation. In Fig. 1.3(a), the pole of the SPDT switch is labeled with **c**. The pole **c** should always be connected to either throw **a** or throw **p**.

Figure 1.3(b) shows the SPDT switch, configured with a MOSFET and diode. This figure indicates that a voltage source is placed between the throw **a** and throw **p**, whilst a current source is loaded between the pole **c** and throw **p**. As will be detailed in Section 2.1.3, the connections illustrated in Fig. 1.3(b) are always required for the proper operation of the SPDT switch.

The MOSFET is turned-on/off by the gate drive signal $v_{GS}$ in Fig. 1.3(b). When the MOSFET is turned-on by the *on signal*, the diode is turned-off because the voltage source forces a negative voltage across the diode. Conversely, when the MOSFET is turned-off by the *off signal*, the current source forces the diode to turn on. Thus, the MOSFET-diode pair functions as an SPDT switch whose connection is controlled by the gate drive signal of the MOSFET. Details about the properties of semiconductor switches, structure of SPDT switch, and operation of the MOSFET-diode pair as an SPDT switch will be given in Section 2.1.3.

If the LC filter provides sufficient filtering, the output voltage nearly becomes a dc waveform corresponding to the average value of $v_X$

$$\boxed{v_O(t) \approx V_O = \bar{v}_X(t) = \frac{T_{on}}{T_s} V_B} \tag{1.3}$$

To maintain $V_O = 12$ V at the presence of the battery voltage variation, the controller adjusts the ratio $T_{on}$ to $T_s$. With a fixed $T_s$, the controller changes $T_{on}$ to meet the condition

$$\frac{T_{on}}{T_s} V_B = 12 \text{ V} \tag{1.4}$$

An illustration is given in Fig. 1.4. With a battery voltage $V_B = 24$ V and switching period $T_s = 10$ $\mu$s, the controller generates $T_{on} = 5$ $\mu$s to produce $V_O = (5 \times 10^{-6}/10 \times 10^{-6})\, 24 = 12$ V. If the battery voltage is increased to

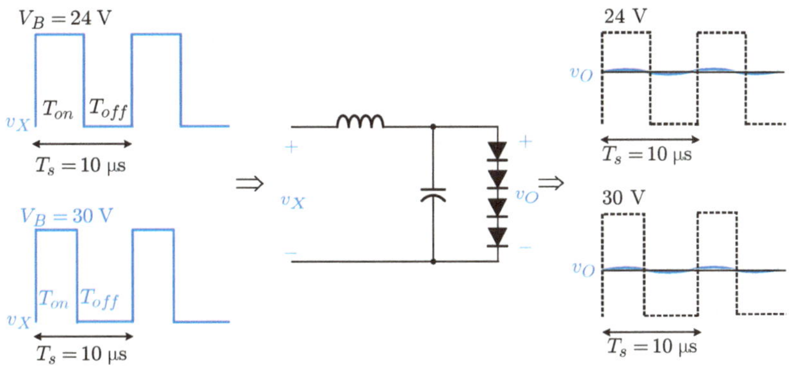


**Figure 1.4** Illustration of output voltage control in Fig. 1.2.

$V_B = 30$ V, the controller reduces $T_{on}$ to 4 $\mu$s to regulate $V_O$ at 12 V: $V_O = (4 \times 10^{-6}/10 \times 10^{-6})\,30 = 12$ V.

Although Fig. 1.1 and Fig. 1.2 both achieve the same goal, a crucial difference exists between them. Figure 1.2 presumes a lossless operation because the SPDT switch and the reactive components in the LC filter do not consume any power. The lossless operation eliminates all the problems associated with the power loss. Because no heat management is required, the circuit could be packaged with a smaller size and lighter weight, thereby making it fully compatible with modern electronic systems.

A more definitive description of the dc-to-dc power conversion is now established as the process of changing the voltage level of a dc source, while eliminating or minimizing power loss. In this perspective, Fig. 1.2 is a typical example of the dc-to-dc power conversion circuit, while the conventional circuit illustrated in Fig. 1.1 is not classified so.

### 1.1.2 PWM Technique

The concept of PWM technique could be envisaged from Fig. 1.4, where the ratio $T_{on}$ to $T_s$ of the SPDT switch is adjusted to keep the output voltage constant. By changing $T_{on}/T_s$ ratio, the pulsewidth of the rectangular voltage waveform is adaptively modulated so that the output voltage remains constant despite the input voltage variation. This control scheme is called the *pulsewidth modulation (PWM) technique* and the dc-to-dc conversion circuit based on the PWM scheme is known as the *PWM dc-to-dc converter*. The PWM dc-to-dc converter is widely adapted to modern industrial and consumer electronics, thereby becoming the most prevailing dc-to-dc power conversion circuit.

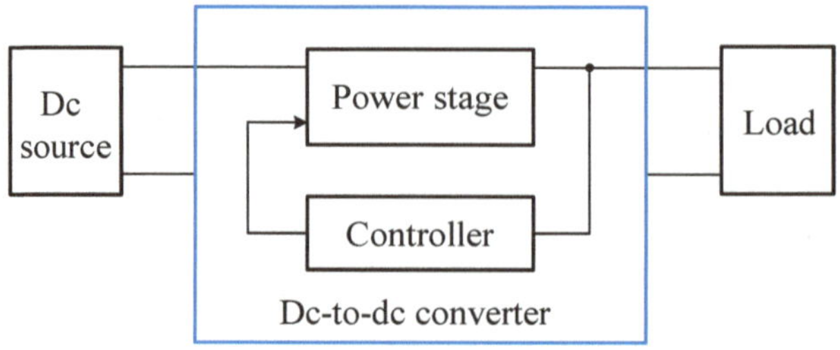


**Figure 1.5** Standalone dc-to-dc power conversion system.

## 1.2 STANDALONE DC-TO-DC POWER CONVERSION SYSTEM

The basic concept illustrated in Fig. 1.2 is generalized into dc-to-dc power conversion systems, whose block diagram representation is shown in Fig. 1.5. The system consists of the dc source, dc-to-dc converter, and load. The dc source provides an arbitrary dc voltage to the dc-to-dc converter. The dc-to-dc converter then converts the level of the given dc voltage into the value required by the load and delivers a rated current to the load. The load is an application system that operates with a fixed dc voltage and consumes electrical power. This section presents characteristic features of the dc source, dc-to-dc converter, and load.

### Dc Source with Non-Ideal Characteristics

The practical dc source falls short of the characteristics of an ideal voltage source in many aspects. First, the voltage level of the dc source could vary with time, as is the case with batteries, fuel cells, and other standalone dc sources. The voltage variation could occur either gradually or abruptly, depending on the characteristics and conditions of the dc source.

Second, a rectified ac source is often used as a substitute for the dc source. For this case, the rectified ac source usually contains a considerable amount of ac components, known as *ac ripple*. In addition, the output of the rectified ac source could be corrupted with various noises. Accordingly, the dc source represents an arbitrary non-ideal source whose voltage could be varied, polluted with ac ripple and noises, and switched from one value to another.

### Dc-to-Dc Converter as Voltage Source

The dc-to-dc converter receives an arbitrary voltage from the non-ideal source and is required to provide a fixed dc voltage for the load. Thus, in addition to altering the voltage level, the dc-to-dc converter should have the capacity of maintaining its output voltage constant at the presence of the variation, ac ripple component, and abrupt change in the input voltage. Ideally, the dc-to-dc converter should function as an *ideal* voltage source, powered from a non-ideal voltage source and programmed to produce the required dc voltage for the load, regardless of the status of the voltage source and load.

Although a practical dc-to-dc converter is more complicated in structure and operation than the circuit in Fig. 1.2, the converter is still recognized as an integration of two functional blocks: the power stage and controller. The power stage alters the level of the input voltage into a desired value using various active and passive circuit components, while the controller provides the necessary signals for the power stage to execute its function.

Dc-to-dc converters come with numerous variations in their power stage configuration and each dc-to-dc converter is differently named after its power stage structure. Despite the wide diversity in structure, all the power stages employ the common electronic components to perform the dc-to-dc power conversion. The power stage utilizes semiconductor devices to implement the function of SPDT switch, energy storage components to perform filtering, and transformers to change the voltage and current levels of circuit variables while transferring electrical energy with galvanic isolation.

The controller also widely varies in structure and functionality. Even so, all the controllers perform the same role of providing the control signals that are required for the power stage to produce a fixed output voltage, regardless of variations in the input voltage and other operational conditions. In PWM dc-to-dc converters, this important function is executed in a closed-loop fashion using the PWM technique. The closed-loop PWM controller uses various analog and digital ICs, as well as discrete circuit components, to generate the required control signals.

### Load as Dynamic Current Sink

The load of a dc-to-dc converter will be an electronic device or system operating with a fixed dc voltage. The load draws the necessary current from the dc-to-dc converter to meet its power requirement. Thus, the load current could fluctuate depending on conditions of the load system. In particular, when high-speed digital systems are connected to a dc-to-dc converter, the current changes could occur frequently and rapidly, including step changes between two different values. Accordingly, the load system presents a dynamic current sink to the dc-to-dc converter, whose current level could change widely and abruptly.

## 1.3 FEATURES AND ISSUES OF PWM DC-TO-DC CONVERTER

The PWM converter is a closed-loop controlled circuitry that operates with the principle of PWM. The PWM dc-to-dc converter is intended to function as an efficient and reliable voltage source, interfacing with the non-ideal dc source and dynamic load system. Accordingly, there are specific features demanded for the dc-to-dc converter and issues involved with implementing such features in a PWM dc-to-dc converter.

## Dc-to-Dc Power Converter Circuits

A dc-to-dc converter is required to accept an arbitrary voltage as the input and to generate a predetermined output voltage. The ratio between the input voltage and output voltage could be either very large or considerably small. Also, a dc-to-dc converter should provide the current needed by the load. Accordingly, there are demanding requirements for the converter power stage in voltage and current ratings, input-to-output voltage conversion ratio, and power handling capacity. In addition, dc-to-dc converters are frequently required to provide galvanic isolation between the source and load. To meet these demands, numerous power stage configurations have been developed, each with a different complexity and functionality. The power stage configuration occupies a large and important portion of the dc-to-dc power conversion technology.

## Dynamic Modeling and Analysis

A dc-to-dc converter should function as a voltage source which holds its output voltage constant at the desired value, irrespective of any possible changes in the input voltage, load current, and other operational conditions. This vital function is achieved by the closed-loop feedback controller operating under the principle of PWM.

It is well known that a closed-loop controlled system becomes unstable if the system is *not* properly designed. An unstable system does not settle into a stationary steady state and keeps on exhibiting a growing or sustained oscillation. On the other hand, a stable system always establishes steady state and stays still thereafter.

Stability is assessed by investigating the dynamic characteristics of the closed-loop controlled system. There are many analytical methods to judge whether a closed-loop controlled system is stable or not. However, these methods are mainly intended for linear time-invariant systems. The PWM dc-to-dc converter falls into the category of the nonlinear time-variant system to which the aforementioned stability analysis methods cannot be directly applied.

The dynamic modeling refers to the analytical process of describing the dynamic characteristics of the nonlinear PWM dc-to-dc converter in a special format to which all the classical analysis methods, originally aimed for linear systems only, could be applied. Therefore, the dynamic modeling allows us to investigate the stability and performance of the nonlinear time-variant PWM dc-to-dc converter using the familiar classical control theory. The dynamic modeling and ensuing analysis using the resultant model are collectively referred to as the *dynamic modeling and analysis*. The dynamic modeling and analysis plays an important role in the PWM power conversion technology and deserves rigorous treatments.

## Dynamic Performance and Control Design

The performance of a dc-to-dc converter will be grouped into two categories in this book series – the static performance and dynamic performance. The *static performance* characterizes the dc-to-dc converter as a static voltage source, which invariably generates a prefixed dc output voltage in steady state. The static performance includes the input-to-output voltage conversion ratio, power handling capacity, output voltage ripple, and other steady-state properties. The static performance is solely determined by the power stage and is irrelevant to the feedback controller.

The second category is the *dynamic performance* which assesses the dc-to-dc converter as a closed-loop controlled dynamic system. Most important dynamic performance is *stability*. The dc-to-dc converter should establish a periodic steady-state operation to produce the desired output voltage. When a certain disturbance is introduced, the converter could temporarily deviate from its steady-state operation. However, the converter should always return to the original periodic operation as the disturbance disappears. This essential feature is possible only when the converter meets the stability criterion.

Another important dynamic performance is the step load response. When a step change occurs in the load current, the output voltage would show a transitional excursion before it returns to its steady-state value. The transitional output voltage response is called the *step load response* in this book series. The step load response is of particular concern when digital logic circuits are employed as the load. Modern logic circuits operate with a very tightly-regulated low voltage, for example $2.1 \pm 0.02$ V, and draw a large pulsating current. These logic circuits naturally and frequently incur substantial step changes in the load current. For this case, the output voltage excursion should be controlled under the regulatory limit, in order to avoid any potential failure of digital logic circuits owing to an excessive transitional overshoot or undershoot in the supply voltage.

The dynamic performance is solely determined by the feedback controller. For a given power stage configuration, the controller should be designed for stability and good dynamic performance. While the controller design is primarily based on the dynamic modeling and analysis of PWM converters, it also requires comprehensive knowledge about the control theory, linear system theory, and feedback compensation design. The controller design will be explored in the second volume of the series.

## 1.4 DC POWER DISTRIBUTION SYSTEMS

The traditional standalone dc power conversion via one single dc-to-dc converter, as illustrated in Fig. 1.5, has advanced to multi-stage parallel-module dc power distribution systems, where a number of dc-to-dc converters and intermediate filter stages are integrated in cascaded and paralleled fashions, in order to power multiple electronic instruments efficiently and reliably. The structure of such power distribution systems is very complex and varied. The complexity and variety of the power distribution systems present new challenges to the analysis and design of individual converters and power distribution systems. This section discusses the issues related with dc power distribution systems.

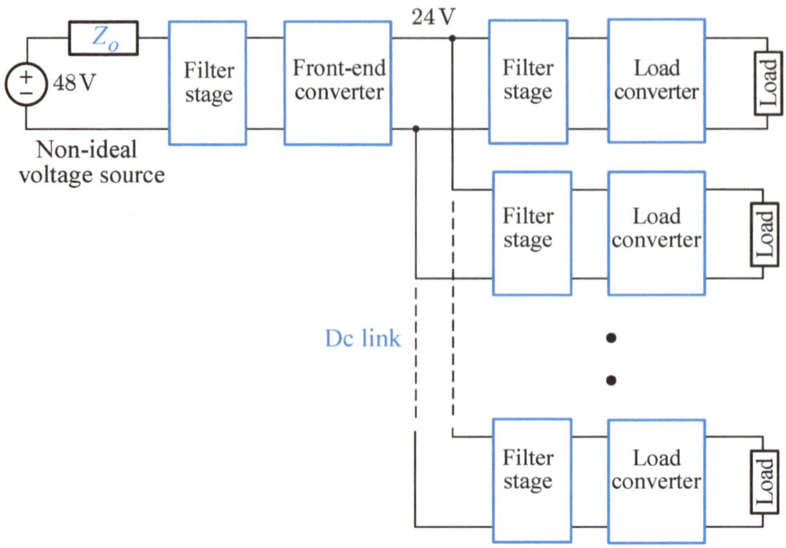


**Figure 1.6** Dc power distribution system for electronic devices and circuits.

### 1.4.1 Structure of Dc Power Distribution Systems

Figure 1.6 depicts a typical dc power distribution system for multiple electronic devices and instruments. The system employs a front-end converter and several load converters in parallel, along with intermediate filter stages. The front-end converter steps down the 48 V source to a regulated 24 V dc link. The 24 V link voltage is then distributed to the load converters through the filter stages. The load converters further alter the voltage level to other values, as required by modern analog and digital electronics.

The input current of a PWM converter is either a dc-biased pulse or dc-biased triangular waveform. If the input current of the load converter is directly drawn from the dc link in its original form, the quality and integrity of the 24 V link voltage will be degraded. When a pulsating input current is extracted from the dc link, the dc link voltage inevitably produces a transitional voltage fluctuation in the shape of the input current of the load converter.

The fluctuation in the dc link voltage will adversely affect the operation and performance of the other downstream converters. To avoid such a situation, a filter stage is placed at the input port of the load converters. The filter stage locally bypasses the fluctuating ac components so that only the smoothly-filtered dc current flows among the dc link and load converters. For the same reason, a filter stage is also placed at the input port of the front-end converter.

The energy source of the power system could be any practical standalone dc voltage source or non-ideal dc voltage obtained by rectifying the utility line. Such practical voltage sources have a finite output impedance, $Z_o(s)$ in Fig. 1.6. In addition, the

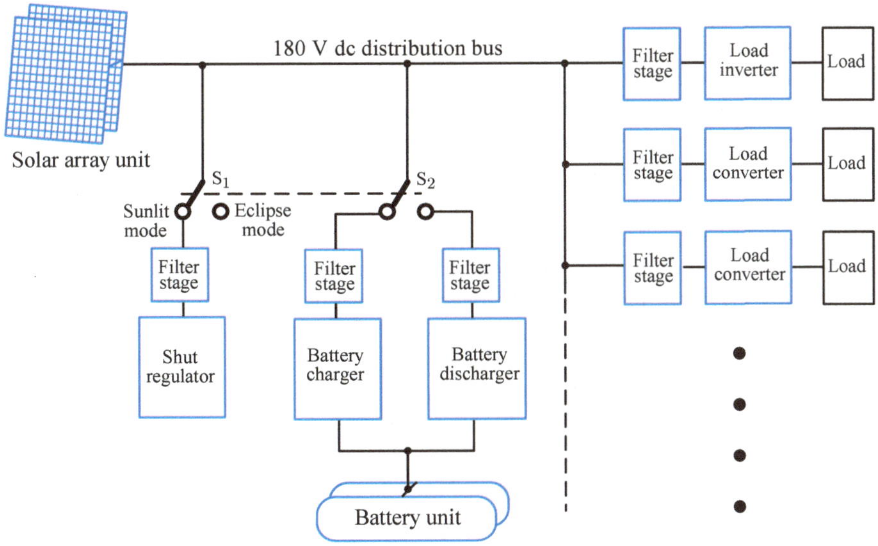


**Figure 1.7** Spacecraft power system.

load converters should support all analog or digital electronics operating with a fixed dc voltage. These loads are hardly ever resistive and present complex load impedance characteristics. The non-resistive loads and non-ideal voltage source complicate the analysis and design of the individual converters and entire dc power distribution system.

### 1.4.2 Space Power Systems

As another example of dc power distribution systems, Fig. 1.7 depicts the architecture of power conditioning systems in earth-orbiting spacecrafts. The prime energy sources of the spacecraft power system are the external solar energy and internal electrical energy of the battery unit. Spacecrafts are usually equipped with the solar array unit, consisting of countless solar cells arranged in a special format, in order to effectively intercept the external solar energy for internal uses.

The solar array unit first alters the solar energy into a dc current, which is converted into the dc voltage by auxiliary circuitry in the solar array unit. The dc voltage then fed to the dc distribution bus. The load converters and inverters, which are placed between the dc distribution bus and payloads, perform the required power conversion and deliver electrical power to payloads in an appropriate form. Each load converter or load inverter is linked to the upstream dc distribution bus via an intermediate filter stage for the reasons explained in the previous section.

The solar array unit produces full power during a sunlit mode, but it generates far less power during an eclipse mode.† Thus, for a balanced power conditioning, some additional functional units are employed. They are the shunt regulator, battery charger, and battery discharger, all connected to the dc distribution bus through a filter stage. The dc distribution bus is regulated at 180 V by the harmonious operation of these functional units. The two synchronized switches, $S_1$ and $S_2$ in Fig. 1.7 are used for a proper power conditioning over the operational period.

In sunlit mode, the solar array unit produces more electrical energy than necessary, thereby resulting in an energy overflow. The switches $S_1$ and $S_2$ are positioned as shown in Fig. 1.7, during the energy-overflow period. A part of surplus energy is fed to the battery charger to recharge the battery unit. The remaining extra energy is thermally dissipated, in a controlled manner, at the shunt regulator. The shunt regulator and battery charger both function as a controlled power consumer and collectively regulate the distribution bus at 180 V during this energy-overflow period.

When the spacecraft enters an eclipse mode, the two synchronized switches, $S_1$ and $S_2$ in Fig. 1.7, alter their connections. Now, the shunt regulator is disabled. At the same time, the battery discharger is activated while the battery charger is deactivated. In this mode, the solar array unit produces less electrical energy than necessary, and the battery discharger pulls out electrical energy from the battery unit to compensate for energy deficit. The battery discharger now regulates the dc distribution bus at 180 V by adjusting the energy influx from the battery unit.

As shown in the system diagram in Fig. 1.7, the intermediate filter stages are employed between the dc distribution bus and all the functional units in the power system. Spacecraft power systems have complex structure/functionality and present challenges in the analysis and design of the individual functional units and entire power conditioning system. In fact, the space power system created considerable momentum and impetus for advancements of modern power electronics technologies.

### 1.4.3 Issues in Dc Power Distribution System Analysis and Design

The analysis and design of PWM converters have long been studied, based on the assumption that the converters are powered from an ideal voltage source and connected to a resistive load. However, neither of these assumptions is valid for PWM converters employed as a functional module in dc power distribution systems. For example, the frond-end converter in Fig. 1.6 sees the combination of the non-ideal voltage source and filter stage as the source subsystem. Likewise, the front-end converter has a complex load subsystem, consisting of all the paralleled filter stages and load converters downstream. Furthermore, the exact structure of the load subsystem is only available after the entire power system is constructed by integrating all the load converters and filter stages. Any usable information about the load subsystem is unavailable in the design and construction stage of the front-end converter. These

†The spacecraft directly faces the sun during a sunlit mode. In an eclipse mode, in which the earth is placed between the sun and spacecraft, the spacecraft travels in the shade and the solar array unit receives only a small fraction of solar energy.

features and many other aspects bring in new challenges to the analysis and design of both the individual converters and entire power distribution system.

1) Use of the existing knowledge: Previous textbooks mostly dealt with standalone converters connected with an ideal voltage source and resistive load. We need to refine and extend the previous work for the converters employed in dc power distribution systems, while making the most and best use of the existing knowledge.

2) Dynamic analysis and control design of converters: The analysis and control of the converters in the power distribution system soon become intractable when the whole system is considered altogether. As a practical alternative, the task could be broken down into two steps. We first analyze and design individual converters as a separate standalone converter, and later incorporate the operational environment in the system in order to evaluate the performance of individual converters in real operation. We should establish theoretical fundamentals and engineering techniques for this two-step approach.

3) Analysis of subsystem interactions and system stability: The dynamics of each converter will be affected by the other converters and filter stages in the power distribution system. The dynamic interactions among subsystems should be correctly incorporated when analyzing the converter dynamics and system stability, in order to avoid any unexpected and irreversible system failures.

4) Line filter design for system stability and performance: The intermediate line filter stages, employed for current filtering to assure the dc link integrity, exert substantial impacts on the system stability and subsystem interactions. For the system stability with high performance, these impacts should be analyzed and incorporated into the line filter design.

5) Design of dc power distribution systems: The new approach should be employed to correctly design the otherwise intractable dc power distribution systems. The design should secure the system stability and subsystem compatibility, without requiring any re-engineering of individual converters.

This book series presents the latest research results about the system analysis and design of practical dc power distribution systems.

## 1.5 SIMULATORS FOR POWER ELECTRONICS CIRCUITS AND SYSTEMS

One efficient approach to learning and teaching power electronics is the use of circuit simulators for illuminations of operation and performance of power converters and power distribution systems. In this section, we overview features and utilities of major commercial simulation software as the circuit/system simulator for nonlinear time-variant power conversion circuits and systems.

This section first discusses six different software in a comparative and contrasting manner. Among the six candidates, we will select three specific software as the platform for illustrative simulations in this book series. We outline the use of the selected software in regard to the scopes and contents of the book series.

1) PSIM®was commercially introduced by Powersim in 2006 and soon emerged as a market-leading simulation tool for power electronics circuits and systems. PSIM adapts the nodal analysis and trapezoidal integration rule as its major numerical algorithms. It provides a schematic capture interface and waveform viewer Simview®.

   PSIM provides a much faster simulation than many SPICE-based circuit simulators, primary due to the use of ideal switch for semiconductor switches. With ideal switches, PSIM is better suited for system level studies rather than device level simulations. PSIM has much simpler interface, compared with other alternatives, and thus provides more intuitive and friendly user interactions.

   Later, PSIM launched a computational method for the frequency-domain transfer function analysis. It has an automated procedure of performing a series of time-domain simulations, while introducing sinusoidal perturbations to a specific circuit variable for a desired frequency range. PSIM post-processes the simulation results through an FFT algorithm in order to extract the frequency response between the perturbed signal and predefined output variable. This computational method has been popularly used to predict the frequency-domain characteristics of power electronics circuits and systems. This feature allows users to perform the frequency-domain analyses using circuit schematics, which themselves serve as the PSIM model.

   Powersim delivers a license-free trial copy of PSIM, which has restrictions on the node number, component count, and data point. Although limited to relatively simple or simplified power electronics circuits, the trial version meets needs of most educational simulations.

2) LTspice®is a high-performance SPICE-based electronic circuit simulator, initially launched by Linear Technology later transferred to Analog Device. It is the most-widely distributed and used SPICE-based software in electronics industries. LTspice provides i) a schematic capture function to enter an electronic schematic diagram for electronic circuits and systems, ii) an enhanced SPICE-based analog circuit simulator, and iii) a waveform viewer to visualize the simulation results.

   LTspice provides various circuit simulation environment, such as the transient analysis, dc analysis, ac analysis, and statistical analysis among many others. It has enhanced and specialized modules that speed up simulations of switched-mode power supplies, which accelerated the acceptance of LTspice in power electronics industries.

   Currently, LTspice®is one of very few license-free circuit simulators that provides the full features of the software without any restrictions. Steady-state operations, transient dynamics, and frequency-domain characteristics of large-scale power distribution systems could be simulated and analyzed using the license-free copy without any limitations on the circuit complexity and components count.

3) PSpice®, standing for Personal Simulation Program with Integrated Circuit Emphasis, is a modified version of the academically-developed SPICE program. PSpice was originally commercialized by MicroSim in 1984. Later, PSpice is further evolved into OrCAD PSpice Designer® with an improved user interface and functionality. Unlike LTspice, PSpice is a proprietary software tool. Yet, PSpice Lite® is available for students as a license-free copy, which comes with full functionality of the software, but with limited computational power and circuit-handling capacity.

4) SIMPLIS®, the acronym of Simulation of Piecewise Linear Systems, was introduced by SIMPLIS Technologies and has been accepted as one leading simulation engine for switched-mode power supply design. SIMPLIS adopted the numerical algorithm based on the piecewise linear analysis technique, in order to cope with computational challenges of switching power systems. Same as SPICE, SIMPLIS works on the component level, yet with a much faster transient analysis of switching circuits. For power electronics systems, the piecewise linear modeling and simulation techniques of SIMPLIS offer the superior convergence behavior over other SPICE-based counterparts. SIMPLIS® offers an advanced computational method for frequency-domain analysis of switching circuits and systems.

5) PLECS®, standing for Piecewise Linear Electrical Circuit Simulation, was introduced by Plexim as a software tool for general circuits and systems simulations. while PLECS is primarily designed for power electronics applications, it can be universally used for any linear or nonlinear dynamic systems. PLECS® is equipped with the multi-domain modeling capacity to include thermal-, magnetic-, and mechanical-domain analyses, as well as power electronics circuits and systems.

6) Matlab/Simulink® is a Matlab-based programming environment, suited for modeling, simulating, and analyzing multi-domain dynamic systems. It is a proprietary software tool, developed by Mathworks. The primary interface is a graphical block-diagramming tool and customizable functional block libraries. It offers an easy integration with the powerful Matlab platform. Matlab/Simulink has been widely used for multi-domain simulations and model-based designs in academia and industries.

We employ PSpice®, PSIM®, and LTspice® as the platform for computer simulations in the three-volume book series. The software will be a vehicle to convey theoretical discussions in the form of graphical illuminations. Most computer models will be prepared for license-free copies of the software so that readers use them as *unrestricted* learning and design aids. Simulation examples will be given with complete model descriptions and detailed discussions about simulation outcomes. Important book contents will be converted into circuit models for selected software. The circuit models include all engineering details so that each model could be immediately converted into working hardware.

The student version PSpice® will be used in presenting theoretical basics and operational details of PWM converters in standalone operation. All programming details about computer models will be provided for the license-free PSpice Lite® so that readers readily reproduce the textbook presentations.

The trial-version PSIM® will be used for time- and frequency-domain study of PWM converters. Time-domain simulations will show the steady-state operation and transient responses. For the frequency-domain study, customized PSIM simulations will be performed to extract frequency responses of PWM converters. The results will be used to validate the frequency-domain analyses and compensation designs. PSIM simulations will be presented as self-contained examples, which not only reinforce theoretical discussions but also demonstrate practical operations of PWM converters.

The license-free LTspice® will be used for simulations of power converter systems or power distribution systems, where the circuit complexity or component count exceed the limit of the trial/educational versions of respective software. LTspice® models will contain all analytical equations so that they will be used as design aids for practicing engineers and educational platforms for motivated students.

## 1.6 THREE-VOLUME BOOK SERIES

This book is the first publication of the three-volume book series that will cover one important branch of power electronics – *PWM Dc-to-Dc Power Conversion Circuits and Systems*. Each volume individually addresses theoretical fundamentals and engineering practices on three distinct areas of the PWM power conversion technology – *converter circuits, dynamics and control, and power system design*. The current book investigates the topological structure and operation of PWM converters. The book series collectively provide the comprehensive knowledge about the analysis, design, and applications of PWM converters and power distribution systems.

### 1.6.1 Volume I: Circuits and Operation of PWM Converters

This first volume deals with power stage components, circuit configurations, and operations of standalone PWM converters. Circuit properties of semiconductor switches and energy storage elements are all discussed as the power stage components of PWM converters. The evolutions of most widely-used PWM converters are described. This book illustrates operations and waveforms of all the important PWM converters via circuit analyses and computer simulations.

***Part I: Power Stage Components and Switching Circuits*** The book begins with discussions about functional basics and circuit properties of the power stage components. This part also presents several practical switching circuits to show operations of the power stage components. The topics covered in this part are as follows:

- Semiconductor switches
- Energy storage and transfer devices
- Switching circuits in practice

We investigate the circuit behavior of semiconductor switches, inductors, capacitors, and transformers, all operating under periodic excitations. The operation of MOSFETs as an active switch, diodes as a passive switch, and MOSFET-diode pairs as a single-pole double-throw (SPDT) switch are first presented. Then, we discuss the basic circuit equations of inductors and capacitors. This part also describes the operation of transformers and introduces a circuit model for practical transformers. Several practical switching circuits are studied to illustrate the functions of the power stage components in real applications.

***Part II: PWM Dc-to-Dc Converter Circuits*** The later part of this book presents topologies and operations of various PWM dc-to-dc converters. The PWM converters are grouped into two categories to address their circuit topologies and operations:

- Non-isolated converter circuits
- Transformer-isolated converter circuits

As the prime example of non-isolated converters, we first investigate the buck converter, which performs the step down dc-to-dc power conversion in the simplest form. The basic principles and functional details of the buck converter are both analyzed. The closed-loop feedback control using PWM scheme is presented. The steady-state and transient responses of the closed-loop controlled buck converter are analytically described and illustrated with computer simulations.

The study is then extended to the boost and buck/boost converters. This part first illustrates the topological origin of the boost and buck/boost converters. The features and operations of the two converters are then discussed in detail. We further examine the topology and operation of Ćuk converter and bidirectional boost converter.

This part also addresses the evolution and proliferation of transformer-isolated converters. Derivations of the flyback converter, forward converters, and bridge-type converters from their respective forerunners are demonstrated. The operations and waveforms of these transformer-isolated converters are theoretically analyzed and graphically presented with illustrative computer simulations.

### 1.6.2 Volume II: Dynamics and Control of PWM Converters

The second volume focuses on two important topics of the PWM converter technology: 1) modeling and dynamic analysis and 2) control design for PWM converters.

***Part I: Modeling, Dynamics, and Performance of PWM Converters*** The first part of this volume starts from the modeling of PWM converters and proceeds with the dynamic analysis of converter power stages. We also discuss the closed-loop performance of PWM converters and presents important techniques to evaluate stability and performance metrics of PWM converters.

- Modeling PWM converters
- Power stage transfer functions
- Dynamic performance of PWM converters

We first develop small-signal models for the power stage, PWM block, and feedback circuit of PWM converters. Later, we build an $s$-domain small-signal model for closed-loop controlled converters by integrating the small-signal models of the individual functional blocks of PWM converters. The $s$-domain model is applicable to all the closed-loop controlled isolated/non-isolated PWM dc-to-dc converters in CCM operation or DCM operation.

This part analyzes power stage transfer functions of the converters power stage. The transfer functions are derived from the $s$-domain small-signal model and later converted into Bode plots, in order to expose the frequency-domain dynamics. Results of this analysis are the bases for the control design and performance evaluation of closed-loop controlled converters.

We explore stability and dynamic performance of PWM converters. We first define the performance metrics for PWM converters as a closed-loop controlled voltage source. We address how the classical analysis techniques, such as Nyquist analysis, loop gain method, and Bode plot, could be adapted to the dynamic analysis of nonlinear time-variant PWM dc-to-dc converters.

***Part II: Control Schemes for PWM Converters*** As the heart of this volume, we investigate control schemes and establish the control design procedures for PWM converters. The two prevalent control schemes, voltage mode control and current mode control, are both treated in detail in three chapters.

- Voltage mode control
- Current mode control I – Functional basics and classical analysis
- Current mode control II – Sampling effects and new design procedures

We first deal with voltage mode control, which employs the output voltage as the sole feedback signal. The design and assessment of voltage mode control are discussed with regard to the frequency- and time-domain performance of the buck converter. Relations between the feedback design parameters and closed-loop performance are analyzed, leading to step-by-step design procedures for voltage mode control. Limitations of voltage mode control are discussed, while addressing the need for current mode control for boost and buck/boost converters.

Next, we study current mode control in great detail, which exploits an additional feedback from the inductor current on top of the output voltage feedback. In current mode control, the inductor current is sampled and held constant for every single switching period, and used as the control variable in the next switching period. Due to this feature, current mode control implants attributes of the sampled-data systems into the converter dynamics. This trait of the control scheme is referred to as *the sampling effects of current mode control.*

Traditionally, current mode control has been analyzed and designed purely using continuous-time $s$-domain techniques, implicitly assuming that the sampling effects only have negligible consequences. This simplified analysis is referred to as the *classical analysis* in the sense that the analysis did not include the sampling effects and was popular in the past. We first investigate current mode control through the classical analysis.

In the early 1990s, a new class of dynamic models for current mode control emerged, which accounts for the sampling effects only using conventional $s$-domain descriptions. These new models, which is called the *new s-domain model for current mode control* in this book, rekindled the interest and attention to the sampling effects of current mode control. We study current mode control using the new $s$-domain model. We address new facets of current mode control which supplement and reinforce the outcomes of the earlier classical analysis. Most importantly, the results of this analysis are used as the baseline to develop alternative control design procedures for current mode control. Several design examples are given to support the validity of the control design procedures.

### 1.6.3 Volume III: Dc Power Distribution Systems

The final volume of the book series is dedicated to the dynamic analysis and system design of dc power distribution systems. The outcomes of the second volume are combined with specific design/analysis techniques to pursue the dynamic analysis, system design, and performance optimization of multi-stage parallel-module dc power distribution systems. As the final outcome, this volume provides the design procedures for dc power distribution systems.

***Part I: Converter Design and System Dynamics*** The first part of this volume deals with the control design and dynamic analysis of individual converters employed in dc power distribution systems. For this task, the results of the previous chapters are combined with an advanced circuit analysis technique, called Middlebrook's extra element theorem. The topics of this part are as follows:

- Uncoupled converter and extra element theorem
- Load-coupled converters and loading effects
- Source-coupled converters and input filter interaction

This part presents the design technique for the PWM converters intended for dc power distribution applications. The design technique does not require any prior knowledge about the source impedance or load impedance of the converter. Later we develop the analysis method that predicts the converter performance at the presence of the actual source and load impedances.

We also study the dynamics and performance of dc-to-dc converters combined with the source impedance and load impedance, in the environment of multi-stage power distribution systems. Conventional circuit analyses are amalgamated with Middlebrook's extra element theorem and asymptotic analysis method, thereby presenting a conceptually new approach to predicting and improving the converter performance and dc link dynamics in dc power distribution systems.

***Part II: Design of Dc Power Distribution Systems*** The final part of this volume presents the design of multi-stage parallel-module dc power distribution systems. We establish the system design procedures for robust stability and programmable performance metrics of power distribution systems. The contents of this part are as follows:

- Power system design approach
- Line filter and source/load impedances of converters
- Design of multi-stage parallel-module dc power distribution systems

We first establish the design approach for multi-stage parallel-module dc power distribution systems, where numerous converters and line filter stages will be repeatedly connected in cascaded and/or paralleled fashions. We break down the big design challenge of the whole power distribution system into smaller design problems of standalone functional units, consisting of an upstream converter, intermediate line filter, and downstream converter.

We will demonstrate that performance of the standalone functional units is governed by the intermediate line filter located between the two converter stages. The line filter dictates the load impedance of the upstream converter. Likewise, the line filter dominates the source impedance for the downstream converter. The load impedance and source impedance decisively affect the performance metrics of the converter stages, as well as the dc link dynamics and quality.

The design problem of the standalone functional unit is further reduced to the design of the intermediate line filter. The line filter is designed to program the performance metrics of the upstream converter and to control the dc link dynamics, while not disturbing the operation of the downstream converter. The downstream converter is programmed to possess desired performance matrices by the ensuing line filter in the next functional unit. The proposed technique only needs the design of line filters, while allowing the independently-engineered converter modules to be harmoniously integrated without any modifications. The validity of the proposed design method will be supported by frequency- and time-domain simulations.

## 1.7 HIGHLIGHTS OF FIRST VOLUME

As the first book of the series, the current volume covers the power stage components, circuit configuration, and operation of PWM dc-to-dc converters. Circuit properties of semiconductor switches and energy storage/transfer elements are all presented. Topologies and operations of most widely-used PWM dc-to-dc converters are discussed.

Chapter 2 deals with the power stage components. This chapter presents the circuit behavior of semiconductor switches, inductors, capacitors, and transformers, all operating under periodic excitations. This chapter first describes the operation of MOSFETs as an active switch, diodes as a passive switch, and MOSFET-diode pairs as a single-pole double-throw (SPDT) switch. Then, this chapter discusses the circuit equations of inductors and capacitors. Chapter 2 also describes the operation of transformers and introduces a circuit model for practical transformers. This chapter finally presents several examples of practical switching circuits, such as solenoid drive circuits and capacitor charging circuits, in order to address the functions of power stage components in real applications.

Chapter 3 presents the simplest PWM dc-to-dc power conversion circuit, known as the buck converter. Theoretical basics and functional details of the buck converter are both presented. This chapter also illustrates circuit analysis techniques that are applicable to all PWM dc-to-dc converters. In addition, Chapter 3 discusses the underlying basics of the PWM technique and closed-loop control of dc-to-dc converters.

Chapter 4 deals with the topology and operation of an important class of PWM dc-to-dc converters. This chapter first introduces two non-isolated converters, boost converter and buck/boost converter. We also study Ćuk converter and bidirectional boost converter. A number of isolated converters, derived from the respective non-isolated converter, are also presented. For each converter, the origin of the circuit topology is first illustrated and the steady-state operation is then investigated using the analysis techniques covered in Chapter 3.

More importantly, this chapter provided systematic explanations about the evolution and proliferation of non-isolated and isolated converter topologies. The boost converter is viewed as a *reverse* buck converter, while the buck/boost converter is derived from the cascaded connection of the buck converter and boost converter. The Ćuk converter is emerged by cascading the boost converter upstream and the buck converter downstream. In addition, the conventional boost converter is modified to the bidirectional boost converter by replacing the passive diode switch with the active MOSFET switch. The operations of Ćuk converter and bidirectional boost converter are discussed in detail.

Addition of a transformer to the non-isolated converters creates a number of new isolated dc-to-dc converters. Development of widely-used forward converters, such as the tertiary-winding reset forward converter, Zener-diode reset forward converter, RCD clamp reset forward converter, and two-switch forward converter, are described. The evolution of the full-bridge converter, half-bridge converter, and push-pull converter are discussed. Operations of the transformer-isolated converters are all illustrated with PSIM®simulations.

## REFERENCES


1. Byungcho Choi, *Pulsewidth Modulated Dc-to-Dc Power Conversion*, 2nd Ed., Wiley, Hoboken, NJ, 2021.
2. R. W. Erickson and D. Maksimović, *Fundamentals of Power Electronics*, 3rd Ed., Springer Nature, 2020.
3. Byungcho Choi, *Dynamics and Control of Switchmode Power Conversions in Distributed Power Systems*, Ph. D. Dissertation, Virginia Polytechnic Institute and State University, Blacksburg, VA, May 1992
4. Syam Kumar Pidaparthy, *Load Impedance Shaping Techniques for Dc Dower Conversion Systems*, Ph D. Dissertation, Kyungpook National University, Daegu, Korea, June 2017.

# CHAPTER 2

# POWER STAGE COMPONENTS AND SWITCHING CIRCUITS



Power stages of PWM dc-to-dc converters contain semiconductor switches, inductors, capacitors, and transformers. This chapter investigates the operations of these power stage components under periodic excitations. The current chapter also presents several circuit theorems that will be used in later chapters for power conversion circuit analyses. Finally, this chapter discusses two practical switching circuits, the solenoid drive circuit and capacitor charging circuit, in order to demonstrate operations of power stage components in practical applications.

## 2.1 SEMICONDUCTOR SWITCHES

Dc-to-dc power conversion circuits extensively use active and passive semiconductor switches. The active switch refers to the three-terminal semiconductor device whose *on/off* status is actively controlled by exciting one of the device terminals. On the

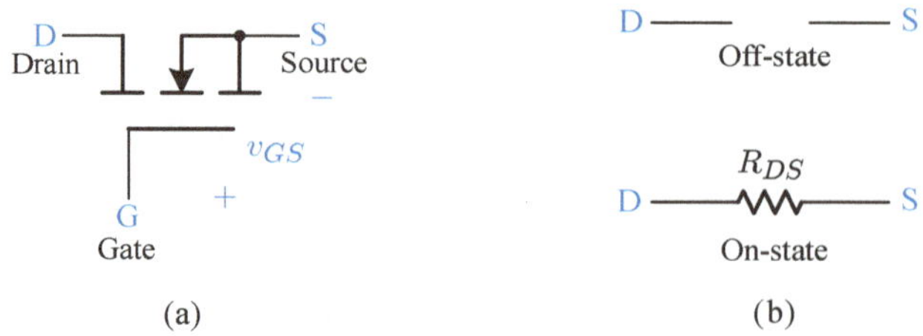


**Figure 2.1** MOSFET. (a) Circuit symbol. (b) Circuit representations.

other hand, the passive switch is the two-terminal device whose on/off status is passively determined by the conditions of the application circuit. The switching action of active and passive switches allows the voltage and current waveforms of the application circuit to be altered, in the format required for the desired dc-to-dc power conversion.

As illustrated in Fig. 1.2 in Section 1.1.1, the dc-to-dc power conversion is performed by converting the dc voltage into a rectangular waveform and filtering the resulting rectangular voltage into another dc voltage. The conversion from the dc input to the intermediate rectangular waveform is performed with the single-pole double-throw (SPDT) switch, which is implemented using a pair of active and passive switches.

There are numerous semiconductor switches available for dc-to-dc conversion circuits. The selection of switching devices depends on both how well the existing devices function and what the application circuits require. In many dc-to-dc conversion circuits, MOSFETs are commonly used for active switches because of their fast switching characteristics compared with other alternatives. For passive switch, fast recovery diodes or Schottky diodes are often selected due to their excellent switching characteristics. This section describes the function of MOSFETs as an active switch, diodes as a passive switch, and MOSFET-diode pairs as an SPDT switch. Emphases are placed on the functional behavior of these devices in dc-to-dc power conversion circuits, rather than on their physical or operational details.

### 2.1.1 MOSFETs

When employed as an active switch, MOSFETs can only have either an off-state or on-state. Figure 2.1 shows the symbol of an n-channel MOSFET and its circuit representations for the off-state and on-state. When the gate drive signal, denoted by $v_{GS}$ in Fig. 2.1(a), is lower than the threshold voltage, the MOSFET is turned-off. At the off-state, the drain-source terminal simply behaves as an open circuit because the conduction channel is not created. When the gate drive signal is higher than the threshold voltage, the conduction channel is created and MOSFET is turned-on. Once turned-on, the drain and source terminals are connected through the resistance of the conduction channel, denoted by $R_{DS}$ in Fig. 2.1(b). The channel resistance $R_{DS}$ is usually so small that a turned-on MOSFET can be viewed as a short circuit.

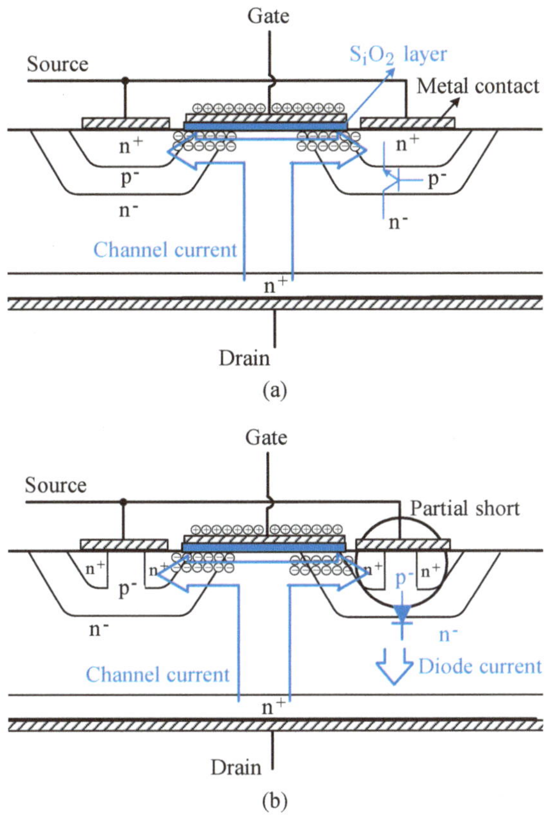


Figure 2.2 Power MOSFET structure. (a) Basic structure. (b) Partial short and body diode.

## Power MOSFETs and Body Diode

Practical MOSFETs for power electronics applications are largely different from general-purpose MOSFETs in structure and characteristics. Most significantly, the power MOSFETs have a built-in anti-parallel body diode, originated from the internal device structure.

Figure 2.2 is the conceptual illustration of an n-channel power MOSFET. The basic structure and current flow of the device are depicted in Fig. 2.2(a). The MOSFET has a vertical structure with the drain terminal at the bottom of the silicon substrate and the source terminal on the top of the device. The source terminal is fabricated by connecting two individual metal contacts, which are built on the top of $n^+$ region of the silicon substrate. The gate terminal is located on the upper surface of the $S_iO_2$ layer that covers the two adjacent $p^-$ regions.

When a rated voltage is applied between the gate and source terminals, a large amount of positive charges are accumulated on the top of the $S_iO_2$ layer. The accumulated positive charges attract the electrons from the two $n^+$ regions and stores them underneath the $S_iO_2$ layer. Consequently, an n-type conduction channel is created between the $n^+$ region and $n^-$ silicon substrate, thus providing a current path from the drain to source terminal. Now, the MOSFET is turned-on and the channel current flows from the bottom to the top of the device. Conversely, when the gate terminal is grounded, the charges stored across the $S_iO_2$ layer will be discharged and the n-type conduction channel disappears. The MOSFET then reestablishes the turn-off state.

Referring to the device structure in Fig. 2.2(a), an npn parasitic BJT is created in the region below the source contact. This parasitic npn BJT could be accidentally turned-on during the turn-off process of the device. During the turn-off transition, the MOSFET could experience a large $\mathrm{d}v/\mathrm{d}t$ across the drain and source terminals, resulting in a substantial internal current flow, $i_{p^- n^+} = C_{eff}\mathrm{d}v/\mathrm{d}t$, where $i_{p^- n^+}$ is the transitional current between the $p^-$ and $n^+$ regions of the parasitic $n^+p^-n^-$ BJT and $C_{eff}$ is the effective capacitance between the drain and source terminals.

The high-resistance path between the source terminal of the MOSFET and base terminal of the npn BJT could induce a considerable voltage drop, which accidentally turns on the base-emitter junction of the npn BJT. To prevent such an erroneous operation, the central portion of the $n^+$ region is partially removed, thus creating a *partial short* between the source contact and base terminal of the parasitic BJT. The device structure with the *partial short* is depicted in Fig. 2.2(b), where the *partial short* is highlighted with a circle.

The *partial short* provides reliable turn-offs with a high $\mathrm{d}v/\mathrm{d}t$ withstanding. However, the *partial short* also produces a $p^-n^-$ diode between the source and drain terminals of the MOSFET. This parasitic diode could deliver the current from the source to drain terminals – in the opposite direction to the channel current – when the $p^-n^-$ diode is forward-biased. This parasitic diode is thus called the *anti-parallel body diode.*

In conventional PWM dc-to-dc power conversion circuits, the potential of drain terminal is usually higher than that of the source terminal. Thus, the body diode is turned off all the time and does not affect the circuit operation. In some applications, on the other hand, the body diode could be employed as a functional diode. For one example, the body diode becomes an important circuit element of resonant power conversion circuits. As another example, the PWM-based bidirectional power conversion circuits often exploit the body diode of MOSFET in order to transfer the power in the reverse direction. Details about this application will be covered in Section 4.5.

## MOSFET Switching

Power MOSFETs invariably function as a switching device in power electronics circuits. The MOSFET switching is simply controlled by the gate signal. Figure 2.3 depicts the MOSFET switching, where the power MOSFET is portrayed as a parallel combination of the MOSFET and its body diode. In Fig. 2.3, the body diode of the MOSFET remains turned-off all the time, due to the presence of the positive voltage source $V_{DS}$.

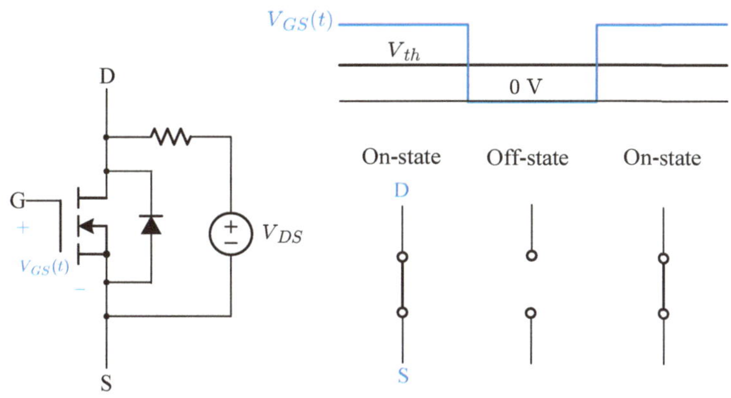


**Figure 2.3** Power MOSFET switching.

The drain and source terminals of the MOSFET presents a short circuit when $v_{GS}$ is larger than the threshold voltage $V_{th}$. Otherwise, the two terminals become an open circuit. In practical applications, the $v_{GS}$ is significantly larger than the threshold voltage to build a sufficiently thick conduction channel for a small $R_{DS}$. This practice is called the *overdrive* of MOSFETs.

### 2.1.2 Diodes

Diodes only presume one of two possible states, the off-state or on-state. Figure 2.4 depicts the symbol of a diode and its circuit representations for off-state and on-state. At an off-state, the diode becomes open-circuited. At an on-state, a practical diode is viewed as a series connection of a voltage source $V_D$, resistor, and *ideal* diode. The voltage source and resistor are necessary to approximate the nonlinear $v - i$ characteristics of the diode to a piecewise linear function. The ideal diode ensures the *unidirectional* current flow from the anode terminal to cathode terminal. The values of the voltage source and resistor are usually negligibly small. Accordingly, the on-state circuit representation is approximated to a short circuit for practical circuit analyses. Yet, the current flows only from the anode terminal to cathode terminal.

#### Diode Switching

The state of a diode is determined by the condition of the application circuit. When the application circuit forces a positive voltage from the anode to cathode terminal, the diode establishes the on-state. Conversely, when the application circuit imposes a negative voltage, the diode is turned-off. As an illustration, Fig. 2.5(a) shows a simple circuit consisting of a diode, resistor, and time-varying voltage source, $v_S$. The diode is turned-off when $v_S$ is negative and turned-on when $v_S$ is positive. The transition from the off-state to on-state occurs at the moment the negative voltage $v_S$ increases to hit the *zero* voltage. The voltage waveforms across the diode and resistor

Figure 2.4 Diode. (a) Circuit symbol. (b) Circuit representations.

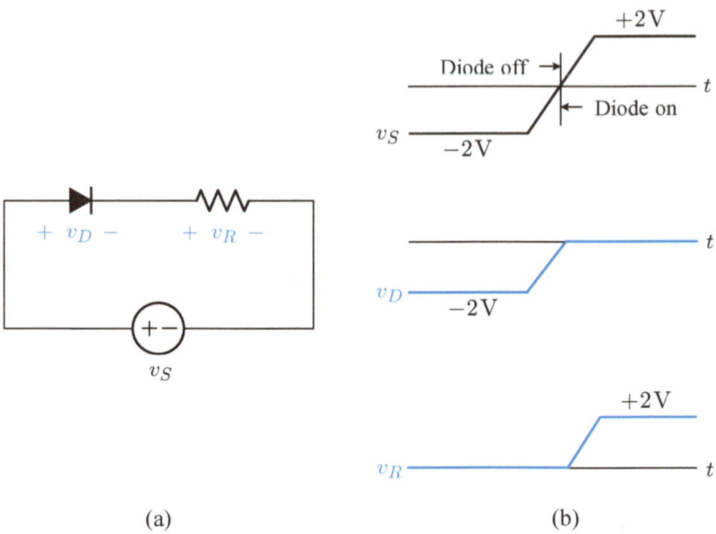


Figure 2.5 Diode switching with voltage source. (a) Circuit diagram. (b) Voltage waveforms.

are determined as shown in Fig. 2.5(b), based on the circuit behavior of the diode at the on-state and off-state.

The state of a diode can also be judged based on the direction of the diode current in the application circuit. When the application circuit forces a *positive* current flow from the anode terminal to cathode terminal, the diode is turned-on and retains the on-state as long as the current remains positive. On the other hand, when the application circuit forces a *negative* current flow from the cathode terminal to anode terminal, the diode becomes turned-off.

Figure 2.6 is a simple circuit which illustrates the diode switching using a time-varying current source, $i_S$. When $i_S$ is positive, the diode is turned-on and carries the entire current. Conversely, when $i_S$ becomes negative, the diode is turned-off and the current diverts to the resistor. The transition from the on-state to off-state occurs at the instant the positive current $i_S$ reduces to the *zero* current.

The voltage and current waveforms shown in Figs. 2.5 and 2.6 frequently emerge in practical switching circuits. Operations of switching circuits produce linearly increasing, decreasing, or zero-crossing voltage/current waveforms, which determine

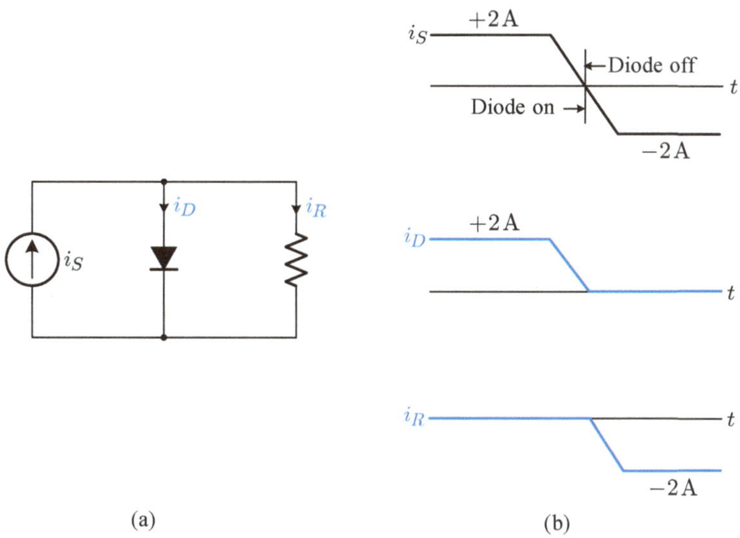


Figure 2.6 Diode switching with current source. (a) Circuit diagram. (b) Current waveforms.

the status of the diodes. Examples of such circuits will be seen very often in this chapter and forthcoming chapters.

### 2.1.3 MOSFET-Diode Pair as SPDT Switch

When a MOSFET and diode are paired together, they function as an SPDT switch in power electronics circuits. This section illustrates the implementation of the SPDT switch with a MOSFET-diode pair. Figure 2.7(a) shows a simple switching circuit where a voltage source is connected to a current source via an SPDT switch. The SPDT switch periodically toggles its position as illustrated in Fig. 2.7(a). Figure 2.7(b) shows two representations of the switching circuit with two different SPDT switch positions: one is at **a** position and the other is at **p** position. Finally, Fig. 2.7(c) depicts an implementation of the SPDT switch using MOSFET and diode.

The MOSFET is turned-on/off by the gate drive signal $v_{GS}$. When the MOSFET is turned-on by the *on signal*, the diode is turned-off because the voltage source forces a negative voltage across the diode. Conversely, when the MOSFET is turned-off by the *off signal*, the current source forces the diode to turn on. The equivalence between Fig. 2.7(a) and Fig. 2.7(c) now becomes apparent. As shown in this example, the SPDT switch is usually constructed using a MOSFET-diode pair.

The switching circuit in Fig. 2.7(a) is the only workable configuration of the SPDT switch, voltage source, and current source. Figure 2.8 shows the three *unworkable* cases, which could encounter the situations of shorting a voltage source, opening a current source, paralleling unequal voltage sources, and/or cascading unequal current sources – all resulting in a catastrophic circuit failure.

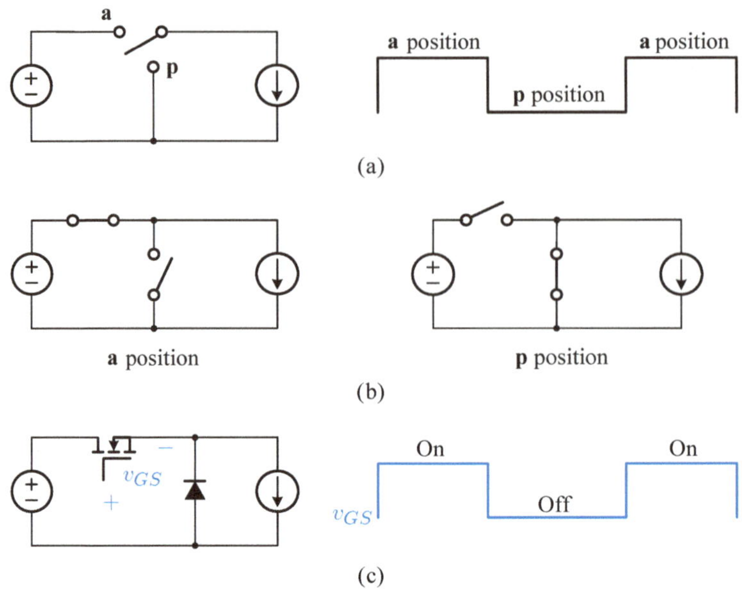


Figure 2.7 Simple switching circuit. (a) Original circuit with SPDT switch. (b) Equivalent circuits. (c) Implementation of SPDT switch using semiconductor switches.

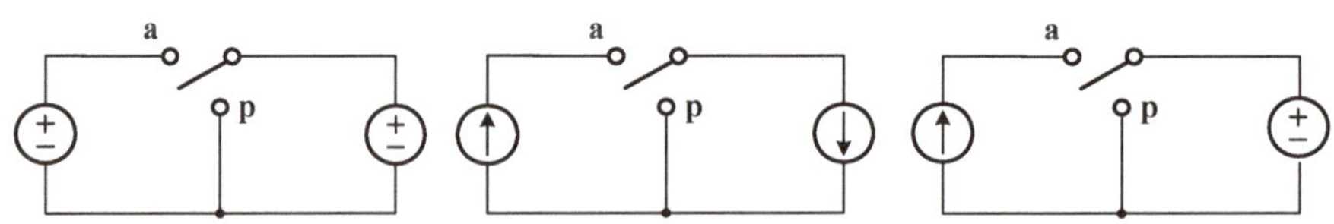


Figure 2.8 Incompatible connections of SPDT switch.

## SPDT Switch in Practical Switching Circuits

The MOSFET and diode are often directly linked to operate as an SPDT switch, as shown in Fig. 2.7(c). Yet, there are other cases where the SPDT switch configuration is not so apparent.

Simple examples are given in Fig. 2.9.† Figure 2.9(a) is a solenoid drive circuit, which will be studied in Section 2.3.1. The MOSFET and diode are connected as an SPDT switch in order to properly operate the solenoid drive circuit. Figure 2.9(b) is a capacitor charging circuit, which is capable of charing a capacitor with an arbitrary dc voltage. Section 2.3.2 will reveal that the MOSFET-diode pair works as an SPDT switch, even though they are located in distance with transformer isolation. In other cases, where multiple MOSFETs and diodes are employed altogether, a specific pair

†Operations of the circuits in Fig. 2.9 will be studied later in this chapter.

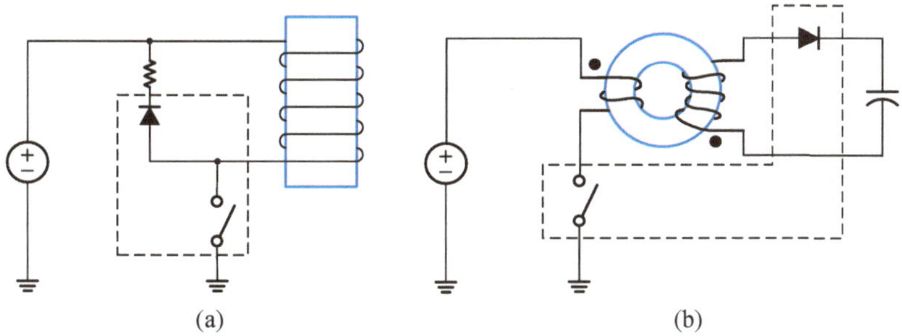

Figure 2.9 SPDT in practical switching circuits. (a) SPDT switch in solenoid drive circuit. (b) SPDT switch in capacitor charging circuit.

of the MOSFET and diode are functionally combined to work as an SPDT switch. Such examples will be covered in Chapter 4.

## 2.2 ENERGY STORAGE AND TRANSFER DEVICES

Dc-to-dc power conversion circuits utilize energy storage devices to perform filtering and energy transfer devices to transfer electrical energy while altering the magnitude of voltage and current waveforms. Energy storage and transfer devices include inductors, capacitors, and transformers. This section analyzes the circuit behavior of such devices under periodic excitations.

### 2.2.1 Inductors

We first examine circuit properties of inductors. In addition to basic circuit equations, the operations of inductors under periodic excitations are discussed. This section also presents important circuit theorems for inductors, which will be used in the remaining part of this book.

#### Circuit Equations

An inductor is typically constructed by winding a copper wire on a magnetic core. When the current $i_L$ flows through the inductor as shown in Fig. 2.10(a), a series of electro-magnetic phenomena occur inside the magnetic core. First, the current passing through the copper winding generates magnetic field intensity $H$. The direction of $i_L$ and $H$ are determined by Ampere's right-hand rule, as illustrated in Fig. 2.10(b). The magnetic field intensity in turn produces magnetic flux density: $B = \mu H$ with $\mu$ representing the permeability of the core. The total magnetic flux across the core is then calculated as $\phi = BS$ where $S$ is the cross-sectional area of the core. Finally, the magnetic flux linkage is given by $\lambda = N\phi$ where $N$ is the number of the copper winding turns. The cause-and-effect of the aforementioned process is summarized

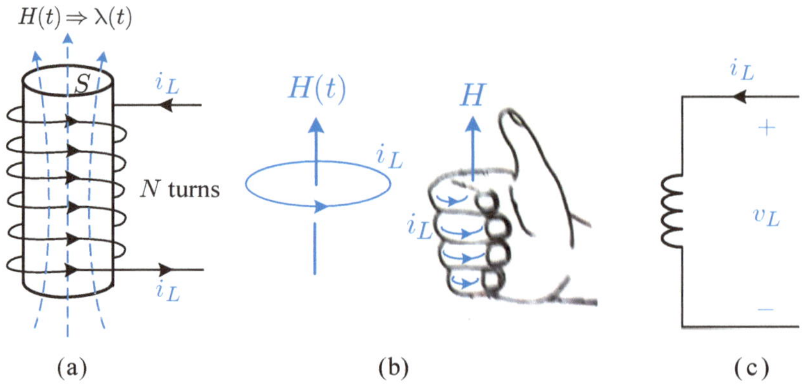


**Figure 2.10** Inductor and its terminal properties. (a) Electro-magnetic process within inductor. (b) Ampere's right-hand rule. (c) Polarity/direction of inductor voltage/current waveform.

as $i_L(t) \Rightarrow H(t) \Rightarrow B(t) \Rightarrow \phi(t) \Rightarrow \lambda(t)$. When the current $i_L$ is considered as the input of the electro-magnetic process, the magnetic flux linkage $\lambda$ becomes the output of the process. The ratio of the output variable, $\lambda$, to the input variable of the process, $i_L$, is defined as the inductance of the inductor

$$L \equiv \frac{\lambda(t)}{i_L(t)} \tag{2.1}$$

According to Faraday's law, the time-varying magnetic flux linkage induces a voltage across the inductor terminals

$$v_L(t) = \frac{\mathrm{d}\lambda(t)}{\mathrm{d}t} \tag{2.2}$$

Using the definition of the inductance, Faraday's law is rewritten as

$$v_L(t) = L\frac{\mathrm{d}i_L(t)}{\mathrm{d}t} \tag{2.3}$$

thus establishing the $i - v$ relationship of the inductor. The polarity of the induced voltage is defined as shown in Fig. 2.10(c).

Equation (2.3) is integrated to yield an alternative circuit equation for the inductor

$$i_L(t) = \frac{1}{L}\int v_L(t)\,\mathrm{d}t \tag{2.4}$$

When a dc voltage $V_S$ is applied across the inductor, the inductor current is given by

$$i_L(t) = \frac{V_S}{L}t \tag{2.5}$$

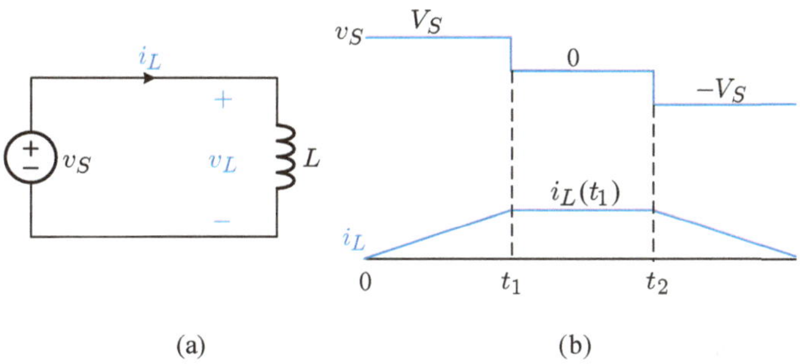


**Figure 2.11** Voltage drive of inductor. (a) Circuit diagram voltage. (b) Inductor voltage and current waveforms.

Expression (2.5) indicates that the inductor current continues to increase linearly without bounds, as long as the dc voltage is present. In practice, however, a real inductor cannot sustain an excessively large current and eventually becomes *saturated.* The saturation of an inductor will be discussed later in this section.

In practical dc switching circuits, inductors are alternatively excited by several dc voltages of different values – this feature is called the *voltage drive* of an inductor. For example, an inductor could be successively exposed to a positive voltage $V_S$, *zero* voltage, or negative voltage, $-V_S$. Figure 2.11 illustrates this conceptual operational scenario. Referring to Fig. 2.11(a), we assume that the voltage source $v_S$ varies as illustrated in the upper plot in Fig. 2.11(b). With the positive voltage $v_S = V_S$ for $0 < t < t_1$, the inductor current increases as

$$i_L(t) = \frac{1}{L}\int v_L(t)\,\mathrm{d}t = \frac{V_S}{L}t$$

When $v_S$ is stepped down to zero at $t = t_1$, the equation (2.3) yields

$$v_L(t) = L\frac{\mathrm{d}i_L(t)}{\mathrm{d}t} = 0 \quad \Rightarrow \quad \frac{\mathrm{d}i_L(t)}{\mathrm{d}t} = 0$$

which indicates that the inductor current remains constant at $i_L(t_1) = (V_S/L)t_1$. Now assume that the inductor voltage is further reduced to $v_S = -V_S$ at $t = t_2$. The inductor current then declines linearly

$$i_L(t) = -\frac{V_S}{L}t$$

from $i_L(t_1)$, as illustrated in Fig. 2.11(b).

From the preceding analysis, we conclude the evolution of the inductor current under the voltage drive condition. With a positive voltage $V_S$, the inductor current linearly increases with the slope $V_S/L$. When the *zero* voltage is applied to the inductor, for example by shorting the inductor terminals, the inductor current remains unchanged and circulates through the shorted path. With a negative voltage, $-V_S$, the inductor current decreases with the slope, $-V_S/L$.

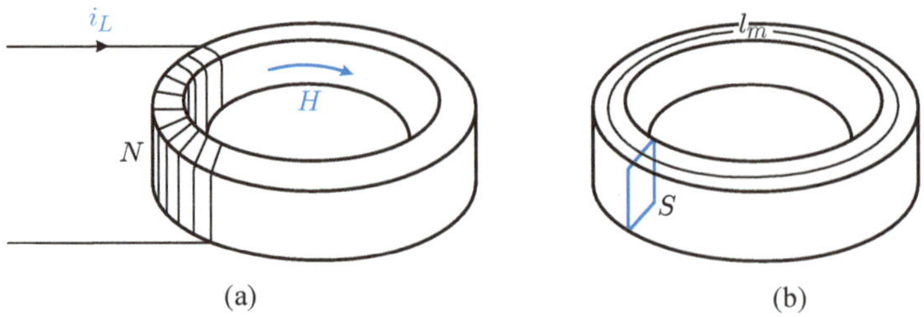


Figure 2.12 Toroidal inductor. (a) Inductor structure. (b) Geometry of toroidal core.

## EXAMPLE 2.1 Inductance of toroidal inductor

This example evaluates the inductance of an inductor fabricated using a toroidal core, as depicted in Fig. 2.12. Referring to Fig. 2.12, the inductance is given by

$$L = \frac{\lambda(t)}{i_L(t)} = \frac{N\phi(t)}{i_L(t)} = \frac{NSB(t)}{i_L(t)} = \frac{\mu_r\mu_o NS}{i_L(t)} H(t) \tag{2.6}$$

where $\mu_r$ is the relative permeability of the core material, $\mu_o$ is the permeability of free space, $N$ denotes the turns of the inductor winding, and $S$ is the cross-sectional area of the core.

Ampere's law leads to the relationship

$$H(t)\, l_m = N\, i_L(t) \qquad \Rightarrow \qquad H(t) = \frac{N\, i_L(t)}{l_m} \tag{2.7}$$

where $l_m$ is the length of magnetic path along the toroidal core. From (2.6) and (2.7), it follows that

$$L = \frac{\mu_r\mu_o NS}{i_L(t)} \frac{N\, i_L(t)}{l_m} \tag{2.8}$$

The final expression for the inductor is given by

$$\boxed{L = \mu_r\mu_o \frac{S}{l_m} N^2} \tag{2.9}$$

## EXAMPLE 2.2 Inductive switching circuit 1

This example illustrates the behavior of the inductor excited by a pulse voltage waveform. Figure 2.13 shows a simple switching circuit, along with its circuit waveforms. Assume that the inductor is initially unenergized and the switch is opened. When the switch is closed at $t = 0$, a dc voltage $V_S$ is applied across the inductor and $i_L$ starts to increase linearly with a slope $V_S/L$. The switch is then opened at $t = t_1$, thereby forcing the inductor current to collapse. The change in the inductor current in turn induces a voltage across the inductor. Equation (2.3) indicates that

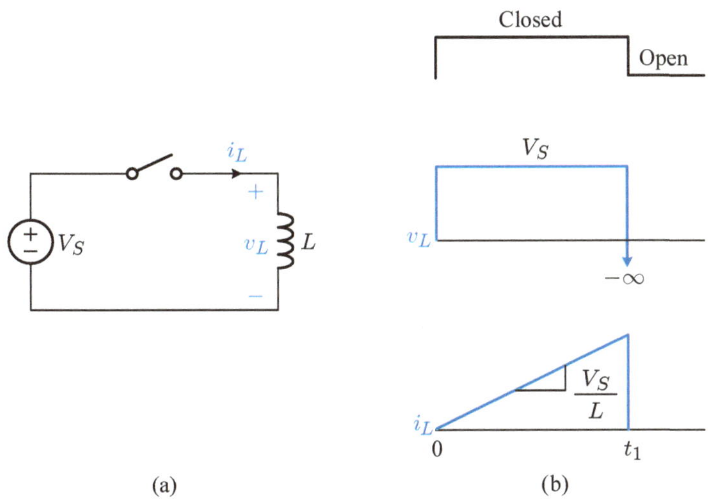


**Figure 2.13** Inductive switching circuit. (a) Circuit diagram. (b) Circuit waveforms.

a negative voltage, whose magnitude is proportional to the decaying slope of the inductor current, will be induced across the inductor.

When the switch is assumed to open instantaneously, the magnitude of the induced voltage is infinite. In practice, however, the magnitude of the negative voltage is finite because a real switch can only react with a finite response time. Even so, the magnitude is still large enough to destroy any practical semiconductor switch.

The behavior of the inductor can also be explained using the energy conservation principle. As the inductor current ramps up, the magnetic energy inside the inductor continues to increase. When the inductor current is collapsed at $t = t_1$, the total energy $E_m = 0.5Li_L^2(t_1)$ is instantaneously released in the form of a voltage spike. As illustrated in this example, a sudden interruption of the inductor current incurs a *destructive* voltage spike and therefore should be avoided, unless the circuit is intentionally designed to operate so.†

## Inductive switching circuit 2 — EXAMPLE 2.3

This example shows the operation of another inductive switching circuit. In the circuit in Fig. 2.14(a), a dc voltage source is connected to an inductor through an active-passive switch pair. The on/off status of the active switch is controlled by the switch drive signal shown in Fig. 2.14(b). As explained earlier, the active-passive switch pair functions as an SPDT switch. When the active switch turns on, the

†The engine plugs in a conventional gasoline car generate the periodic voltage spikes in the manner similar to Fig. 2.13 to activate the traction chain.

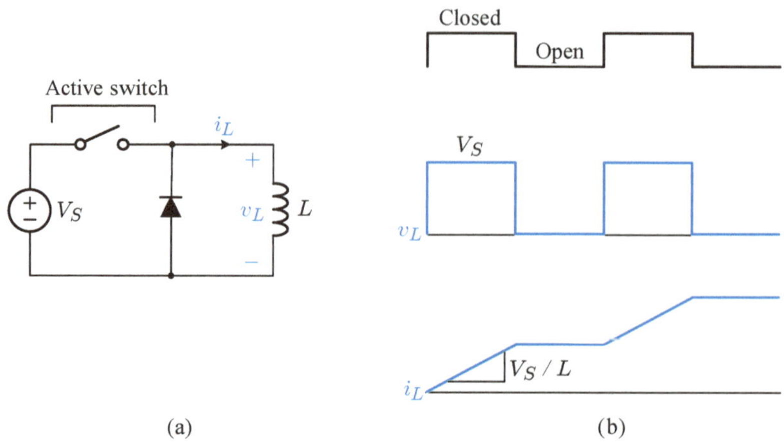


Figure 2.14 Inductive switching circuit. (a) Circuit diagram. (b) Circuit waveforms.

passive switch is turned-off; conversely, when the active switch turns off, the passive switch is turned-on. With the active switch turned-on, the dc voltage $V_S$ is applied across the inductor $L$ and the inductor current thus ramps up with a slope $V_S/L$.

When the active switch turns off, the diode is turned-on because the current-carrying inductor behaves as a current source. Once the diode is turned-on, the voltage across the inductor becomes zero and is clamped at zero thereafter. Referring to (2.3), the inductor current remains unchanged when its voltage stays constant; the inductor current thus maintains the value that has developed at the instant the switch is just turned-off. The fixed inductor current now circulates the loop formed by the inductor and the shorted diode. When the switch is turned-on in the next operational cycle, the inductor current again starts to increase. By repeating this process, the inductor current proceeds as depicted in Fig. 2.14(b).

## Inductor as Current Source

In practical switching circuits, an inductor often functions as a transitional current source. This feature is illustrated in Fig. 2.15.

In the left circuit diagram in Fig. 2.15(a), the switch $S_1$ is closed and $S_2$ is opened $t = 0$, while $S_3$ was opened. A linearly increasing current is then developed along the path through $V_S$, $S_1$, and $L$

$$i_L(t) = \frac{V_S}{L}t$$

At $t = t_1$, $S_1$ is opened and $S_2$ is closed at the same instant, while $S_3$ remains open. The inductor carries a constant current

$$I_L = i_L(t_1) = \frac{V_S}{L}t_1$$

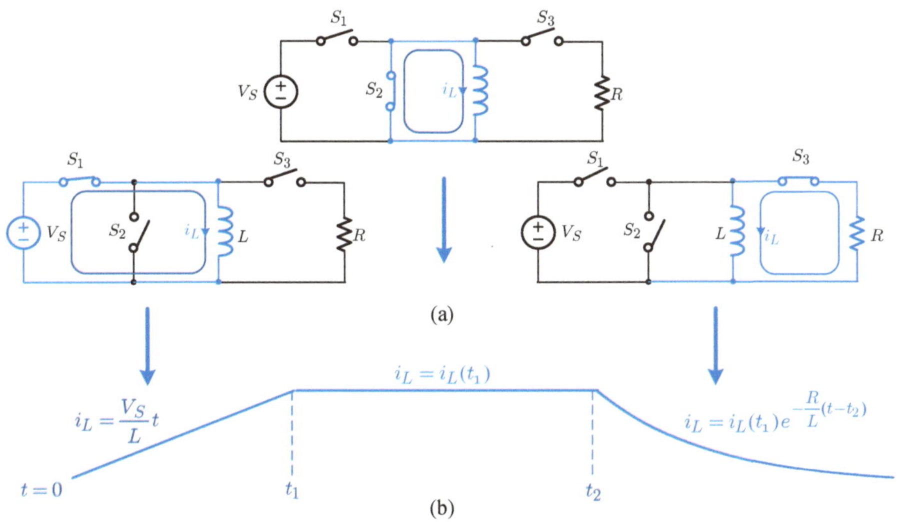


**Figure 2.15** Inductor as transitional current source. (a) Circuit diagram. (b) Inductor current.

circulating the path through $L$ and $S_2$, as shown in the middle plot. Now, assume that the current-carrying inductor is connected to a resistor $R$ at $t = t_2$, by turning-on the switch $S_3$ and simultaneously turning off $S_2$, while $S_1$ remains open. The inductor current now decays exponentially

$$i_L(t) = i_L(t_1)\, e^{-\frac{R}{L}(t - t_2)}$$

until the inductor current vanishes. The inductor current is shown in Fig. 2.15(b). It becomes apparent that the current-carrying inductor behaves as a transitional current source for the time period $t_1 < t < t_2$.

## Saturation of Inductor

Inductors invariably *saturate* when they are forced to carry an excessively large current. Figure 2.16 illustrates the relationship between the inductor current $i_L$ and magnetic flux linkage $\lambda$. As $i_L$ is increased from zero, so is $\lambda$. However, the linear relationship between $i_L$ and $\lambda$ is valid only within a limited range. When the inductor current exceeds the critical values, $\pm\, i_{L\,crit}$ in Fig. 2.16, the magnetic flux linkage remains the same at $\pm\, \lambda_{sat}$, despite any further increase in the inductor current; this phenomenon is referred to as the magnetic saturation.

Because the slope of the $i_L$–$\lambda$ curve represents the inductance, $L = \Delta\lambda(t)/\Delta i_L(t)$, the saturation implies a *zero* inductance for the inductor. Thus, the saturation causes a profound impact on the circuit operation. For example, when a dc voltage $V_S$ is applied to an inductance $L$, the current increases toward $i_{L\,crit}$ by the equation

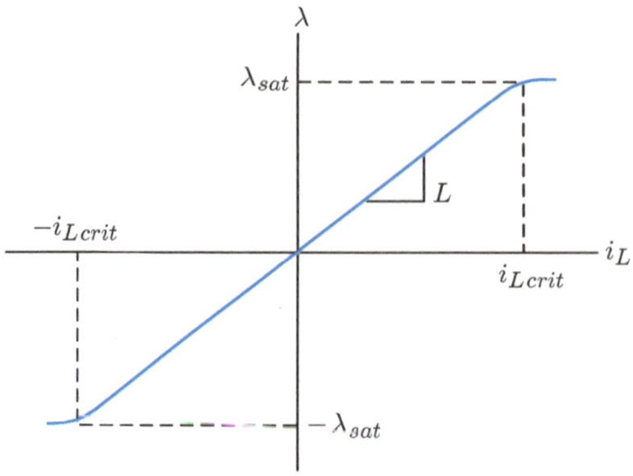


Figure 2.16 $i_L$–$\lambda$ characteristics of inductor.

$i_L(t) = (V_S/L)\,t$. When the inductor current reaches $i_{L\,crit}$, magnetic saturation occurs. Upon magnetic saturation, the inductor current shoots up boundlessly because the inductance becomes *zero* at the instant of saturation. A real circuit cannot support such an excessive inductor current, and magnetic saturation therefore should be avoided to prevent a catastrophic failure of the circuit.

Magnetic saturation is attributed to the properties of the core material used for the inductor. Numerous magnetic dipoles† exist inside the core material. When no external current is present, the magnetic dipoles are randomly oriented. For this case, the core material does not exhibit any magnetic properties because the effects of individual magnetic dipoles are canceled each other.

When the inductor current begins to flow, the magnetic field is developed inside the core and some magnetic dipoles start to align in parallel with the magnetic field, thereby generating additional magnetic flux linkage; this effect is known as the magnetic induction. As the current continues to grow, more magnetic dipoles align with the magnetic field, thereby increasing the magnetic flux linkage. When the inductor current reaches the critical value $i_{L\,crit}$, all the magnetic dipoles line up to the direction of the magnetic field and the magnetic flux linkage attains its maximum value, $\lambda_{sat}$. The core now becomes saturated and the magnetic flux linkage remains at $\lambda_{sat}$ even if the inductor current is further increased. The inductance of the saturated inductor is zero because the slope of the $i_L - \lambda$ curve is zero.

### Flux Linkage Balance Condition or Volt-Sec Balance Condition

In most switching circuits, inductors are excited in such a way that the magnetic flux linkage increases during one part of a switching period and then decreases during the other part of the switching period. The *flux linkage balance condition* asserts that the increase in the flux linkage, $\Delta\lambda_{inc}$, in one switching period should be equal to the

†Very small conceptual magnets that are assumed to exist inside magnetic materials.

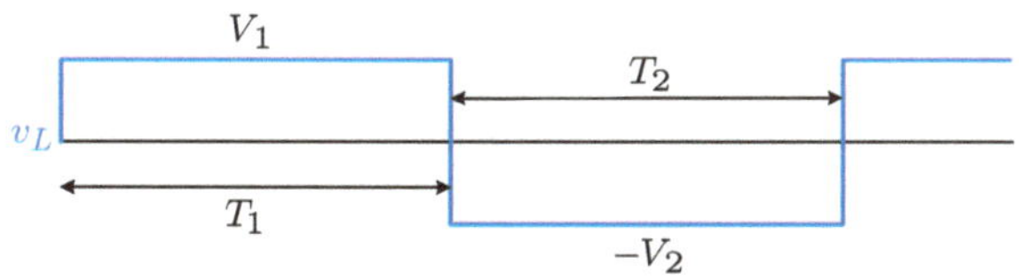


**Figure 2.17** Volt-sec balance condition: $V_1 T_1 = V_2 T_2$.

decrease in the flux linkage, $\Delta\lambda_{dec}$, in that switching period

$$\boxed{\Delta\lambda_{inc} = \Delta\lambda_{dec}} \tag{2.10}$$

or equivalently the net change in the magnetic flux linkage over one switching period must be zero.

The justification for the flux linkage balance condition is self-explanatory. If the net change in magnetic flux linkage within each switching period is *not* balanced at zero, the magnetic flux linkage will continue to grow toward the positive direction or negative direction, eventually encountering the magnetic saturation illustrated in Fig. 2.16. Therefore, all the inductors in a workable inductive switching circuit should satisfy the flux linkage balance condition. More precisely, any workable inductive switching circuit is to be designed in such a structure that all the inductors in the circuit shall meet the flux linkage balance condition. For such cases, we safely presume the flux linkage balance condition for inductors.

The flux linkage balance condition is paraphrased into an alternative form, which is more convenient for circuit analysis purposes. Faraday's law states that

$$v_L(t) = \frac{\mathrm{d}\lambda(t)}{\mathrm{d}t} \tag{2.11}$$

where $v_L$ is the inductor winding voltage and $\lambda$ denotes the magnetic flux linkage across the inductor winding. Equation (2.11) is rearranged as

$$\mathrm{d}\lambda(t) = v_L(t)\mathrm{d}t \tag{2.12}$$

to express the flux linkage change as a function of the inductor voltage. Now assume that the inductor is excited by a rectangular voltage waveform in Fig. 2.17. The positive voltage $V_1$ increases the magnetic flux linkage, whereas the negative voltage $-V_2$ decreases the flux linkage. The flux linkage increase during $T_1$ is given by

$$\Delta\lambda_{inc} = V_1 T_1 \tag{2.13}$$

and the flux linkage decrease during $T_2$ becomes

$$\Delta\lambda_{dec} = V_2 T_2 \tag{2.14}$$

By equating (2.13) and (2.14) based on the flux linkage balance condition, it follows that

$$\boxed{\Delta\lambda_{inc} = \Delta\lambda_{dec} \quad \Rightarrow \quad V_1 T_1 = V_2 T_2} \tag{2.15}$$

which states that the product of the voltage and time interval during which the inductor voltage is positive should be equal to that calculated when the inductor voltage is negative, or equivalently *the average value of the inductor voltage over one switching period must be zero.* This principle is referred to as the *volt-sec balance condition.* The volt-sec balance condition is generalized as follows. The average value of the inductor voltage calculated over integer multiples of the switching period is zero. Furthermore, the average value of the inductor voltage in general can be considered to be zero, assuming that the averaging is performed over a sufficiently longer period than the switching period.

A properly designed inductive switching circuit will establish the steady-state equilibrium where all the circuit variables are settled into the values that satisfy the volt-sec balance condition on the inductors. Thus, the volt-sec balance condition can be used as a circuit theorem in evaluating steady-state values of circuit waveforms.

Some circuits that violate the volt-sec balance condition can be conceptually devised, for instance Example 2.3 where the inductor current continues to increase as time elapses. However, such circuits eventually destroy themselves by an over-current condition that triggers the inductor saturation. Thus, all practical switching circuits are to be engineered in such a manner that all the inductors shall meet the volt-sec balance condition in steady state.

### Freewheeling Path and Freewheeling Diode

As illustrated in Example 2.2, the inductor current should *not* be discontinued abruptly because a sudden interruption of the inductor current generates a destructive high voltage spike. Therefore, the circuit path, which is connected to an inductor and to be open-circuited during the circuit operation, should always be accompanied by an alternative circuit path. The alternative path shall be operative only provided that the original inductor current path is broken. This alternative path for the inductor current is called the *freewheeling path.* The freewheeling path is usually constructed with a diode, which is normally turned-off and only becomes turned-on when the original inductor current path is disrupted. The diode employed to provide a freewheeling path is called a *freewheeling diode.*

#### EXAMPLE 2.4 Inductive switching circuit with freewheeling path

This example illustrates the operation of an inductive switching circuit with a freewheeling path. Figure 2.18 shows the circuit diagram and waveforms of an inductive switching circuit. If the freewheeling diode located in the middle of Fig. 2.18(a) is not present, a high voltage spike will be generated when the switch is opened. The freewheeling diode prevents such a voltage spike. When the switch is turned-off, the freewheeling diode conducts, thereby creating a freewheeling path for the inductor current.

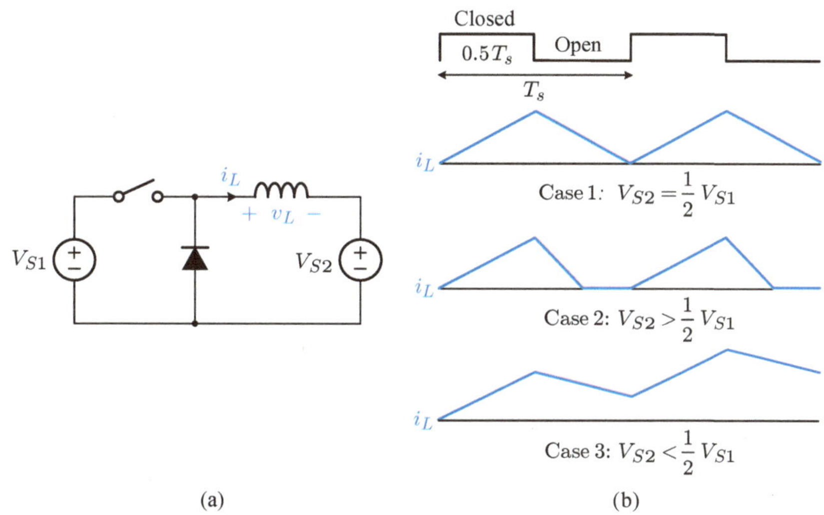


**Figure 2.18** Inductive switching circuit with freewheeling path. (a) Circuit diagram. (b) Inductor current waveforms.

When the switch is turned-on, the inductor current increases according to the equation

$$i_L(t) = \frac{v_L}{L}t = \frac{V_{S1} - V_{S2}}{L}t \tag{2.16}$$

On the other hand, when the switch is turned-off and the freewheeling diode conducts, the inductor current decreases as

$$i_L(t) = -\frac{V_{S2}}{L}t \tag{2.17}$$

Now, it is assumed that the active switch is turned-on for half the switching period and turned-off for the remaining switching period, as shown in Fig. 2.18(b). With this assumption, the following three different cases are considered:

- Case 1: $V_{S2} = \frac{1}{2}V_{S1}$
- Case 2: $V_{S2} > \frac{1}{2}V_{S1}$
- Case 3: $V_{S2} < \frac{1}{2}V_{S1}$

The inductor current waveforms $i_L$ for these three cases are shown in Fig. 2.18(b). For Case 1, the increasing slope of the inductor current is identical to the decreasing slope, resulting in a periodic triangular waveform. For Case 2, the falling rate is faster than the rising rate, as such, the inductor current $i_L$ is reduced to zero before the onset of the next operational period. When $i_L$ is reduced to zero, the diode turns off and remains off for the remaining switching period, because the diode cannot carry the current in the reverse direction.

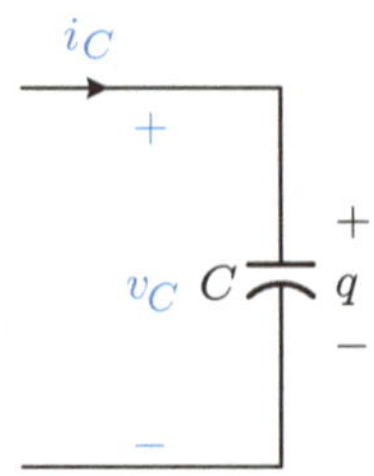


Figure 2.19 Capacitor and polarity/direction of terminal voltage/current waveforms.

Finally, for Case 3, the rate of the inductor current falling is slower than the rate of the inductor current rising. At each switching period, the final value of the inductor current will be larger than the initial value. Accordingly, the inductor current will respond as shown in the bottom plot in Fig. 2.18(b). Case 3 violates the flux linkage balance condition and the circuit never reaches steady state. This circuit will eventually encounter the inductor saturation and ensuing catastrophic failure. Readers are urged to sketch the inductor voltage waveforms for the three cases, in order to confirm the compliance with or violation of the volt-sec balance condition on the inductor, particularly to understand how Case 2 meets the volt-sec balance condition.

### 2.2.2 Capacitors

Along with inductors, capacitors are widely used in dc-to-dc power conversion circuits as an energy storage component. This section discusses circuit properties of capacitors and investigates their operation under periodic excitations. The current section also presents important circuit theorems for capacitors.

#### Circuit Equations

Capacitors are typically fabricated by placing a pair of conductor plates in parallel and filling the gap with a dielectric material. When a voltage $v_C$ appears across the parallel plates, the electric charge is accumulated inside the dielectric material, as shown in Fig. 2.19. The capacitance of a capacitor is defined as the ratio of the accumulated electric charge $q$ to the voltage $v_C$ across the capacitor

$$\boxed{C \equiv \frac{q(t)}{v_C(t)}} \tag{2.18}$$

On the other hand, the current through the capacitor is defined as

$$i_C(t) \equiv \frac{\mathrm{d}q(t)}{\mathrm{d}t} \tag{2.19}$$

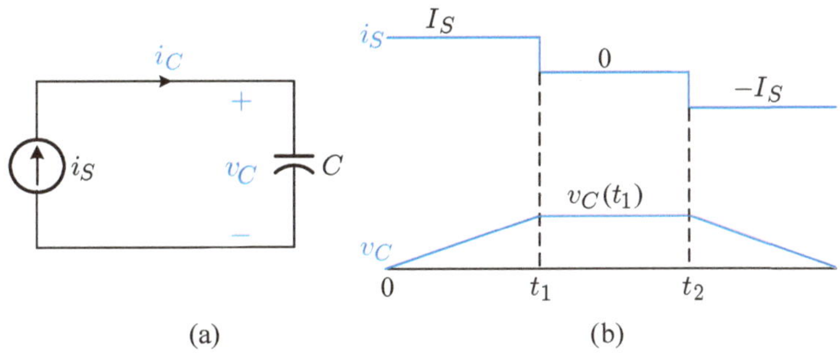


**Figure 2.20** Current drive of capacitor. (a) Circuit diagram. (b) Capacitor current and voltage.

Using the definition of the capacitance in (2.18), (2.19) is rearranged as

$$\boxed{i_C(t) = C\frac{\mathrm{d}v_C(t)}{\mathrm{d}t}} \tag{2.20}$$

leading to the $v-i$ relationship for capacitors. The polarity/direction of the voltage/current waveforms associated with a capacitor is defined as shown in Fig. 2.19. Equation (2.20) is integrated to yield an alternative circuit equation for capacitors

$$\boxed{v_C(t) = \frac{1}{C}\int i_C(t)\,\mathrm{d}t} \tag{2.21}$$

When a capacitor is connected to a dc current $I_S$, the voltage across the capacitor rises linearly

$$\boxed{v_C(t) = \frac{I_S}{C}t} \tag{2.22}$$

Equation (2.22) indicates that the capacitor voltage continues to increase boundlessly at the presence of the dc current. However, a practical capacitor will be damaged or destroyed when the capacitor voltage is increased excessively.

The capacitors in the switching circuits are excited by a dc current source that could vary among a positive current $I_S$, *zero* current, and negative current, $-I_S$. This feature is referred to as the *current drive* of a capacitor. Figure 2.20 exemplifies the capacitor waveforms under the current drive.

As shown in the upper plot of Fig. 2.20(b), a positive current $I_S$ is applied to the capacitor at $t = 0$. The capacitor voltage then increases as

$$v_C(t) = \frac{1}{C}\int i_C(t)\,\mathrm{d}t = \frac{I_S}{C}t$$

When the $i_S$ is stepped down to zero at $t = t_1$, the expression (2.20) indicates that

$$i_C(t) = C\frac{\mathrm{d}v_C(t)}{\mathrm{d}t} = 0 \quad \Rightarrow \quad \frac{\mathrm{d}v_C(t)}{\mathrm{d}t} = 0$$

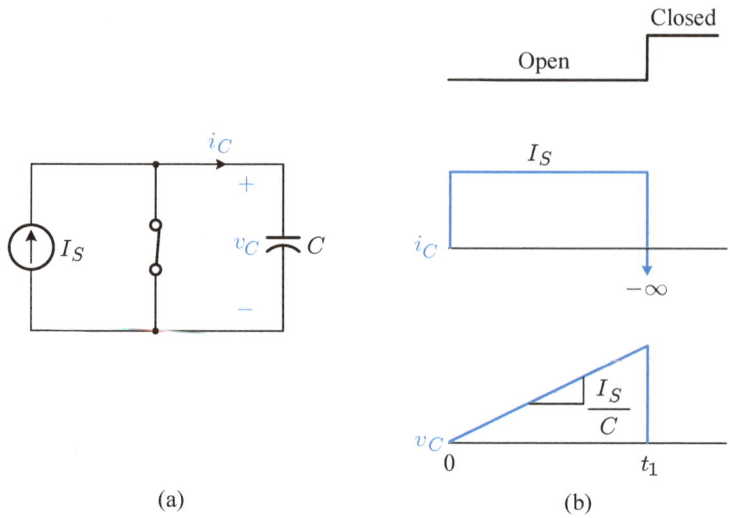


Figure 2.21 Capacitive switching circuit. (a) Circuit diagram. (b) Circuit waveforms.

and the capacitor voltage is fixed at $v_C(t_1) = (I_S/C)t_1$. Now we assume that the inductor current is further stepped down to $i_S = -I_S$ at $t = t_2$. The capacitor voltage then decreases with the equation

$$v_C(t) = -\frac{I_S}{C}t$$

as illustrated in Fig. 2.20(b).

## EXAMPLE 2.5 Capacitive switching circuit 1

This example illustrates the operation of a simple capacitive switching circuit. Figure 2.21(a) shows a capacitive switching circuit where an ideal current source is connected to an uncharged capacitor. The switch in the middle of the circuit is initially closed. When the switch is opened at $t = 0$, the dc current $I_S$ flows into the capacitance $C$, thereby raising the capacitor voltage as $v_C(t) = (I_S/C)t$. When the switch is closed at $t = t_1$, the capacitor voltage is forced to collapse. A sudden collapse of the capacitor voltage in turn induces an infinitely large current according to (2.20): $i_C = C\,\mathrm{d}v_C/\mathrm{d}t = -\infty$.

Certainly, a real semiconductor switch cannot carry such a large current and will be permanently damaged. In terms of the energy conservation principle, the energy accumulated in the capacitor at $t = t_1$, $E_e = 0.5Cv_C^2(t_1)$, is instantaneously released when the switch is closed. This instantaneous energy discharge will destroy the semiconductor switch.

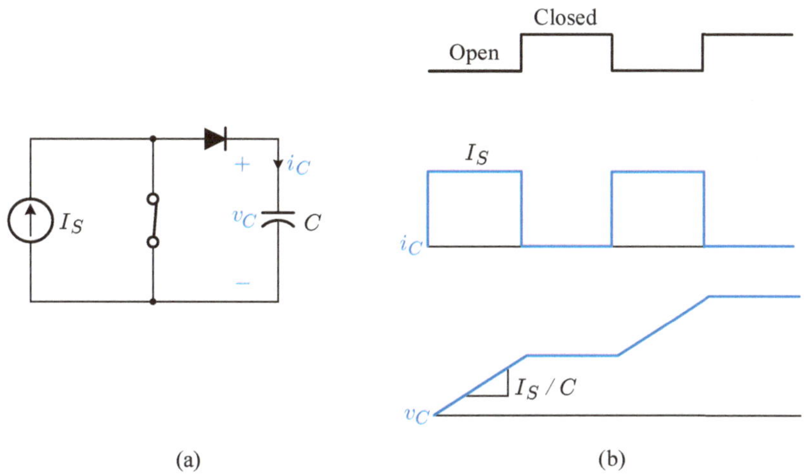


**Figure 2.22** Capacitive switching circuit. (a) Circuit diagram. (b) Circuit waveforms.

## Capacitive switching circuit 2 — EXAMPLE 2.6

The operation of another capacitive switching circuit is discussed here. Figure 2.22(a) shows a capacitive switching circuit in which the current source is connected to a capacitor through a diode. The switch in the middle of the circuit is initially closed. When the switch is opened, the current source turns on the diode and charges the capacitor, thereby raising the capacitor voltage by the equation $v_C(t) = (I_S/C)t$. When the switch is closed, the diode is reverse-biased by the elevated capacitor voltage. The capacitor voltage remains constant because the capacitor is isolated from the current source by the turned-off diode. The diode, which isolates the charged capacitor from the closed switch, is called an *isolation diode*. When the switch is opened in the next switching period, the capacitor voltage starts to rise again. By repeating this process, the circuit produces the waveforms shown in Fig. 2.22(b).

### Capacitor as Voltage Source

It is easy to conceive that a charged capacitor behaves as a transitional voltage source. A simple illustration is given in Fig. 2.23. In the left circuit diagram in Fig. 2.23(a), the switch $S_1$ is opened and $S_2$ is closed at $t = 0$, while $S_3$ was opened. The capacitor is then charged by $I_S$ to develop the voltage across its terminals

$$v_C(t) = \frac{I_S}{C}t$$

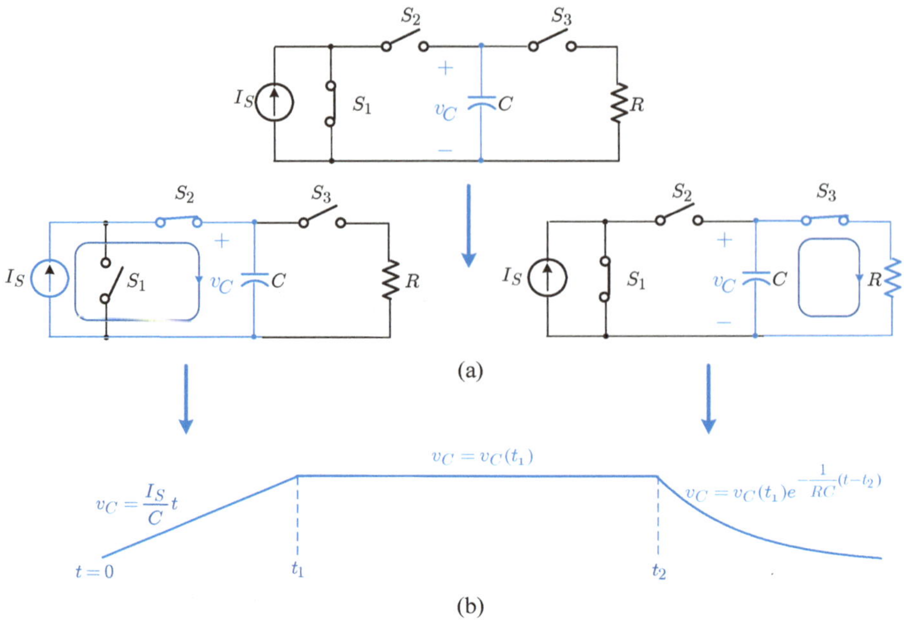


**Figure 2.23** Capacitor as transitional voltage source. (a) Circuit diagram. (b) Capacitor voltage current.

At $t = t_1$, $S_1$ is closed and $S_2$ is opened, while $S_3$ remains open. The isolated capacitor holds a constant voltage

$$V_C = v_C(t_1) = \frac{I_S}{C}t_1$$

and becomes a transitional voltage source. When the charged capacitor is connected to a resistor $R$ at $t = t_2$ by turning-on the switch $S_3$, while the other two switches remain unchanged, the capacitor voltage decays

$$v_C(t) = v_C(t_1)\; e^{-\dfrac{1}{RC}(t-t_2)}$$

until the stored charge is all depleted.

## Insulation Breakdown

As the capacitor voltage continues to rise, it eventually reaches the critical value that breaks the insulation of the dielectric material inside the capacitor. This insulation breakdown causes a permanent damage to the capacitor and application circuit. Accordingly, all capacitors are specified with the highest voltage they can withstand. The circuit shown in Fig. 2.22 is not practical because the capacitor voltage will eventually rise to the critical value that triggers insulation breakdown.

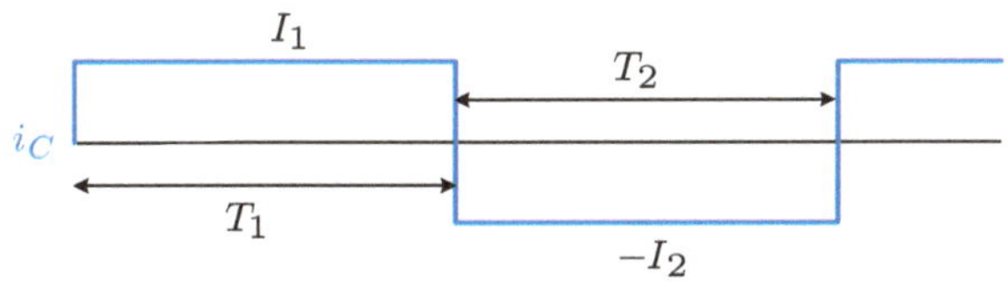


**Figure 2.24** Amp-sec balance condition: $I_1T_1 = I_2T_2$.

## Charge Balance Condition or Amp-Sec Balance Condition

The *charge balance condition* states that the net change in the charge accumulation in a capacitor should be balanced at zero for each switching period

$$\boxed{\Delta q_{inc} = \Delta q_{dec}} \tag{2.23}$$

where $\Delta q_{inc}$ is the charge increase during one switching period and $\Delta q_{dec}$ is the charge decrease in that switching period. Same as the flux linkage balance condition, the charge balance condition is a prerequisite to assume steady-state operation for a capacitive switching circuit.

Because the incremental charge in a capacitor, $\Delta q$, is given by the product of the capacitance $C$ and incremental capacitor voltage, $\Delta v_C$

$$\Delta q = C\Delta v_C \tag{2.24}$$

the violation of the charge balance condition implies that the capacitor voltage will continue to rise in the positive or negative polarity until the capacitor encounters insulation breakdown. Because this is unacceptable to any application circuits, all capacitors should meet the charge balance condition.

The charge balance condition is converted into an alternative form using the definition of the current

$$i_C(t) = \frac{\mathrm{d}q(t)}{\mathrm{d}t} \tag{2.25}$$

where $i_C$ is the capacitor current. Equation (2.25) is written as

$$\mathrm{d}q(t) = i_C(t)\mathrm{d}t \tag{2.26}$$

to describe the charge balance condition as a function of the capacitor current, $i_C$.

When a capacitor is excited by a periodic rectangular current shown in Fig. 2.24, the charge balance condition implies

$$\boxed{I_1T_1 = I_2T_2} \tag{2.27}$$

because the positive current $I_1$ accumulates the electric charge, whereas the negative current $-I_2$ depletes the electric charge. Equation (2.27) is referred to as the *amp-sec balance condition* as it places a constraint on the product of the magnitude and the period of the capacitor current.

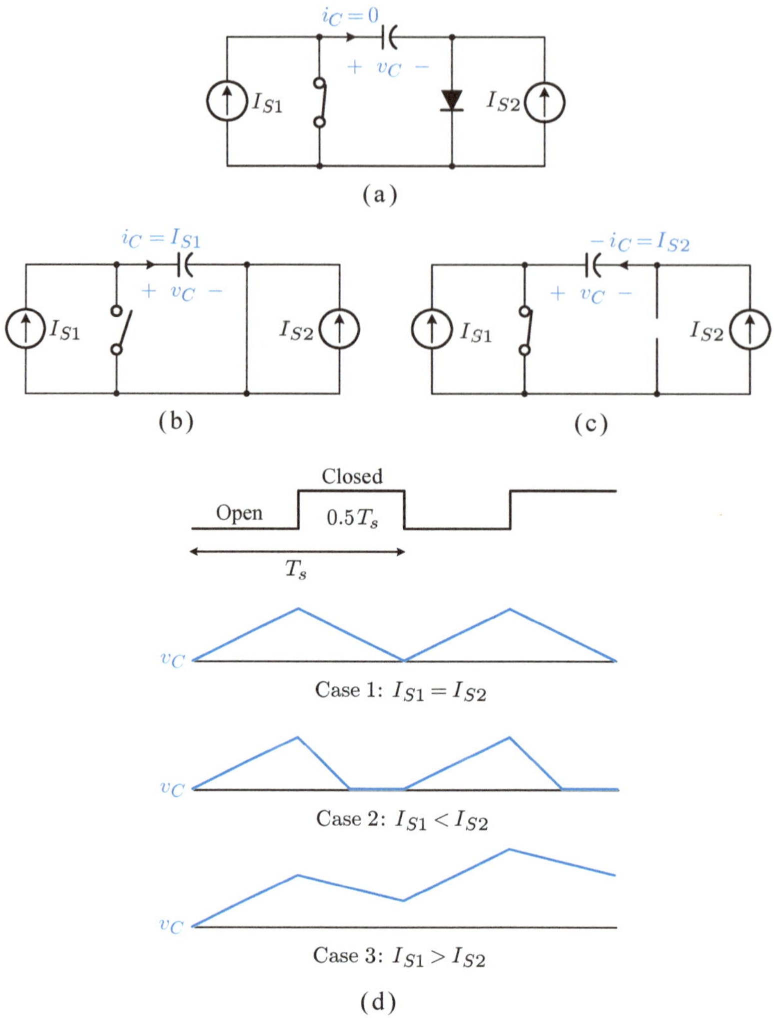


**Figure 2.25** Capacitive switching circuit. (a) Circuit diagram. (b) Current flow with switch open. (c) Current flow with switch closed. (d) Capacitor voltage waveforms.

As an extension of the charge balance condition, the average value of the capacitor current in general is considered zero. This generalized charge balance condition could simplify the analysis of practical capacitive switching circuits, under the assumption that the circuits are all configured in such a way that all capacitors could meet the amp-sec balance condition.

## Capacitive switching circuit 3 EXAMPLE 2.7

This example illustrates the operation of another capacitive switching circuit. The circuit configuration and waveforms of the switching circuit are shown in Fig. 2.25. Figure 2.25(a) shows the condition of the circuit before operation. The circuit consists of two current sources, active switch, diode, and capacitor. Initially, the capacitor is uncharged and the active switch is closed. Under this situation, the current $I_{S1}$ flows through the active switch while $I_{S2}$ circulates via the diode. For this case, the capacitor current $i_C$ is zero, as shown in Fig. 2.25(a), and the circuit remains stationary.

The circuit begins its operation by first opening the active switch and later closing it, as shown in Figs. 2.25(b) and 2.25(c), and repeating the aforementioned operations periodically. When the active switch is first opened in Fig. 2.25(b), $I_{S1}$ flows into the capacitor while $I_{S2}$ still runs through the diode. Accordingly, the capacitor voltage increases linearly: $v_C(t) = (I_{S1}/C)t$.

Now, the active switch is closed as shown in Fig. 2.25(c). The current source $I_{S1}$ diverts its flow from the capacitor to the shorted active switch. Concurrently, the elevated capacitor voltage $v_C$ is applied to the diode in the negative polarity. The diode is then turned-off and $I_{S2}$ runs into the capacitor in the *reverse* direction. The reverse current $I_{S2}$ brings down the capacitor voltage linearly: $v_C(t) = (-I_{S2}/C)t$. The preceding two operations are repeated with an equal time length for each operational period, as shown in Fig. 2.25(d). The following three cases are considered under this operational scenario:

- Case 1: $I_{S1} = I_{S2}$
- Case 2: $I_{S1} < I_{S2}$
- Case 3: $I_{S1} > I_{S2}$

The capacitor voltage waveforms $v_C$ for these three cases are shown in Fig. 2.25(d). For Case 1, the capacitor voltage $v_C$ becomes a symmetric triangular waveform with the identical rising and falling slopes. For Case 2, the falling slope is steeper than the rising slope. Thus, $v_C$ reduces to zero before the onset of the next operational period. When $v_C$ becomes zero, the diode is turned-on. Under this condition, $I_{S2}$ redirects its path to the diode and $v_C$ is thus held zero for the remaining operational period; namely, the circuit now returns to the stationary condition of Fig. 2.25(a). In the ensuing operational periods, the circuit repeats the structure of Fig. 2.25(a), (b), and (c) in sequence for periodic operation.

For Case 3, the rate of the capacitor voltage decreasing is slower than the rate of the capacitor voltage increasing. Accordingly, the capacitor voltage $v_C$ will increase cycle-by-cycle, as illustrated in Fig. 2.25(d). In fact, Case 3 breaks the charge balance condition and the circuit never reaches steady state. It would be informative to sketch the capacitor current waveform $i_C$, in order to comprehend how each of the three cases does or doesn't meet the amp-sec balance condition.

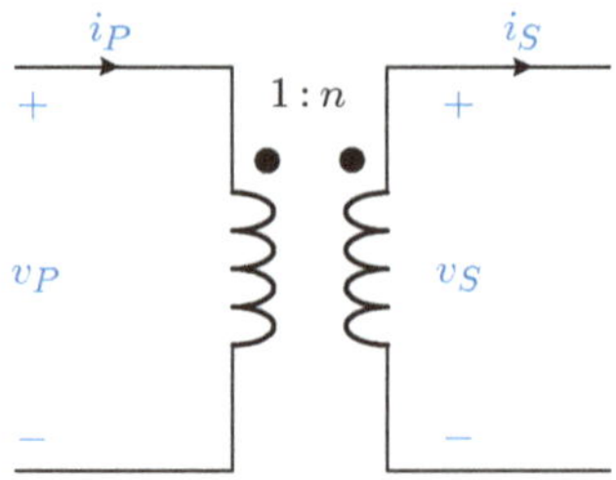


**Figure 2.26** Symbol and polarity/direction of voltage/current waveforms of ideal transformer.

### 2.2.3 Transformers

Transformers are widely used in dc-to-dc power conversion circuits to change the levels of voltage and current waveforms while transferring electrical energy. Although the electro-magnetic process inside the transformer is rather involved, the external circuit properties can readily be described by a circuit model. This section discusses the circuit model for practical transformers and illustrates its use.

First, the concept of the *ideal* transformer is discussed. Then, the electro-magnetic process of practical transformers is described using the equations of the ideal transformer and inductor, leading to the circuit model of practical transformers.

#### Ideal Transformer

The ideal transformer is a conceptual device that satisfies predefined relationships among its circuit variables. Figure 2.26 shows the symbol and polarity/direction of terminal voltage/current waveforms of an ideal transformer. The following terms are defined for the circuit variables and parameter associated with the ideal transformer:

- $v_P(t)$: primary voltage — $i_P(t)$: primary current
- $v_S(t)$: secondary voltage — $i_S(t)$: secondary current
- $n$: turns ratio

The symbol *dot* • in Fig. 2.26 is associated with the definitions for the polarity/direction of the primary and secondary voltage/current waveforms. In other words, the polarity and direction of the circuit variables are determined in reference to the position of •. Details about • will be given in the next section which deals with non-ideal practical transformers. The turns ratio $n$ is the key parameter that defines the relationships between the terminal voltage and current waveforms of the ideal transformer

$$\boxed{v_S(t) = n\, v_P(t)} \tag{2.28}$$

$$\boxed{i_P(t) = n\, i_S(t)} \tag{2.29}$$

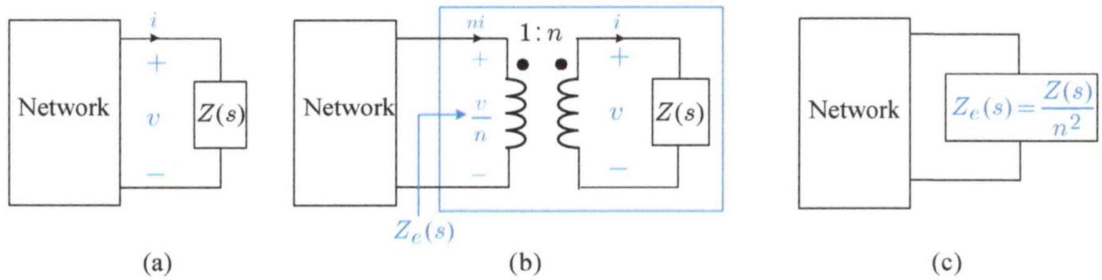


**Figure 2.27** Ideal transformer and impedance scaling property. (a) Circuit element $Z(s)$. (b) Impedance element preceded by ideal transformer. (c) Scaled impedance element $Z_e(s) = Z(s)/n^2$.

The ideal transformer is an imaginary device whose terminal circuit variables are governed by (2.28) and (2.29). The relations (2.28) and (2.29) are always effective regardless of the nature or magnitude of the voltage and current variables. For example, if a dc value $v_P = V_1$ is applied as the primary voltage, the secondary voltage is also a dc, given by $v_S = nV_1$, for any operational circumstances.

## Impedance Scaling Property

The relations (2.28) and (2.29) carry fundamental properties of the ideal transformer. One important property is shown in Fig. 2.27 that describes the *impedance scaling property*. In Fig. 2.27(a), a certain network is connected to a circuit element that presents the impedance $Z(s) = v/i$. Now, a $1 : n$ ideal transformer is inserted between the network and the impedance element $Z(s) = v/i$, as shown in Fig. 2.27(b). We consider the connection of the ideal transformer and the impedance element $Z(s)$ as a new composite element, as highlighted with a box in Fig. 2.27(b). The expressions of the circuit variables around the 1:n ideal transformer are determined as shown in Fig. 2.27(b) based on (2.28) and (2.29). The composite element in the box now represents an impedance, $Z_e(s)$, which is given by

$$Z_e(s) = \frac{v/n}{ni} = \frac{1}{n^2}Z(s) \tag{2.30}$$

where $n$ is the turns ratio of the ideal transformer and $Z(s)$ is the impedance of the standalone circuit element. The combination of the ideal transformer and the circuit element in Fig. 2.27(b) is redrawn as a single impedance element of $Z_e(s) = Z(s)/n^2$ in Fig. 2.27(c). The relation depicted in Fig. 2.27 is referred to as the impedance scaling property.

The relation in (2.30) can be used to infer other important properties of the ideal transformer. Figure 2.28 shows such examples. Figure 2.28(a) is the case where the secondary port of the ideal transformer is shorted so $Z(s) = 0$. From (2.30), it can be inferred that *shorting* the secondary port of the ideal transformer is the same as *shorting* the primary port, because the scaled impedance $Z_e(s)$ becomes null, $Z_e(s) = 0/n^2 = 0$, for this case. The case with the secondary port opened is

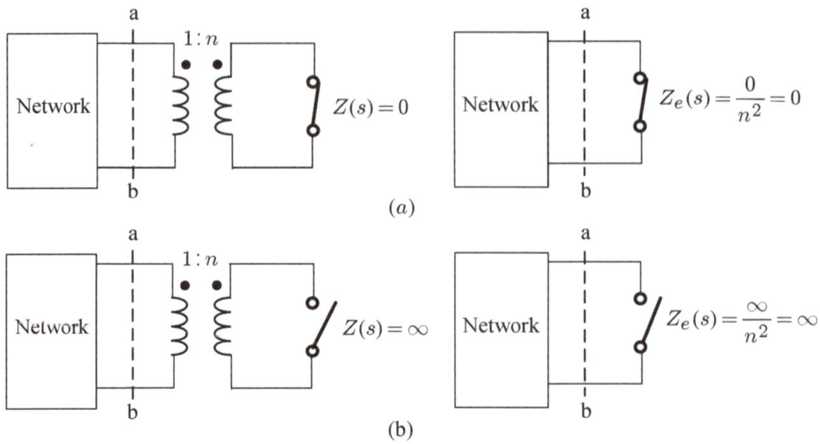


**Figure 2.28** Special cases of (2.30). (a) Case with $Z_e(s) = 0$. (b) Case with $Z_e(s) = \infty$.

shown in Fig. 2.28(b). Here, the a − b port is opened since $Z_e(s) = \infty/n^2 = \infty$. The situations shown in Fig. 2.28 are stated as follows. *When one port of an ideal transformer is shorted or opened, the other port becomes the same.*

Although defined as a conceptual device, the ideal transformer plays a central role in characterizing the circuit properties of practical transformers. In fact, the ideal transformer is conceived as a means of describing the circuit properties of non-ideal practical transformers. Now, we start the discussions about practical transformers.

## Practical Transformers

Practical transformers are fabricated by wrapping two or more copper windings around a magnetic core. Figure 2.29(a) shows a simplified structure of a two-winding transformer built on a toroidal magnetic core.† If one winding is referred to as the primary winding, the other winding becomes the secondary winding. To determine the terminal circuit characteristics of the practical transformer, it is necessary to know *how the primary and secondary windings are wound.* In other words, the polarity/direction of the terminal circuit waveforms depends on the internal pattern of the transformer windings. The following *dot convention* has been used to specify the winding pattern of transformers.

Dot Convention: For both the primary and secondary windings, one end of the winding is marked with a *dot* • as a means of indicating the *polarity* of the transformer windings. The location of • is determined as follows. When the primary current flows into the *dotted* end of the winding and the secondary current

†In practical transformers. the two windings are not separated as shown in Fig. 2.29(a). The two windings are alternatively and evenly wound all around the magnetic core, in order to maximize the magnetic coupling between them.

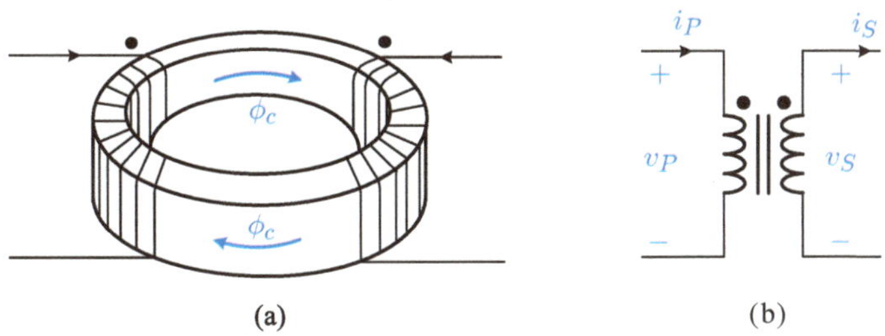


**Figure 2.29** Structure, symbol, and polarity of practical transformer. (a) Structure and dot convention. (b) Symbol and polarity of practical transformer.

also enters the dotted end of the winding, two magnetic fluxes, generated by the two currents running into the respective windings, are directed to be additive, as illustrated in Fig. 2.29(a). Namely, the primary and secondary currents flowing into the winding terminals marked • produce a mutually additive magnetic flux. Readers are urged to examine Fig. 2.29(a) using Ampere's right-hand rule shown in Fig. 2.10(b), in order to confirm the dot convention, pattern of the transformer windings, and direction of the magnetic flux, all illustrated in the figure.

The operation of the transformer is based on the coupling between the primary winding and the secondary winding via the electro-magnetic induction. According to Lenz's law, the electro-magnetic induction occurs in such a way that the induced magnetic flux, produced as the outcome of the induction, opposes the first magnetic flux that initiated the induction process. In conjunction with the dot convention, Lenz's law is reiterated as follows. *When the initiating current enters the dotted end of the primary winding, the induced current should leave from the dotted end of the secondary winding.* The two magnetic fluxes, the initiating flux and the induced flux, then oppose each other, thereby complying with Lenz's law.

Figure 2.29(b) shows the symbol of a practical transformer together with the polarity/direction of its terminal voltage/current waveforms. The directions of the terminal currents are defined by Lenz's law and dot convention.† The polarities of the winding voltages accord with the directions of the winding currents. The polarity/direction of the terminal circuit variables, along with the location of • in Fig. 2.29(b), is consistent with that of the ideal transformer in Fig. 2.26. The vertical bars between the primary and secondary windings signify the presence of a magnetic core in the practical transformer. The vertical bars also symbolically differentiate the practical transformer from the ideal transformer.

The polarity of the transformer can be defined in the other way around. In other words, the position of • on each winding can be swapped, along with the reversal

† Accordingly, the directions of the winding current in Fig. 2.29(b) and Fig. 2.29(c) are different as illustrated.

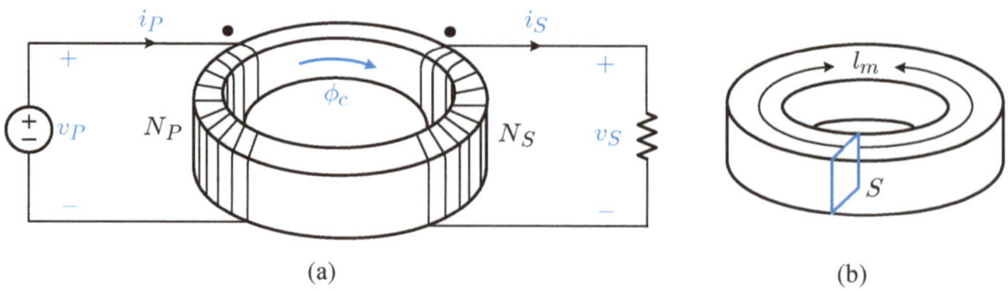


**Figure 2.30** Simple circuit using practical transformer. (a) Circuit configuration. (b) Geometry of toroidal core.

of the polarity/direction of voltage/current waveforms associated with each winding. This, of course, does not alter the circuit properties of the transformer.

## Circuit Model for Practical Transformers

Unlike to the cases of inductors and capacitors, it is not straightforward to directly set up the circuit equations for practical transformers. One alternative approach is to develop a specific circuit model, which describes the electrical behavior of transformers using a combination of standard circuit components. In this section, we develop a circuit model for practical transformers by combining an inductor and ideal transformer. Once the circuit model is obtained, the equations for practical transformers can readily be written using the circuit equations of the inductor and ideal transformer.

A circuit model for practical transformers can be created based on Faraday's law, Ampere's law, and properties of the magnetic core. Figure 2.30 shows the configuration used in developing a circuit model for the practical transformer. To simplify the model derivation, a *perfect* magnetic coupling is presumed. Namely, the common magnetic flux, $\phi_c$ in Fig. 2.30(a), entirely passes through both the primary and secondary windings without any leakage component.

According to Faraday's law, the terminal voltage of each winding is given by

$$v_P(t) = \frac{\mathrm{d}\lambda_P(t)}{\mathrm{d}t} = N_P \frac{\mathrm{d}\phi_c(t)}{\mathrm{d}t} \tag{2.31}$$

$$v_S(t) = \frac{\mathrm{d}\lambda_S(t)}{\mathrm{d}t} = N_S \frac{\mathrm{d}\phi_c(t)}{\mathrm{d}t} \tag{2.32}$$

where $\lambda_P$ and $\lambda_S$ represent the magnetic flux linkages across the primary and secondary windings due to the common magnetic flux $\phi_c$, while $N_P$ and $N_S$ denote the number of the primary and secondary winding turns. From (2.31) and (2.32), the relationship between the terminal voltages is given as

$$\boxed{v_S(t) = \frac{N_S}{N_P} v_P(t)} \tag{2.33}$$

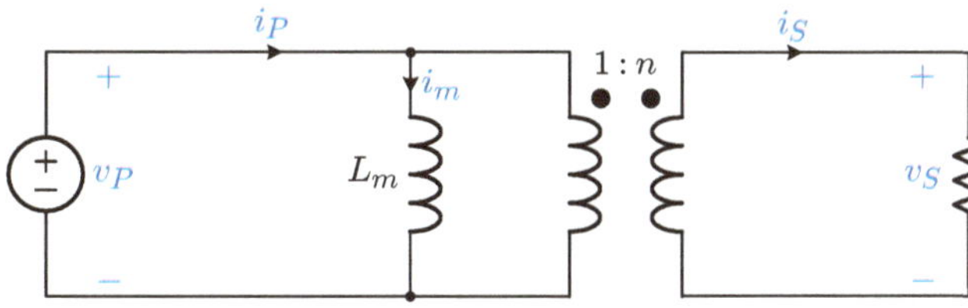


**Figure 2.31** Equivalent circuit representation of Fig. 2.30.

The connection between the terminal currents is formulated by applying Ampere's law along the magnetic path of the transformer

$$N_P\, i_P(t) - N_S\, i_S(t) = H_c l_m \tag{2.34}$$

where $H_c$ is the magnetic field intensity associated with $\phi_c$, and $l_m$ is the length of the magnetic path. The – sign in the second term of the left-hand side of (2.34) indicates that $i_S$ is coming out of the • terminal, while $i_P$ is entering the • terminal. Equation (2.34) is rearranged as

$$\boxed{i_P(t) = \frac{N_S}{N_P} i_S(t) + \frac{H_c l_m}{N_P}} \tag{2.35}$$

The last term in (2.35) is referred to as the magnetizing current

$$i_m(t) \equiv \frac{H_c l_m}{N_P} \tag{2.36}$$

The magnetic flux linkage across the primary winding is given by

$$\lambda_P(t) = N_P \phi_c(t) = N_P \mu_r\, \mu_o H_c S \tag{2.37}$$

with $\mu_r$ the relative permeability of the core material, $\mu_o$ the permeability of free space, and $S$ the cross-sectional area of the core. The inductance, associated with the magnetizing current of (2.36) and magnetic flux linkage of (2.37), is now defined as

$$\boxed{L_m = \frac{\lambda_P(t)}{i_m(t)} = \frac{N_P \mu_r\, \mu_o H_c S}{\dfrac{H_c l_m}{N_P}} = \mu_r\, \mu_o \frac{S}{l_m} N_P^2} \tag{2.38}$$

This inductance is termed as the magnetizing inductance because it is associated with the magnetizing current. Equation (2.38) indicates that the magnetizing inductance in fact corresponds to the inductance that is evaluated at the primary side of the transformer with the secondary winding removed or unaccounted for, as can be confirmed from Example 2.1.

The circuit behavior of Fig. 2.30, given by (2.33) and (2.35), is represented by the circuit model in which the practical transformer is replaced with an ideal transformer plus magnetizing inductance. Figure 2.31 shows such a circuit model, where the turns ratio of the ideal transformer is determined as

$$\boxed{n = \frac{N_S}{N_P}} \tag{2.39}$$

and the magnetizing inductance $L_m$ is given by (2.38). The validity of Fig. 2.31 is confirmed by showing that the circuit model reproduces the original equations of (2.33) and (2.35). Referring to Fig. 2.31 and the circuit equation of the $1:n$ ideal transformer, the relationship between the terminal voltages is given as

$$v_S(t) = nv_P(t) \tag{2.40}$$

and the terminal currents are related as

$$i_P(t) = ni_S(t) + i_m(t) \tag{2.41}$$

The magnetizing current $i_m$ is evaluated as

$$i_m(t) = \frac{\lambda_P(t)}{L_m} = \frac{N_P\,\mu_r\,\mu_o H_c S}{\mu_r\,\mu_o \dfrac{S}{l_m} N_P^2} = \frac{H_c l_m}{N_P} \tag{2.42}$$

Now, the current equation (2.41) becomes

$$i_P(t) = \frac{N_S}{N_P} i_S(t) + \frac{H_c\,l_m}{N_P} \tag{2.43}$$

which is the original current equation of (2.35).

The preceding model derivation is summarized in Fig. 2.32. Figure 2.32(a) shows the symbol and Fig. 2.32(b) depicts the circuit model of a practical transformer. The turns ratio of the ideal transformer is given by

$$\boxed{n = \frac{N_S}{N_P}} \tag{2.44}$$

and the magnetizing inductance is determined as

$$\boxed{L_m = \mu_r\,\mu_o \frac{S}{l_m} N_P^2} \tag{2.45}$$

The magnetizing inductance represents the *non-ideal* property of practical transformers. Thus, it is a good engineering practice to maximize the magnetizing inductance while complying with other design constraints. The larger the magnetizing inductance, the closer the practical transformer to the ideal transformer. The magnetizing inductance becomes boundlessly large when the permeability of the core material is assumed to be infinite. For such a case, the magnetizing inductance disappears and the circuit model reduces to the ideal transformer.

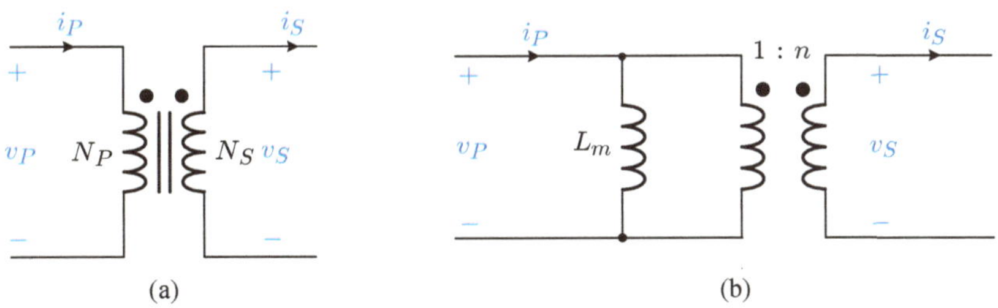


**Figure 2.32** Symbol and circuit model for practical transformer. (a) Symbol. (b) Circuit model.

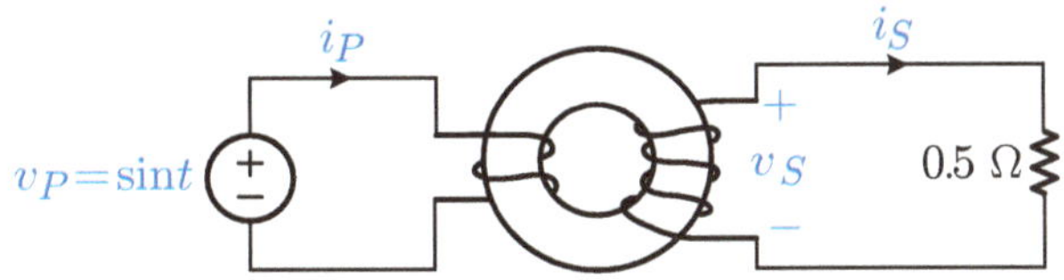


**Figure 2.33** Simple transformer circuit.

Figure 2.32(b) is the simplest circuit model for practical transformers. The model becomes more complex when the imperfect magnetic coupling, parasitic circuit components, and detailed core properties are incorporated. However, the operations of most dc-to-dc power conversion circuits can duly be described using the transformer model in Fig. 2.32(b).

## Simple transformer circuit EXAMPLE 2.8

The operation of a two-winding ideal transformer in a simple circuit is shown in this example. Referring to Fig. 2.33, assume the permeability of the core material is infinite and the transformer thus behaves as an ideal transformer. A sinusoidal voltage $v_P(t) = \sin t$ is applied to the primary winding of the two-winding transformer, whose second winding is terminated with a 0.5 Ω resistor. *From the structure and turns of the windings,* the secondary winding voltage is given by $v_S(t) = -2\sin t$ and the secondary current is $i_S(t) = -2\sin t/0.5 = -4\sin t$. The primary current becomes $i_P(t) = -2 \cdot i_S(t) = (-2)(-)4\sin t = 8\sin t$.

## Simple circuit with practical transformer EXAMPLE 2.9

This example illustrates the use of the circuit model for practical transformers. Figure 2.34(a) shows a simple circuit in which a practical two-winding transformer is connected to a voltage source $v_P$ and current source $i_S$. The waveforms for the voltage source $v_P$ and current source $i_S$ are shown in Fig. 2.34(a). The transformer

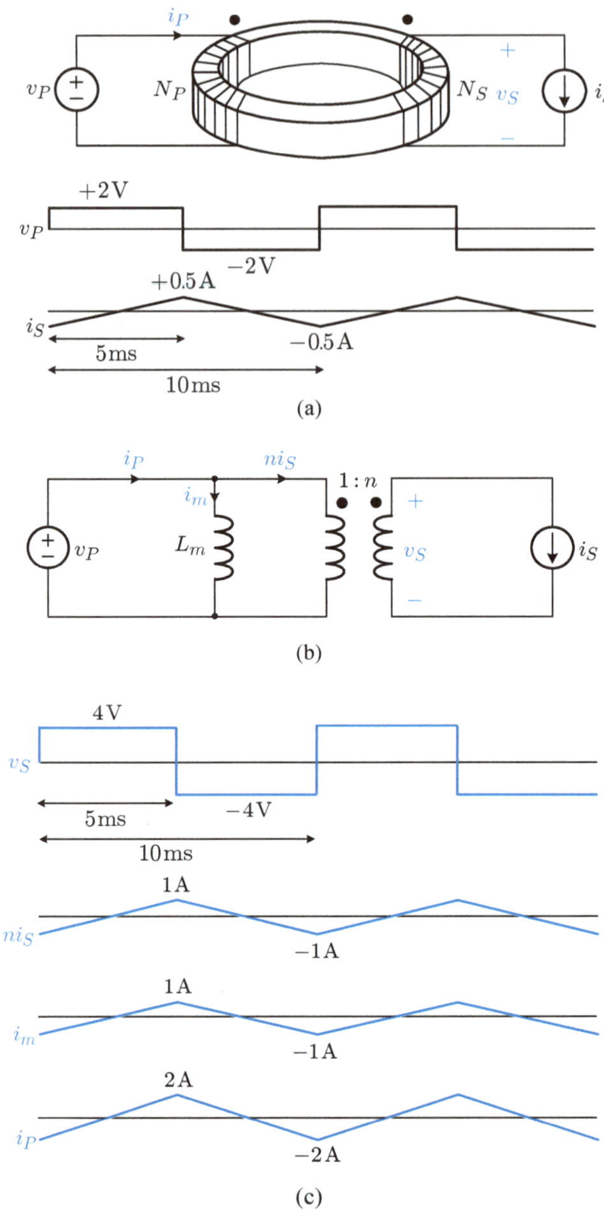


**Figure 2.34** Simple transformer circuit. (a) Circuit configuration. (b) Circuit model. (c) Circuit waveforms.

is built using a toroidal core with $\mu_r = 5000$, $S = 1\ \text{cm}^2$, and $l_m = 4\pi\times10^{-1}$ cm. The turns of the primary winding is $N_P = 10$ and that of the secondary winding is $N_S = 20$. Figure 2.34(b) shows the circuit model of Fig. 2.34(a). The magnetizing inductance is evaluated as

$$\begin{aligned} L_m &= \mu_r\,\mu_o \frac{S}{l_m} N_P^2 \\ &= 5\times10^3\,4\pi\times10^{-7}\,\frac{10^{-4}}{4\pi\times10^{-3}}\,10^2 = 5\ \text{mH} \end{aligned}$$

and the turns ratio is

$$n = \frac{N_S}{N_P} = \frac{20}{10} = 2$$

The circuit variables are evaluated as

$$v_S(t) = n v_P(t) = 2\,v_P(t) \tag{2.46}$$

and

$$i_P(t) = n i_S(t) + i_m(t) = 2\,i_S(t) + i_m(t) \tag{2.47}$$

with

$$i_m(t) = \frac{1}{L_m}\int v_P(t)\,\mathrm{d}t \tag{2.48}$$

The average value of $i_m$ is assumed zero in this example. Figure 2.34(c) shows the steps of constructing $v_S$ and $i_P$, based on the expressions (2.46) through (2.48) along with $v_P$ and $i_S$ waveforms given in Fig. 2.34(a).

## 2.3 SWITCHING CIRCUITS IN PRACTICE

Switching circuits, constructed with the power stage components discussed in this chapter, have been used in many industrial and consumer electronics. In such applications, semiconductor switches and energy storage/transfer elements are appropriately combined to operate as a single practical switching circuit. This section presents a couple of such examples: the solenoid drive circuit and capacitor charging circuit.

### 2.3.1 Solenoid Drive Circuits

An inductor fabricated by winding a copper coil around an iron core is called a solenoid. The solenoid is often used as an actuator in industrial applications. When a solenoid is connected to a voltage source by turning on a semiconductor switch, a linearly-increasing current is established in the solenoid inductance and energy is thus accumulated inside the solenoid. Although some energy is dissipated during the operation as an actuator, most of the accumulated energy will still remain in the solenoid after operation. The solenoid drive circuit must be properly designed so that the remnant energy is safely removed from the solenoid inductance without damaging the semiconductor switch. In this section, different solenoid drive circuits are analyzed focusing on their efficiency. To simplify ensuing discussions, the energy

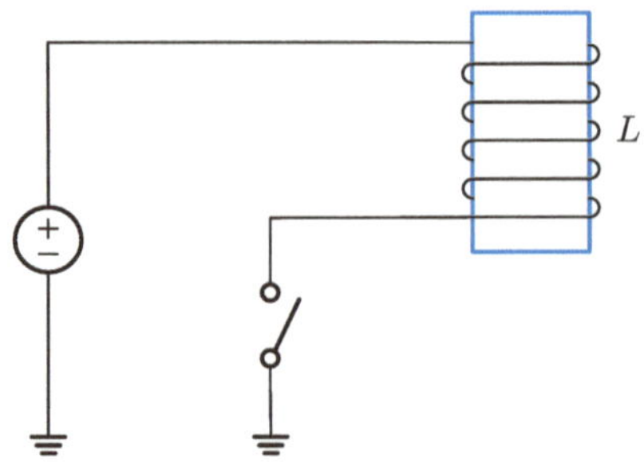


Figure 2.35 Conceptual solenoid drive scheme.

consumed during the actuator operation is assumed negligible and the solenoid is thus represented by a pure inductor.

A conceptual solenoid drive circuit is shown in Fig. 2.35. It becomes immediately apparent that this drive circuit is *not* workable due to the lack of the freewheeling path. When the switch is closed, the inductor current increases linearly, thereby storing energy in the solenoid. When the switch is opened, the inductor current suddenly loses its path and the stored energy is instantaneously released in an uncontrolled manner. This abrupt energy discharge, occurring in the form of a high voltage spike, would destroy the semiconductor switch. This section discusses two different solenoid drive circuits that operate without damaging the semiconductor switch.

## Dissipative Solenoid Drive Circuit

Figure 2.36(a) shows the first workable solenoid drive circuit where a freewheeling path, consisting of a diode and resistor, is employed in parallel with the solenoid. When the switch is closed, the diode is reverse-biased and the freewheeling path does not affect the circuit operation. On the other hand, when the switch is opened, the diode conducts to activate a freewheeling path for the inductor current. This prevents the instantaneous energy discharge and protects the switch. Detailed circuit operations are as follows.

Energy Build-up Period: When the switch is closed, the freewheeling diode is reverse-biased and the circuit is represented by Fig. 2.36(b). The inductor current increases linearly

$$\boxed{i_L(t) = \frac{V_S}{L}t} \tag{2.49}$$

thereby piling up energy in the solenoid inductance.

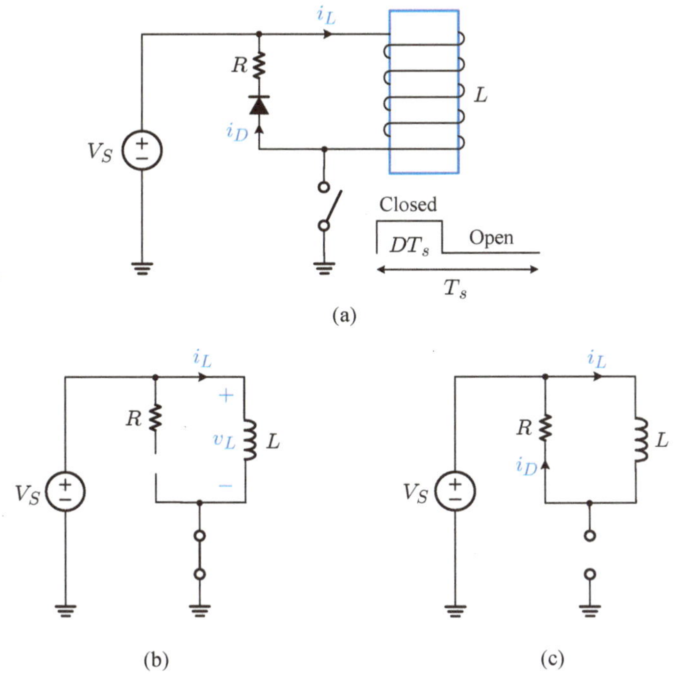


**Figure 2.36** Dissipative solenoid drive circuit. (a) Circuit diagram. (b) Energy build-up period. (c) Energy removal period.

Energy Removal Period: When the switch is opened at $t = DT_s$,† the diode is turned-on and the freewheeling path is created. The equivalent circuit in this period is shown in Fig. 2.36(c). The freewheeling current is given by

$$i_D(t) = i_L(t) = i_L(DT_s)\, e^{-\frac{R}{L}t} \tag{2.50}$$

As the freewheeling current circulates through the diode, resistor, and solenoid, the energy stored in the solenoid inductance is gradually dissipated at the resistor. As the freewheeling current reduces to a negligible value, the stored energy is practically all removed by dissipation. Although the solenoid drive circuit functions properly without damaging the switch, efficiency of the drive circuit will be low because the stored energy is dissipated in the resistor. This solenoid drive circuit is referred to as the dissipative solenoid drive circuit.

† $D$ denotes the ratio of the switch-on period to the operational period, and is called the duty ratio of the switch.

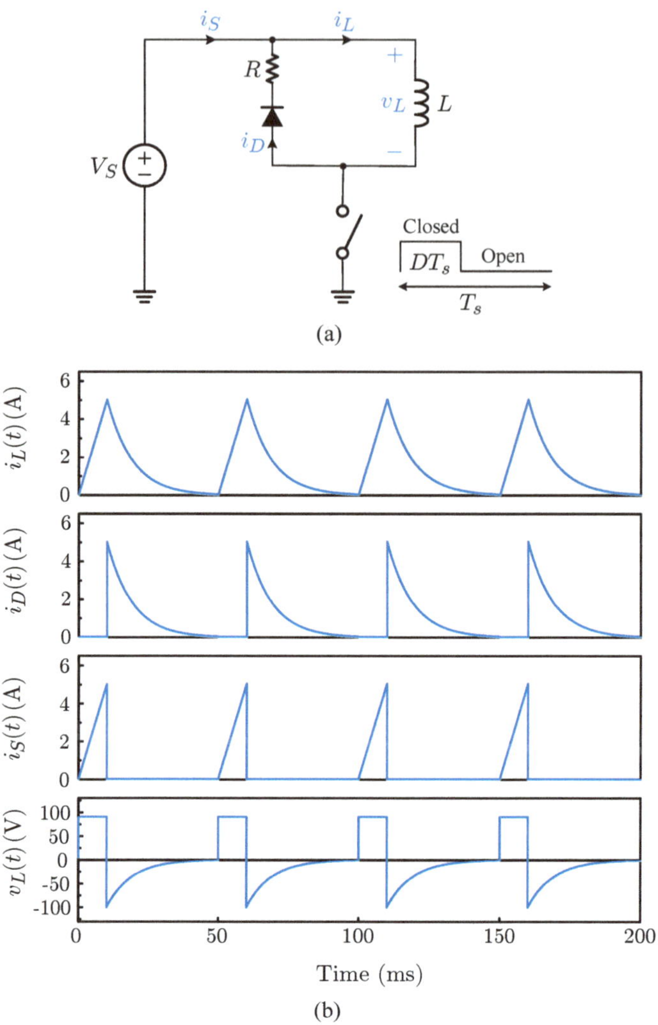


Figure 2.37 Dissipative solenoid drive circuit. (a) Circuit diagram. (b) Circuit waveforms.

## EXAMPLE 2.10 Dissipative solenoid drive circuit

This example illustrates the operation of the dissipative solenoid drive circuit using PSpice®simulations. Figure 2.37(a) is the circuit diagram of a dissipative solenoid drive circuit. Figure 2.37(b) illustrates the major circuit waveforms with $V_S = 90\,\text{V}$, $L = 180\,\text{mH}$, $R = 20\,\Omega$, $T_s = 50\,\text{ms}$, and $D = 0.2$. When the switch is closed, the inductor current grows linearly and reaches the peak value at $t = DT_s = 0.2 \cdot 50 \times$

$10^{-3}$

$$i_{L\,peak} = \frac{V_S}{L} DT_s = \frac{90}{180\times10^{-3}} 0.2 \cdot 50\times10^{-3} = 5\text{ A}$$

The total energy

$$E_m = \frac{1}{2} L i^2_{Lpeak} = \frac{1}{2} 180\times10^{-3}\, 5^2 = 2.25\text{ J}$$

is stored in the inductor.

When the switch is opened, the energy removal period starts and the inductor current $i_L$ freewheels through the loop created by the diode, resistor, and solenoid. The diode current $i_D$ decays exponentially while dissipating energy at the resistor. When the stored energy is all dissipated in the resistor, $i_D$ reduces to zero. The operation of the circuit is repeated four times in Fig. 2.37. The waveforms of the source current $i_S$ and inductor voltage $v_L$ show details of the circuit operation.

## PSIM® Dissipative solenoid drive circuit — EXAMPLE 2.11

In this example, we employ PSIM®software to perform time-domain simulations of the dissipative solenoid drive circuit studied in Example 2.10. The model for and outcomes of the PSIM simulation are displayed in Table 2.1. The format of Table 2.1 will be used throughout this book for all forthcoming PSIM simulations.

The PSIM model is shown at the top of Table 2.1. The PSIM model closely resembles the circuit diagram of the drive circuit. As one customized functional circuit module, the PWM block is included in the PSIM model. The operation of the PWM block is illustrated at the bottom of the table. The PSIM model is self-explanatory and easy to figure out. The PSIM model is stored in the **PSIM Model Library** under the file name **Solenoid_Drive_Circuit_01**, as indicated in the first row of the table. This referencing pattern will be adopted to all PSIM simulations.

The simulation results are shown in the middle of the table. The operational conditions of the drive circuit are identical to the case of Example 2.10. The circuit waveforms are an exact copy of Fig. 2.37.

The operation of the PWM block are shown at the bottom of the table. The PWM block is configured with an RS flip-flop, op-amp comparator, and internal pulse signal, $v_{pulse}$. The leading edge of $v_{pulse}$ triggers the turn-on of MOSFET. The frequency, magnitude, and duty cycle of $v_{pulse}$ are appropriately programmed to produce a narrow pulse train for the periodic MOSFET turning-on.

The PWM block requires two external signals, the ramp signal $v_{ramp}$ and control signal $v_{con}$, to produce the pulsewidth modulated MOSFET gate drive signal, $v_q$ in Table 2.1. The ramp signal $v_{ramp}$ is constructed using the programmable voltage source with a predefined magnitude, frequency, slopes, and dc offset voltage.† The ramp signal is synchronized with the internal pulse signal $v_{pulse}$ at the same frequency for the constant frequency PWM scheme.

The control signal $v_{con}$ is the input signal that determines the duty ratio of $v_q$. In this example, $v_{con}$ is fixed at a constant value. Thus, the duty ratio of $v_q$ is also fixed

†The internal pulse signal $v_{pulse}$ and the ramp signal $v_{ramp}$ should have the same frequency for a synchronized constant-frequency operation.

**Table 2.1** Dissipative Solenoid Drive Circuit

PSIM model: Solenoid_Drive_Circuit_01

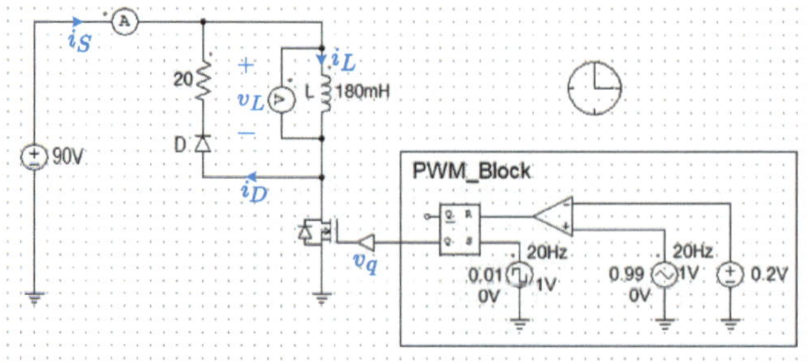


Simulation outcomes

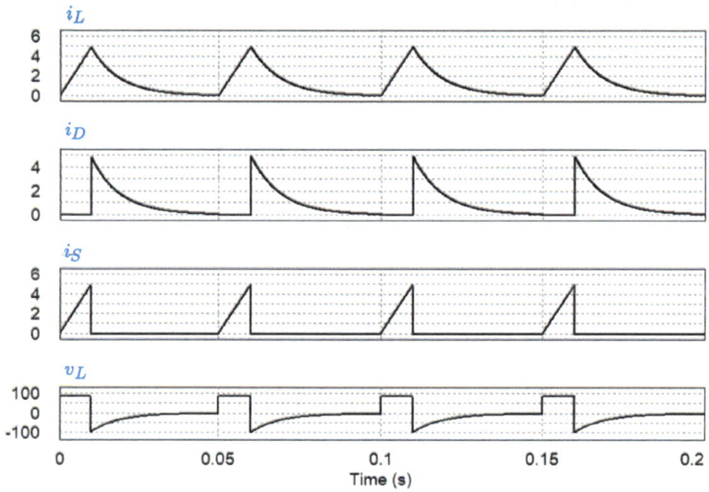


PWM block

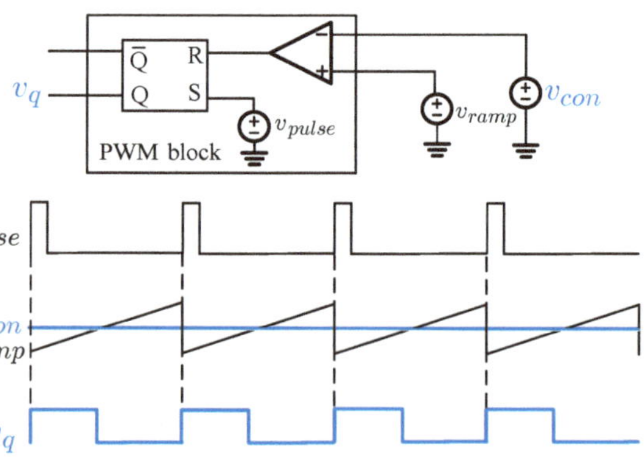

and the PWM block operates in an open-loop fashion. The PWM block generates the gate drive signal with the prefixed duty ratio at the given frequency.

It is apparent that the PWM block will operate in the closed-loop fashion, provided that $v_{con}$ is fed from the output of the application circuit. Thus, the PWM block will be used in all forthcoming open-loop and closed-loop PSIM simulations.

## PSIM® Alternative solenoid drive circuit EXAMPLE 2.12

This example introduces another dissipative solenoid drive circuit. Table 2.2 shows the PSIM model and operation of the alternative drive circuit. In this circuit, the 20 Ω resistor in Example 2.11 is replaced with a Zener diode $D_z$ whose breakdown voltage is $V_z = 90$ V. The other operational conditions remain the same.

When the MOSFET switch is turned on, the drive circuit follows the energy build-up operation of Examples 2.10 and 2.11. When the switch is turned off, the elevated inductor current turns on the regular diode $D$ and initiates the breakdown of Zener diode $D_z$. The inductor current now freewheels through the diode, Zener diode, and solenoid inductor. This operational period constitutes the energy removal period, in which the energy in the solenoid inductance is dissipated at the Zener diode.

During the energy removal period, Zener diode behaves as a constant voltage source of $V_z = 90$V. The breakdown voltage $V_z = 90$ V appears across the solenoid inductor in the negative polarity and the inductor current thus linearly decreases from the peak value $i_{L\,peak} = i_L(DT_s)$ with the slope $-V_z/L$.

When the declining inductor current reduces to zero, the regular diode $D$ turns off and remains off until the onset of the next energy removal period. The current and voltage waveforms in Table 2.2 confirm the operation of the circuit. The waveforms also validate the volt-sec balance condition on the solenoid inductance and a controlled removal of the stored energy.

### Non-Dissipative Solenoid Drive Circuit

Efficiency of the solenoid drive circuit will be improved if the energy stored in the solenoid inductance is recovered by the drive circuit, rather than wasted in the circuit. Figure 2.38(a) shows a non-dissipative solenoid drive circuit. The circuit employs a pair of synchronized switches and two diodes. The synchronized switches are turned-on and turned-off simultaneously, while the two diodes are used to create a freewheeling path for the inductor current. By turning on the two synchronized switches, a linearly-increasing inductor current is produced. When the synchronized switches are turned-off, the inductor current turns on the two diodes and flows through them. Now, the energy stored in the solenoid inductance is transferred back to the voltage source. Details of the drive circuit are as follows.

Energy Build-up Period: When the synchronized switches are closed, the two diodes are individually reverse-biased by $V_S$. The equivalent circuit in this period is shown in Fig. 2.38(b). During this period, the solenoid inductor current increases

**Table 2.2** Alternative Solenoid Drive Circuit

PSIM model: Solenoid_Drive_Circuit_02

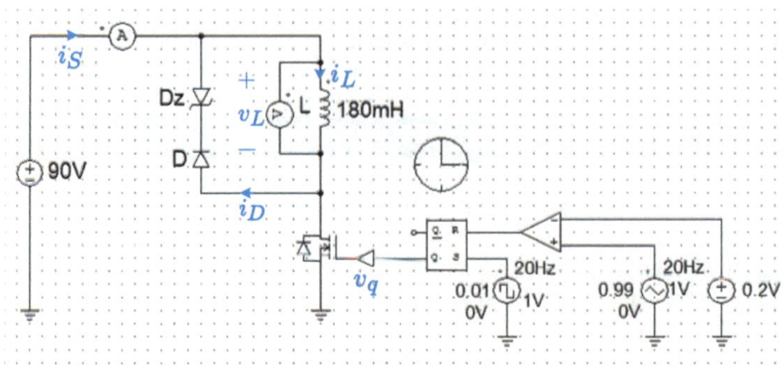


Simulation outcomes

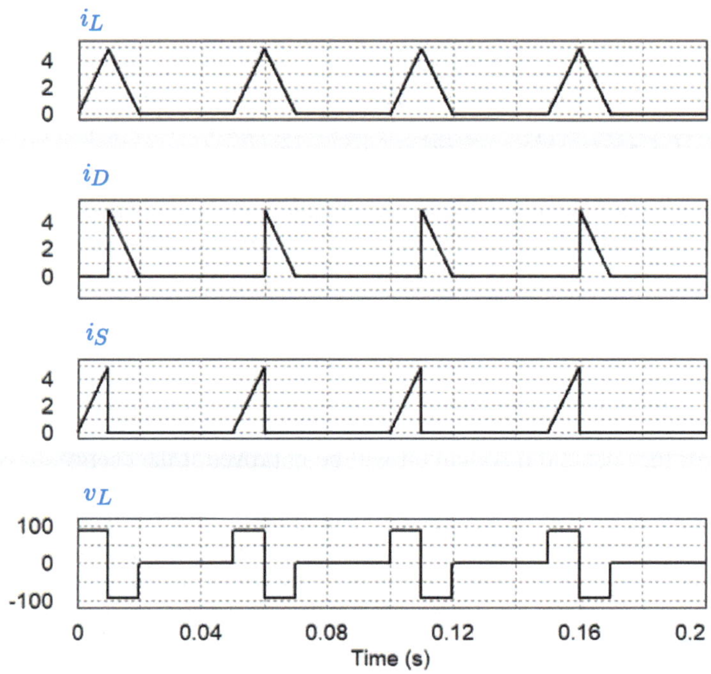


linearly

$$\boxed{i_L(t) = \frac{V_S}{L}t} \tag{2.51}$$

and energy is delivered from the source to solenoid.

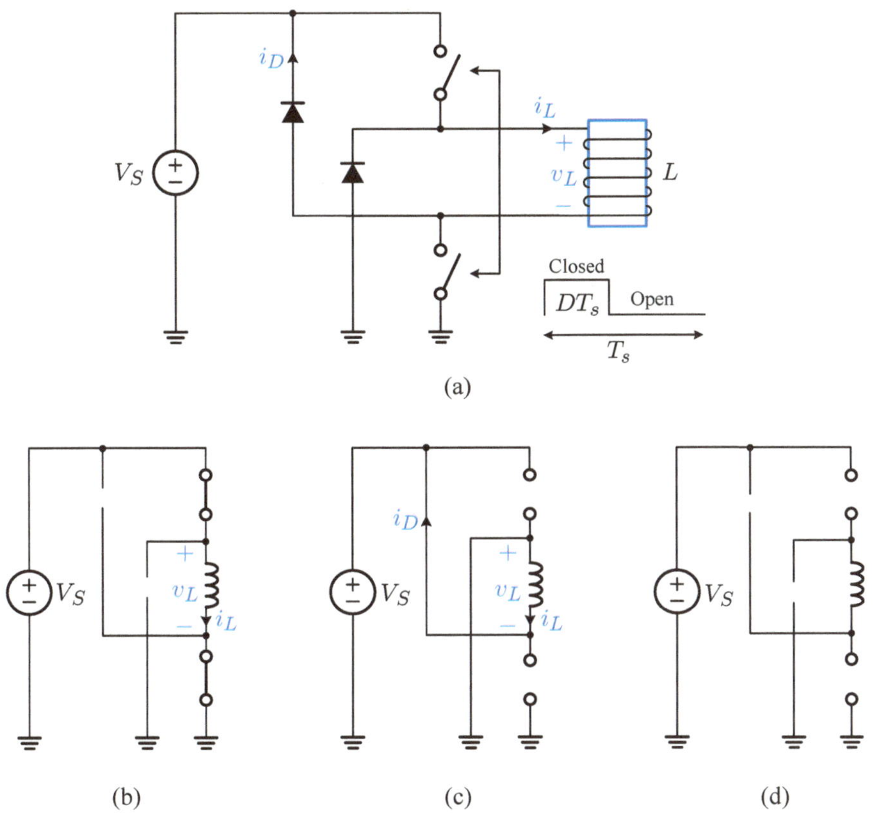


**Figure 2.38** Non-dissipative solenoid drive circuit. (a) Circuit diagram. (b) Energy build-up period. (c) Energy recovery period. (d) Recess period.

Energy Recovery Period: When the switches are opened at $t = DT_s$, the two diodes conduct together and a freewheeling path is created as shown in Fig. 2.38(c). The solenoid voltage now becomes $-V_S$ and the freewheeling current ramps down from the peak with a slope $-V_S/L$. The declining inductor current becomes the diode current, $i_L(t) = i_D(t)$. During this period, the freewheeling current runs into the voltage source and the energy stored in the solenoid inductance is transferred back to the voltage source. The stored energy is thus recovered by the drive circuit, rather than dissipated in the circuit.

Recess Period: When the freewheeling diode current is reduced to zero, the two diodes are turned-off and remain off thereafter. Because the solenoid is isolated from the source, all the circuit variables are zero, as shown in Fig. 2.38(d).

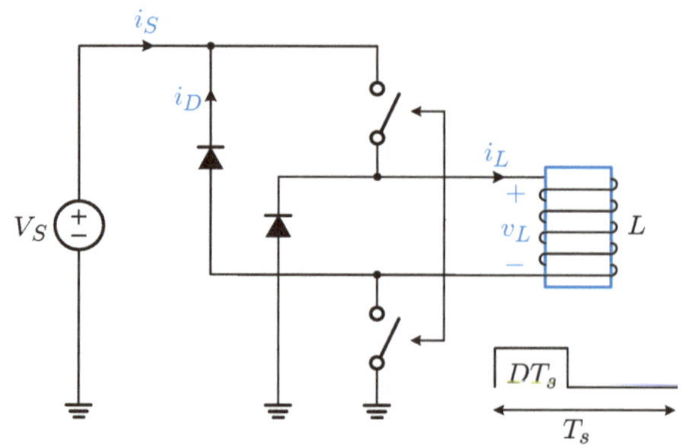


(a)

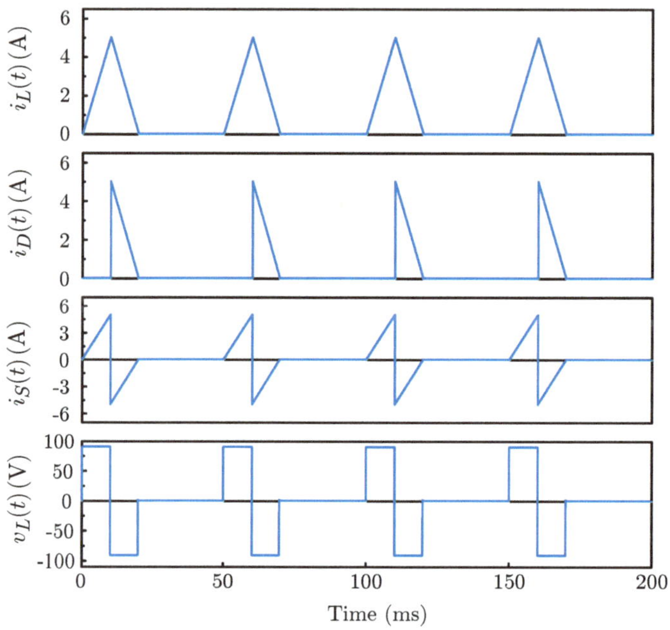


(b)

**Figure 2.39** Non-dissipative solenoid drive circuit. (a) Circuit diagram. (b) Circuit waveforms.

## EXAMPLE 2.13 Non-dissipative solenoid drive circuit

This example illustrates the operation of the non-dissipative solenoid drive circuit. Figure 2.39 shows the circuit diagram and simulated waveforms of a non-dissipative solenoid drive circuit. The operational conditions and circuit parameters are $V_S = 90\,\text{V}$, $L = 180\,\text{mH}$, $T_s = 50\,\text{ms}$, and $D = 0.2$.

Referring to Fig. 2.39(b), the circuit operation is explained as follows. During the energy build-up period, the circuit waveforms are the same as the previous dissipative drive circuit. However, during the energy recovery period, the inductor current $i_L$ flows back to the source through the freewheeling diodes, thereby returning the accumulated inductor energy to the source. The source current $i_S$ becomes negative in this period: $i_S = -i_D = -i_L$. The negative current signifies the retrieval of the energy transferred to the solenoid inductor during the energy build-up period.

## PSIM® Non-dissipative solenoid drive circuit — EXAMPLE 2.14

Table 2.3 displays the PSIM model and simulation results of the non-dissipative solenoid drive circuit, studied in Example 2.13. The PSIM model combines the solenoid drive circuit and the PWM block discussed in the previous examples. The output of the PWM block is commonly fed to the gates of the two synchronized MOSFET switches. The simulation outcomes copy those of Example 2.13.

## PSIM® Composite solenoid drive circuit — EXAMPLE 2.15

This example introduces a composite solenoid drive circuit which is constructed as follows. The non-dissipative solenoid drive circuit in Example 2.14 is integrated with the dissipative solenoid drive circuit covered in Example 2.12. The resulting circuit is simplified to yield a composite solenoid drive circuit. The construction of the composite drive circuit is illustrated in Fig. 2.40.

The non-dissipative and dissipative solenoid drive circuits are first cascaded as shown in Fig. 2.40(a). We assume that the three MOSFET switches operate in synchronization. Then, the redundant switch $Q_3$ is removed and the drive circuit is rearranged into the composite drive circuit in Fig. 2.40(b). Referring to Fig. 2.40(b), the operation of the solenoid drive circuit is described as follows.

**1)** When the switches $Q_1$ and $Q_2$ are closed, the three diodes $D_1$, $D_2$, and $D_3$ are all reverse-biased. The voltage source $V_S$ is then connected to the series combination of the two inductors $L_1$ and $L_2$. Accordingly, a linearly rising current $i_L$ flows through the loop formed by $V_S$, $Q_1$, $L_1$, $Q_2$, and $L_2$

$$i_L(t) = \frac{V_S}{L_1 + L_2} t$$

At $t = DT_s$, the inductor current attains the peak value

$$i_{L\,peak} = \frac{V_S}{L_1 + L_2} DT_s$$

and accumulates the energy $E_{m1} = 0.5 L_1 i^2_{L\,peak}$ in the upper solenoid and $E_{m2} = 0.5 L_2 i^2_{L\,peak}$ in the lower solenoid.

**2)** When the two MOSFETs are turned off at $t = DT_s$, the four regular/Zener diodes, $D_1$-$D_2$ in the upstream non-dissipative circuit and $D_3$-$D_z$ in the downstream dissipative circuit, are all turned on. The inductor current $i_L$, which was passing through both the $L_1$ and $L_2$, now behaves as two separated cur-

**Table 2.3** Non-Dissipative Solenoid Drive Circuit

| PSIM model: Solenoid_Drive_Circuit_03 |
| --- |
| 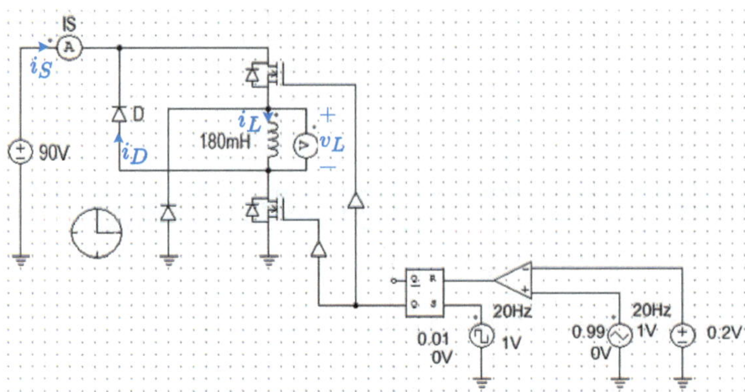  |
| Simulation outcomes |
| 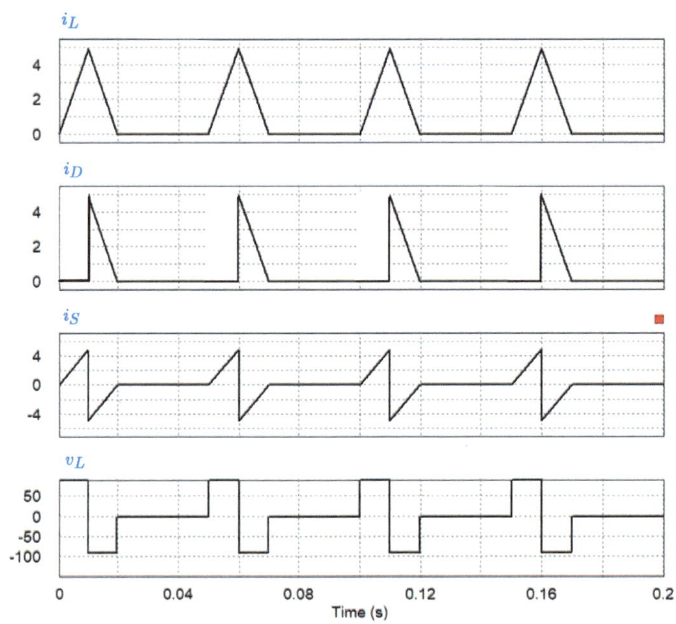  |

rent streams: one stream circulating along Path A in Fig. 2.40(b) and the other freewheeling through Path B. In this period, the input voltage $V_S$ is applied to $L_1$ in the negative polarity and the breakdown voltage $V_z$ of the Zener diode $D_z$ is connected across $L_2$ also in the negative polarity. Thus, $i_{L1}$ linearly decreases with the slope $-V_S/L_1$ and $i_{L2}$ does the same with the slope $-V_z/L_2$.

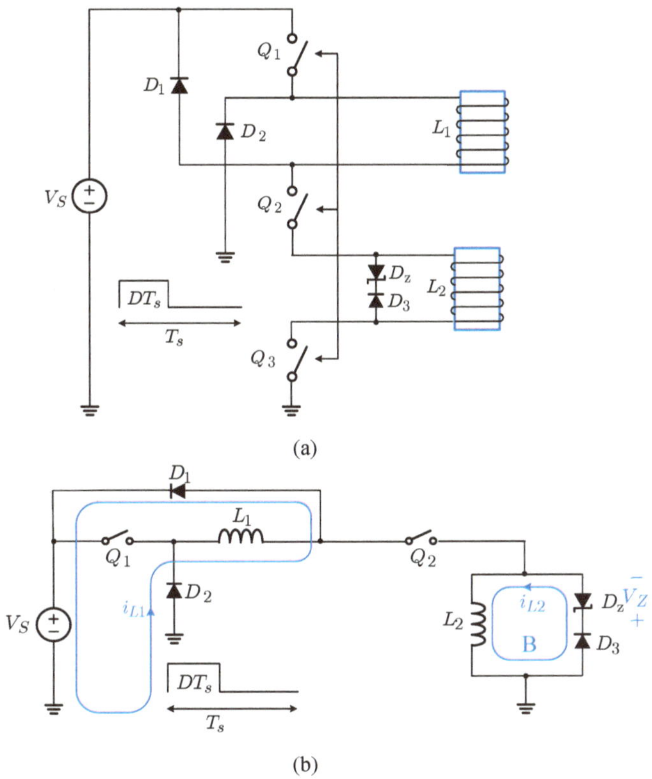


**Figure 2.40** Composite solenoid drive circuit. (a) Cascade connection of two solenoid drive circuits. (b) Composite solenoid drive circuit.

The energy stored in $L_1$ is recovered by the input voltage source, yet the energy in $L_2$ is dissipated at the Zener diode in a controlled manner.

3) As the two current streams in the two solenoids reduce to zero, the respective freewheeling diodes turn off and stay off for the remaining part of the operational period.

The PSIM model and simulation of the composite drive circuit are shown in Table 2.4. The operational conditions of Example 2.14 are adopted for the non-dissipative circuit, while those of Example 2.12 are used for the dissipative circuit. Detailed circuit analyses and interpretations of the simulation results are left as an exercise for readers.

### 2.3.2 Capacitor Charging Circuit

As another example of industrial applications, Fig. 2.41(a) shows a switching circuit consisting of a practical transformer, active switch, diode, and capacitor. Here, the

**Table 2.4** Composite Solenoid Drive Circuit

PSIM model: Solenoid_Drive_Circuit_04

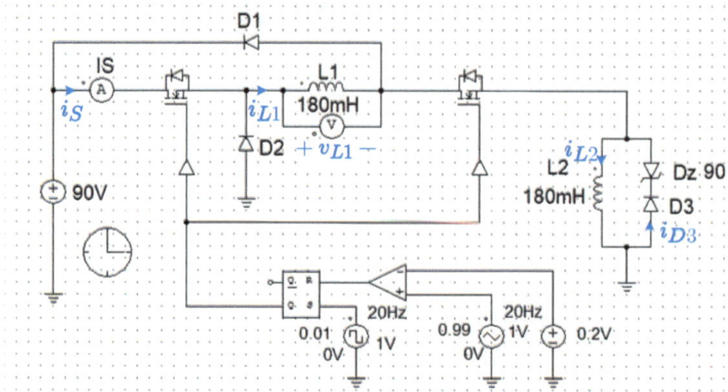


Simulation outcomes

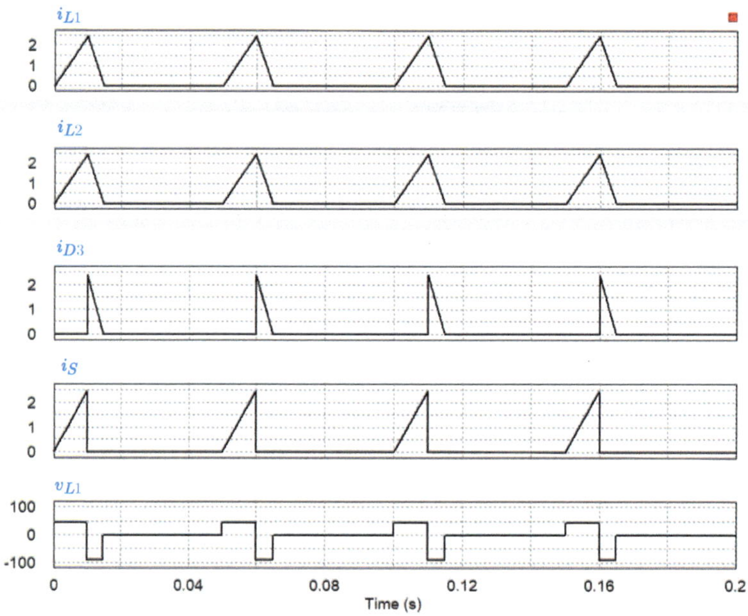


transformer windings are arranged such that • appears at the opposite terminals of the primary and secondary windings. This circuit is capable of charging the capacitor to a desired voltage level. A simple application will be a high voltage generator, which has been used in flash lamp driving circuitry for hand-held cameras. Figure 2.41(b) shows the circuit model of Fig. 2.41(a), where the practical transformer is replaced

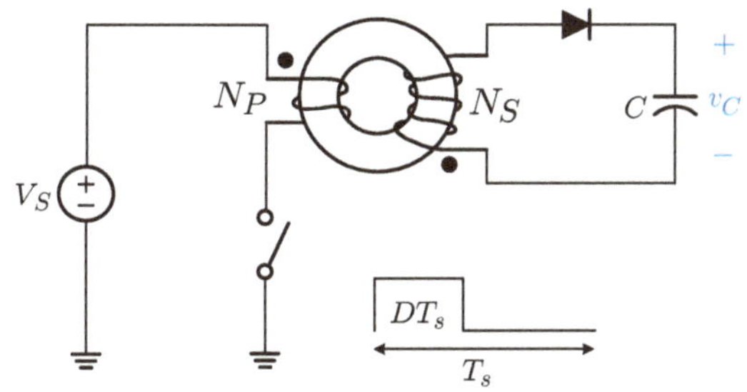


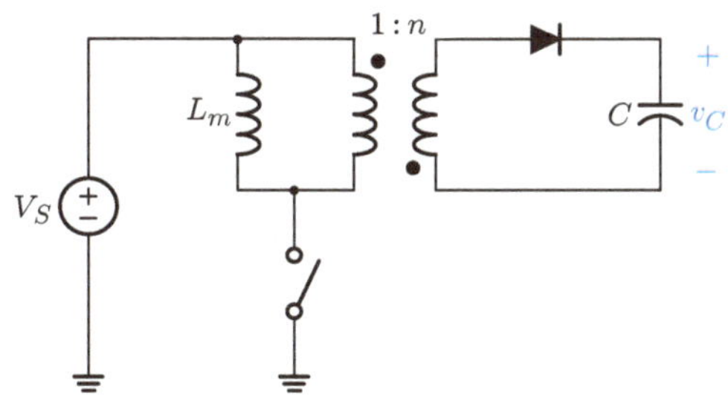


Figure 2.41 Capacitor charging circuit and its circuit model. (a) Capacitor charging circuit. (b) Circuit model: $L_m$ is the magnetizing inductance and $n = N_S/N_P$ is the turns ratio of transformer.

with a $1 : n$ ideal transformer and magnetizing inductance $L_m$. The operation of the capacitor charging circuit is explained using the circuit model in Fig. 2.41(b).

Energy Build-up Period: When the switch is closed, the energy build-up period starts. The circuit model for this period is shown in Fig. 2.42(a). The source voltage $V_S$ is applied across the magnetizing inductance and primary winding of the ideal transformer. Due to the polarity of the transformer windings, the diode is turned-off by the reverse voltage of $nV_S + v_C$, where $v_C$ is the voltage across the capacitor. A linearly-increasing current flows through the magnetizing inductance and active switch

$$i_m(t) = i_Q(t) = \frac{V_S}{L_m}t \tag{2.52}$$

In this period, the primary winding does not carry any current because the secondary winding is open-circuited; for this case, the primary and secondary winding currents are both zero to meet the current equation of the $1 : n$ ideal transformer. With the increasing magnetizing current, energy is piled up in the magnetizing inductance of the transformer.

Energy Transfer Period: When the switch is opened at $t = DT_s$, the energy transfer period begins. As shown in Fig. 2.42(b), the magnetizing current now diverts

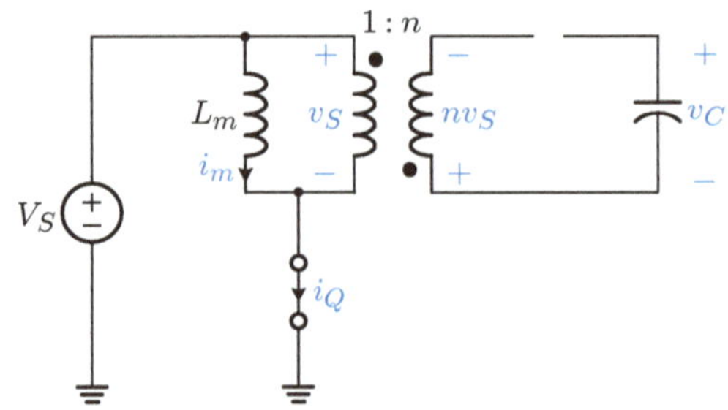


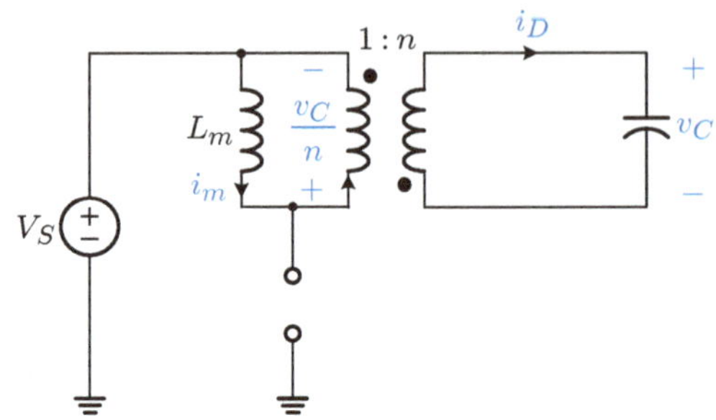


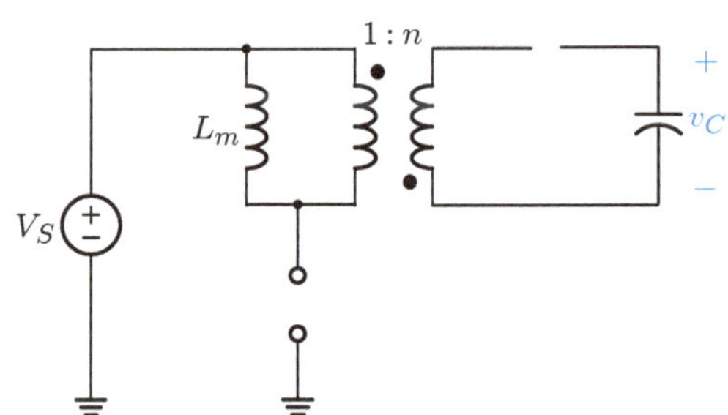


Figure 2.42 Operation of capacitor charging circuit. (a) Energy build-up period. (b) Energy transfer period. (c) Recess period.

from the active switch to the primary winding of the ideal transformer, and circulates through the loop formed by the magnetizing inductance and the primary winding. This current in turn forces the diode to conduct, thus complying with the current equation of the $1 : n$ ideal transformer. Under this situation, the capacitor voltage is reflected to the primary winding and applied across the magnetizing inductance in the *negative* polarity. The magnetizing current now declines from the peak value of $i_m(DT_s) = V_S/L_m\, DT_s$ with the equation

$$i_m(t) = i_m(DT_s) - \frac{1}{L_m}\int \frac{1}{n} v_C(t)\, \mathrm{d}t \tag{2.53}$$

As the magnetizing current $i_m$ continues to decrease, the accumulated energy is transferred to the capacitor, thereby increasing the capacitor voltage. The diode current is determined by the circuit equation of the $1 : n$ ideal transformer

$$i_D(t) = \frac{1}{n} i_m(t) \tag{2.54}$$

Recess Period: When $i_m$ is reduced to zero, so is the diode current and the accumulated energy is completely transferred to the capacitor. Now, the diode is turned-off and the capacitor voltage is held constant. The circuit model in this recess period is shown in Fig. 2.42(c).

By repeating the above operational cycle, the charge storage in the capacitor will be gradually increased, along with the elevation in the capacitor voltage.

## Capacitor charging circuit — EXAMPLE 2.16

Operational details of the capacitor charging circuit are illustrated in this example. Figure 2.43 shows the circuit diagram and simulated waveforms of a capacitor charging circuit with $V_S = 24\,\mathrm{V}$, $L_m = 120\,\mu\mathrm{H}$, $C = 5\,\mu\mathrm{F}$, $n = 1$, $T_s = 100\,\mu\mathrm{s}$, and $D = 0.4$.

Figure 2.43(b) shows the major waveforms of the capacitor charging circuit. The circuit operation is explained as follows. The magnetizing current $i_m$, which is identical to the switch current $i_Q$ during the energy build-up period, increases up to the peak value

$$i_{m\,peak} = \frac{V_S}{L_m} DT_s = \frac{24}{120\times 10^{-6}} 0.4 \cdot 100\times 10^{-6} = 8\ \mathrm{A}$$

thereby accumulating the total energy

$$E_m = \frac{1}{2} L_m i_{m\,peak}^2 = \frac{1}{2} 120 \times 10^{-6}\, 8^2 = 3.84\,\mathrm{mJ} \tag{2.55}$$

in the magnetizing inductance of the transformer. This energy is delivered to the capacitor by the diode current during the energy transfer period. Thus, $E_m$ represents the amount of the energy transfer from the voltage source to capacitor during one operational period. The capacitor voltage $v_C$ continues to increase during the energy transfer period in which the diode current $i_D$ exists. When $i_D$ reduces to zero, the recess period starts and $v_C$ is held constant until the next energy transfer period.

Figure 2.43(b) reveals important features of the circuit operation. First, although the magnetizing current $i_m$ increases linearly during the energy build-up period, it decays in a nonlinear fashion in the energy transfer period. The voltage across the

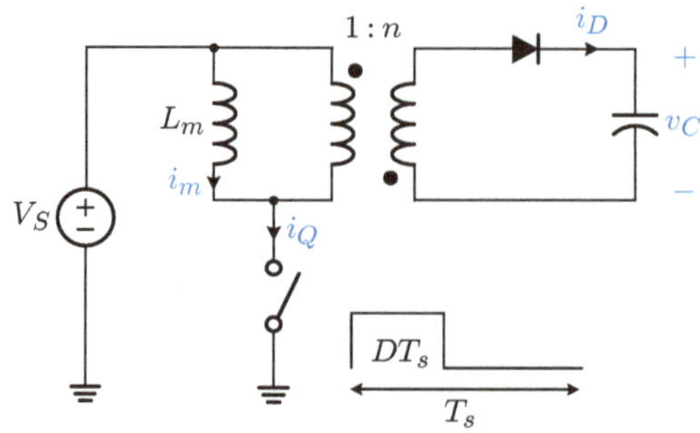


(a)

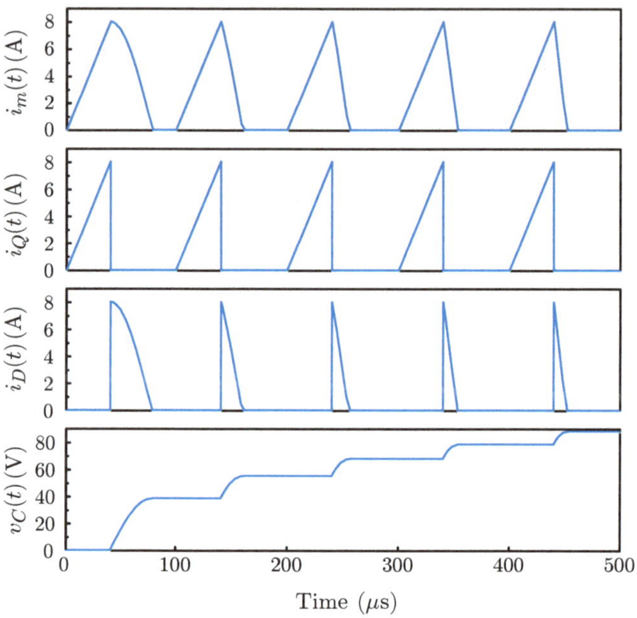


(b)

**Figure 2.43** Capacitor charging circuit. (a) Circuit diagram. (b) Circuit waveforms.

magnetizing inductance is a constant $V_S$ during the energy build-up period. However, in the energy transfer period, the time-varying capacitor voltage is reflected through the ideal transformer and applied to the magnetizing inductor in the negative polarity. Accordingly, the magnetizing current declines in a nonlinear manner.

Second, the duration of the energy transfer period becomes shorter as the operational cycle proceeds. The decaying slope of the magnetizing current is proportional to the magnitude of the capacitor voltage, which increases cycle-by-cycle. Consequently, the rate of the energy transfer becomes progressively faster, resulting in a continuous reduction in the energy transfer period.

Finally, the incremental growth in the capacitor voltage becomes smaller as the operational cycle proceeds. For each operational period, a fixed amount of energy is transferred to the capacitor, resulting in the cycle-by-cycle growth in the capacitor voltage. Because the energy stored in the capacitor is a quadratic function of the capacitor voltage, the increase in the capacitor voltage becomes smaller in proportion to the initial voltage at each operational cycle. More precisely, the following relationship is written from the circuit operation

$$\frac{1}{2}C(v_C(k-1)+\Delta v_C(k))^2 - \frac{1}{2}Cv_C^2(k-1) = E_m \tag{2.56}$$

where $v_C(k-1)$ is the initial capacitor voltage at the $k^{\text{th}}$ period and $\Delta v_C(k)$ is the increase in the capacitor voltage during the $k^{\text{th}}$ period, and $E_m$ is the amount of the energy transfer in one operational period. From (2.56), the capacitor voltage increase in Fig. 2.43(b) is given by

$$\Delta v_C(k) = \sqrt{v_C^2(k-1) + \frac{2E_m}{C}} - v_C(k-1) \tag{2.57}$$

which is a decreasing function of the initial capacitor voltage $v_C(k-1)$.

The capacitor voltage at the end of the $k^{\text{th}}$ operational period, $v_C(kT_s)$, is found from the energy balance relationship

$$kE_m = \frac{1}{2}Cv_C^2(kT_s) \quad \Rightarrow \quad v_C(kT_s) = \sqrt{\frac{2E_m}{C}k} \tag{2.58}$$

where $E_m$ given by (2.55). For example, the capacitor voltage at the end of the 5$^{\text{th}}$ operational period is given by

$$v_C(5\,T_s) = v_C(500\,\mu\text{s}) = \sqrt{\frac{2\cdot 3.84\times 10^{-3}}{5\times 10^{-6}}5} = 87.6\text{ V}$$

Figure 2.43(b) validates this analysis.

Figure 2.44 shows the capacitor voltage $v_C$ for a longer time period. The time interval required to charge the capacitor at a fixed $V_C$ is determined as follows. First, the required number of the charing operation, $k$, is found from

$$\frac{1}{2}CV_C^2 = kE_m \quad \Rightarrow \quad k = \frac{1}{2}CV_C^2\frac{1}{E_m} \tag{2.59}$$

Now, the total time required for charging $V_C$ is given by

$$T_{charge} = kT_s = \underbrace{\frac{1}{2}CV_C^2\frac{1}{E_m}}_{k}\,T_s \tag{2.60}$$

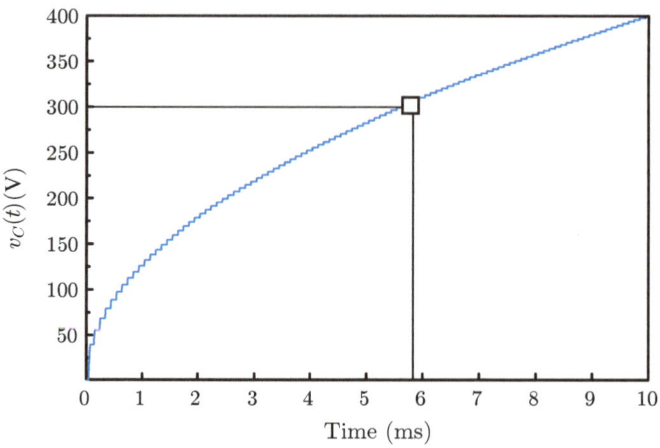


**Figure 2.44** Capacitor voltage waveform.

where $T_s$ is the operational period. As an example, the time needed to charge $V_C = 300$ V is given by

$$\begin{aligned} T_{charge} &= \frac{1}{2}CV_C^2\frac{1}{E_m}T_s \\ &= \frac{1}{2}5\times10^{-6}\,300^2\frac{1}{3.84\times10^{-3}}100\times10^{-6} \\ &= 5.86\text{ ms} \end{aligned}$$

Figure 2.44 shows a close agreement with this analysis.

## EXAMPLE 2.17 PSIM® Capacitor charging circuit

This example illustrates the PSIM simulation of the capacitor charging circuit discussed in Example 2.16. Table 2.5 shows the PSIM model and simulated capacitor voltage. The PSIM model is configured to create the environment that imitates the experimental set-up for measuring the capacitor voltage.

- PSIM model includes the circuit block, labeled as the **voltage display** that emulates the function of the voltage probe and oscilloscope in the experimental set-up.
- The simulation is performed under the condition specified in the **simulation control** dialogue box. This set-up is needed to monitor the capacitor voltage under the *free run* condition that displays the trajectory of the capacitor voltage with respect to time advance.

**Table 2.5** Capacitor Charging Circuit

PSIM model: Capacitor_Charging_ Circuit_01

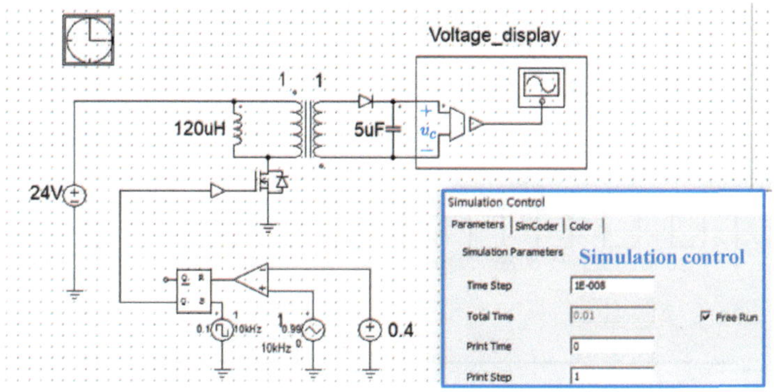


Simulation outcomes

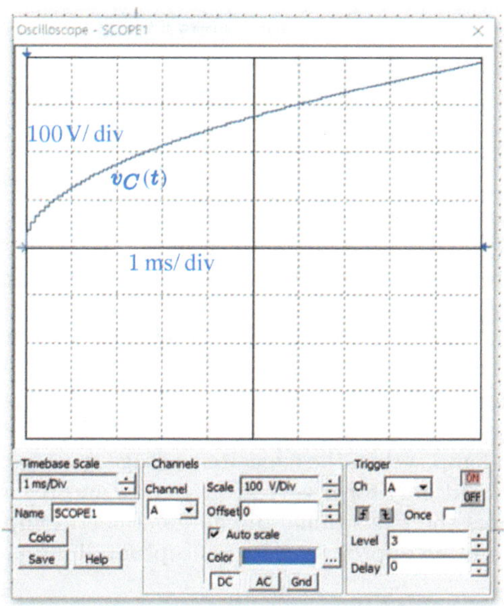


## PSIM® Capacitor charging circuit application EXAMPLE 2.18

This example introduces an important application of the capacitor charging circuit. Figure 2.45 is a functional diagram of the circuit. As shown in Fig. 2.45(a), a capacitor charging circuit is built using a practical transformer and high-voltage capacitor. A zenon-filled flash lamp is connected across the high-voltage capacitor. The flash lamp normally presents an infinitely large resistance and behaves as an open circuit. However, when the terminal voltage exceeds a prefixed threshold value, zenon gases

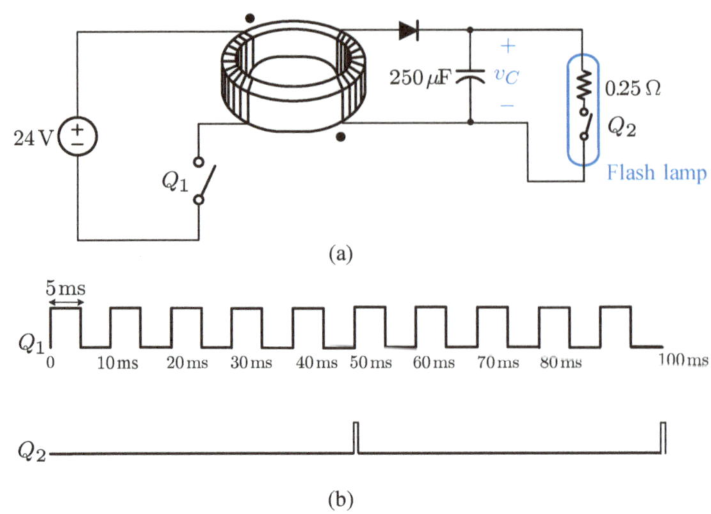


**Figure 2.45** Application example of capacitor charging circuit. (a) Capacitor charging circuit for industrial applications. (b) Switch drive signals.

become ionized and the resistance of the flash lamp suddenly drops to a very small value, thus presenting a nearly short-circuit condition. At this moment, the energy stored in the high-voltage capacitor instantly rushes into the flash lamp in the form of current spike and the flash lamp emits an intensive light-wave pulse. The resulting light-wave pulse is optically post-processed into a laser beam pulse. Typical application examples of the circuit include battery-operated laser pulse generators for military† and industrial uses.

The flash lamp might be modeled as a 0.25 Ω resister in series with a functional switch, $Q_2$ in Fig. 2.45(a). The switch is initially opened to yield the open-circuit condition for the flash lamp. When the capacitor voltage reaches the threshold value, the switch becomes closed only for a very short period and returns to the off-state. The switch $Q_2$ emulates the transitional discharge of energy in the form of current spike. Figure 2.45(a) will be appropriate for conceptual illuminations of the system operation.

As shown in Fig. 2.45(a), the system employ two switches, $Q_1$ and $Q_2$. The main switch $Q_1$ enables the system to accumulate energy into the high-voltage capacitor, thus resulting in a cycle-by-cycle increase in the capacitor voltage. On the other hind, the auxiliary switch $Q_2$ triggers an instantaneous dumping the accumulated energy into the flash lamp, in the form of current spike. Figure 2.45(b) shows the timing diagram of the $Q_1$ and $Q_2$, which are programmed to result in the following periodic operation.

†A sophisticated version of Fig. 2.45 is being employed as the main power supply for high-power pulsed laser generating systems. The first author of this book spent his early twenties in developing and deploying the laser generators for the laser-guided ballistic weapon systems, known as Smart Bombs.

**Table 2.6** Application of Capacitor Charging Circuit

PSIM model: Capacitor_Charging_Circuit_02

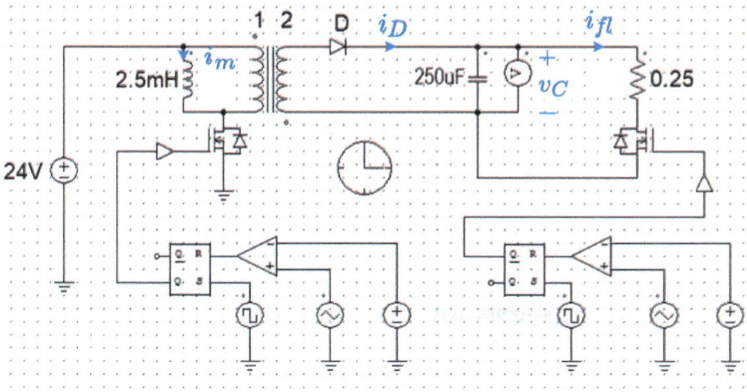


Simulation outcomes

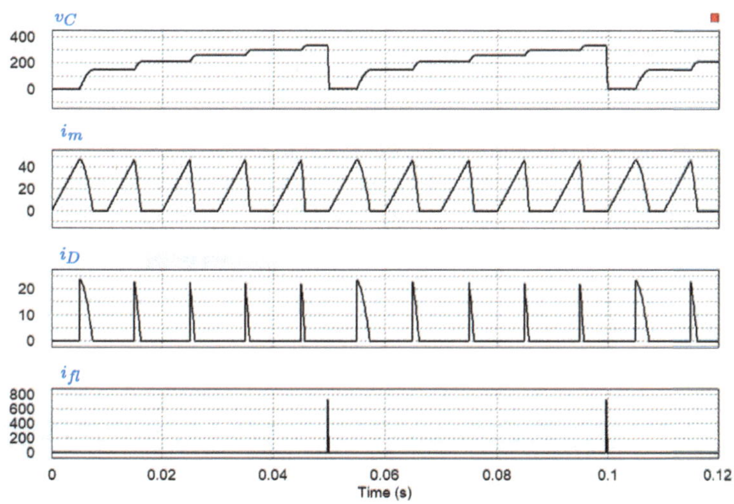


**1)** The capacitor voltage $v_C$ reaches the threshold value of the flash lamp after five consecutive capacitor charing operations.

**2)** At the end of the fifth operational period, $t = 50$ ms, the auxiliary switch $Q_2$ is turned on for a very short period only. Turning-on $Q_2$ emulates the nearly short-circuit condition of the flash lamp.

**3)** Immediately after $t = 50$ ms, the system resumes the steps of 1) and 2) for the repetitive operations.

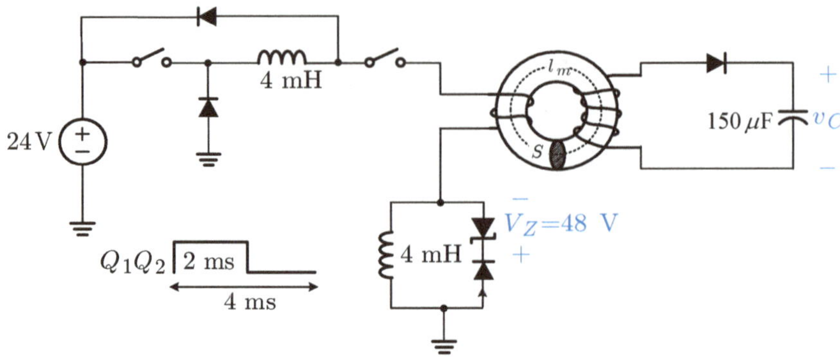


**Figure 2.46** Composite switching circuit.

We assume the transformer is built using a toroidal core with $\mu_r = 5000$, $S = 2\ \text{cm}^2$, and $l_m = 4\pi\times10^{-1}$ cm. The turns of the primary winding are 5 and that of the secondary winding are 10.

The PSIM model of the application circuit is shown in Table 2.6. The two PWM blocks in the model are programmed to produce the switch drive signals shown in Fig. 2.45(b). Table 2.5 shows the PSIM simulation results for $0 < t < 120$ ms. The current waveform $i_{fl}$, at the bottom of the simulation outcomes, produces the periodic high-current spikes that trigger the generation of the laser pulses. Readers are urged to validate the circuit operation and PSIM simulations based on the operational principles of the circuit.

## EXAMPLE 2.19 PSIM® Composite switching circuit

In this example, we consider the switching circuit shown in Fig. 2.46. The circuit combines the composite solenoid drive circuit in Example 2.15 and the capacitor charging circuit of Example 2.17. The transformer is built by winding $N_P = 2$ and $N_S = 4$ turns of copper coils around a toroidal core. The magnetizing inductance of the transformer is 2 mH.

The operation of Fig. 2.46 could be readily understood from the earlier discussions about Examples 2.15 and 2.17. Using the PSIM model in Table 2.7, computer simulations are performed with the conditions specified in the PSIM model. The simulation results are shown in Table 2.7, which displays the inductor current, diode currents, and capacitor voltage of the composite switching circuit. Theoretical validations of the simulation outcomes will be a good summary of this section.

**Table 2.7** Composite Switching Circuit

PSIM model: Composite_Switching_Circuit_01

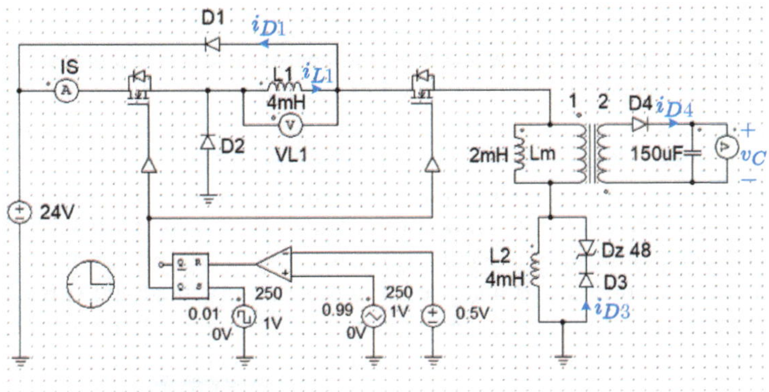


Simulation outcomes

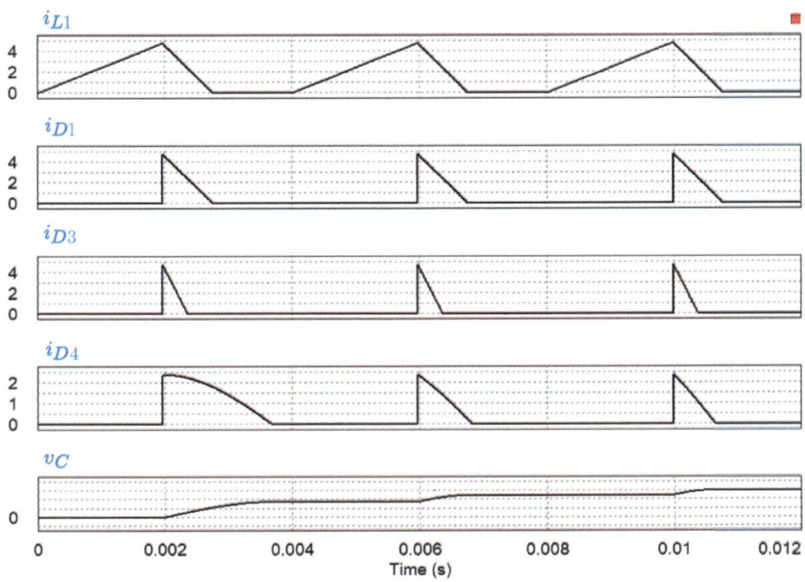


## 2.4 CHAPTER SUMMARY

This chapter investigated the circuit behavior of power stage circuit components. Operations of semiconductor switches, inductors, capacitors, and transformers were presented with practical examples. One essential functional component in dc-to-dc power conversion circuits is the single-pole double-throw (SPDT) switch, which toggles its connection periodically. In all the switching circuits covered in this chapter and other power conversion circuits to be studied in later chapters, the SPDT switch is implemented using the MOSFET-diode pair. In particular, in the capacitor charging circuit in Section 2.3.2, the MOSFET and diode collectively function as an SPDT switch even though they are physically separated by the transformer.

The energy storage and transfer devices include inductors, capacitors, and transformers. The freewheeling path and flux linkage balance condition are critical to understanding operations of inductive switching circuits. For capacitive switching circuits, the charge balance condition can be used to analyze the circuit waveforms. The practical transformer is modeled as a combination of the ideal transformer and magnetizing inductance. The operation of the practical transformer is then analyzed using the circuit equations of the ideal transformer and inductor.

This chapter introduced and analyzed a number of practical switching circuits, including the solenoid drive circuit and capacitor charging circuit, along with several variations and combinations of them. All analytical outcomes are validated using illustrative computer simulations with PSpice® and PSIM® software.

## REFERENCES

1. R. D. Middlebrook and S. Ćuk, *Advances in Switch-Mode Power Conversion*, TESLAco, Pasadena, CA, 1983.
2. Byungcho Choi, *Pulsewidth Modulated Dc-to-Dc Power Conversion*, $2^{nd}$ Ed., Wiley, Hoboken, NJ, 2021.

## PROBLEMS

**2.1** Assume that the inductor and capacitor in Fig. P2.1 are initially unenergized.

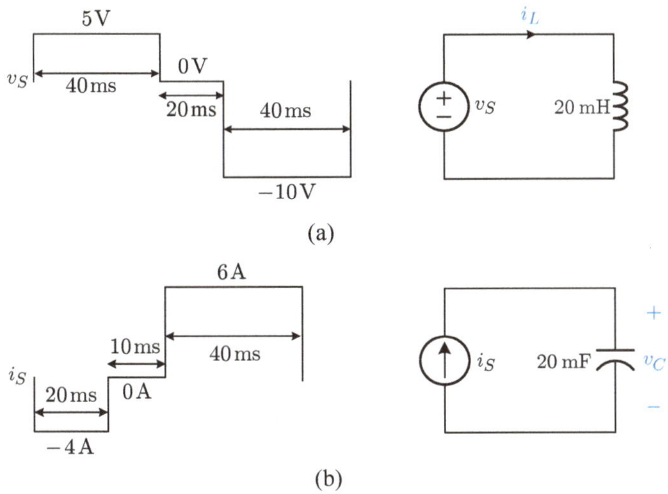


Fig. P2.1

a) For Circuit (a), sketch the inductor current $i_L$ for $0 < t < 100$ ms.
b) For Circuit (b), sketch the capacitor voltage $v_C$ for $0 < t < 70$ ms.

**2.2*** A switch drive signal shown in Fig. P2.2 is commonly applied to the following three circuits.

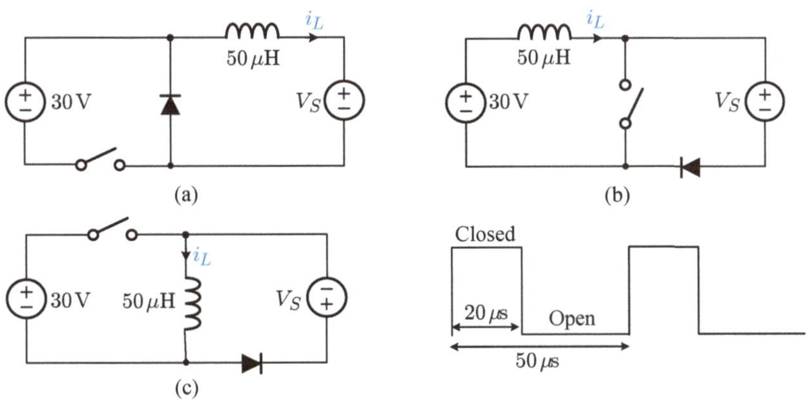


Fig. P2.2

a) For each circuit, determine the magnitude of $V_S$ such that the inductor current $i_L$ settles into a continuous periodic triangular waveform.

b) Assume the inductor is initially unenergized. Using the $V_S$ values found in a), sketch $i_L$ of each circuit for the two operational periods. Show the maximum value of $i_L$.

**2.3*** Shown in Fig. P2.3 are four different switching circuits along with their respective inductor voltage or capacitor current waveform.

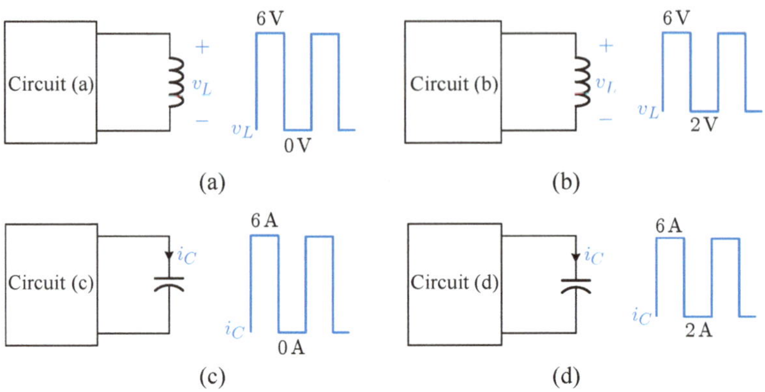


Fig. P2.3

a) Construct Circuit (a) using a 4 V voltage source, 2 V voltage source, active switch, and diode so that the circuit generates the given inductor voltage waveform $v_L$.
b) Build Circuit (b) using the same circuit components as those of a) so that the circuit produces the given inductor voltage waveform $v_L$.
c) Construct Circuit (c) using a 4 A current source, 2 A current source, active switch, and diode so that the circuit generates the given capacitor current waveform $i_C$.
d) Build Circuit (d) using the same circuit components as those of c) so that the circuit produces the given capacitor current waveform $i_C$.

**2.4*** Figure P2.4 shows a switching circuit along with the switch drive signal and $i_L$–$\lambda$ curve of the inductor $L$.
a) Find the inductance of $L$.
b) Assume $V_X = 16$ V and sketch $i_L$ for the first two operational periods. Show the peak value of $i_L$.
c) Find the time instant at which the inductor would saturate.
d) Find the minimum value for the $V_X$ that avoids the inductor saturation.

**2.5*** Consider the circuits shown in Fig. P2.5 and answer the questions.
a) For Circuit (a), assume that the inductor is initially unenergized. Sketch $i_L$ waveform for the first two operational periods for the following two cases.
i) $V_{S1} = 4$ V and $V_{S2} = 6$ V

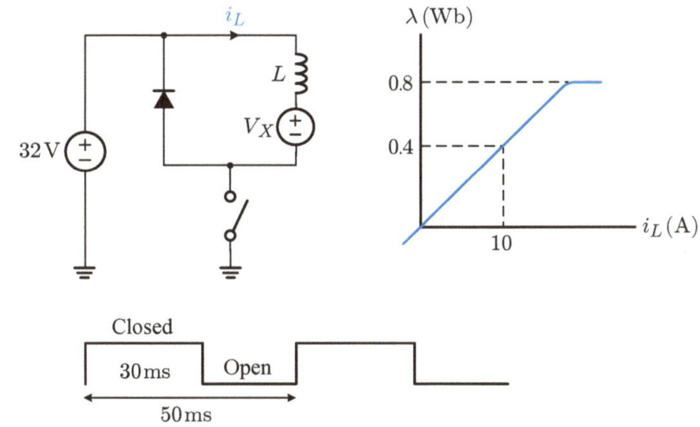


Fig. P2.4

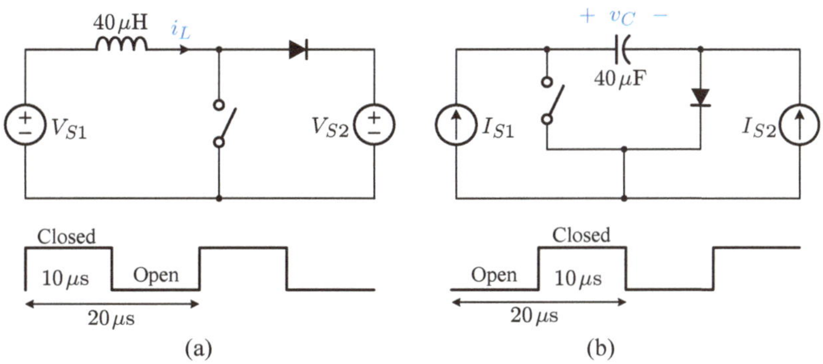


Fig. P2.5

ii) $V_{S1} = 2$ V and $V_{S2} = 6$ V
Show the maximum and minimum values of $i_L$.

b) Assume that the capacitor is initially uncharged in Circuit (b). Draw $v_C$ waveform for the first two operational periods for the following two cases.
i) $I_{S1} = 4$ A and $I_{S2} = 2$ A
ii) $I_{S1} = 2$ A and $I_{S2} = 4$ A
Show the maximum and minimum values of $v_C$.

**2.6** In the circuit in Fig. P2.6, the SPDT switch holds **x** position for $DT_s$ and holds **y** position for $(1 - D)T_s$.

a) For Circuits (a), (b), and (c) find the expression of $v_2/v_1$.

b) For Circuit (d), find the relationship of $i_2/i_1$.

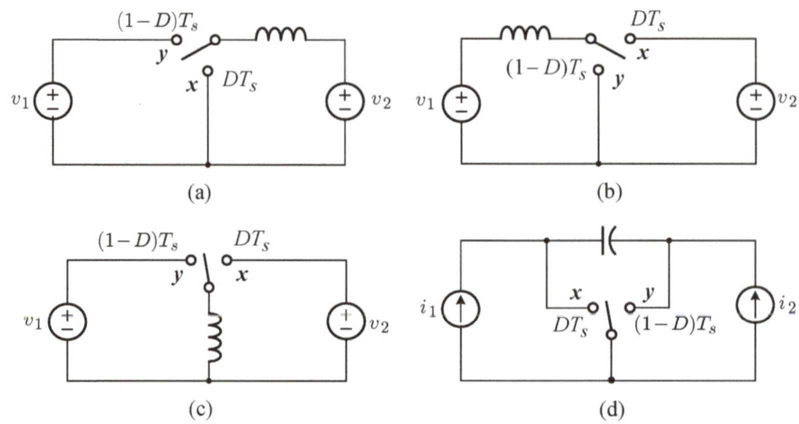

Fig. P2.6

**2.7** In the four switching circuits shown in Fig. P2.7, the SPDT switch holds position **a** for $DT_s$ and position **p** for $(1 - D)T_s$. Answer the following questions.

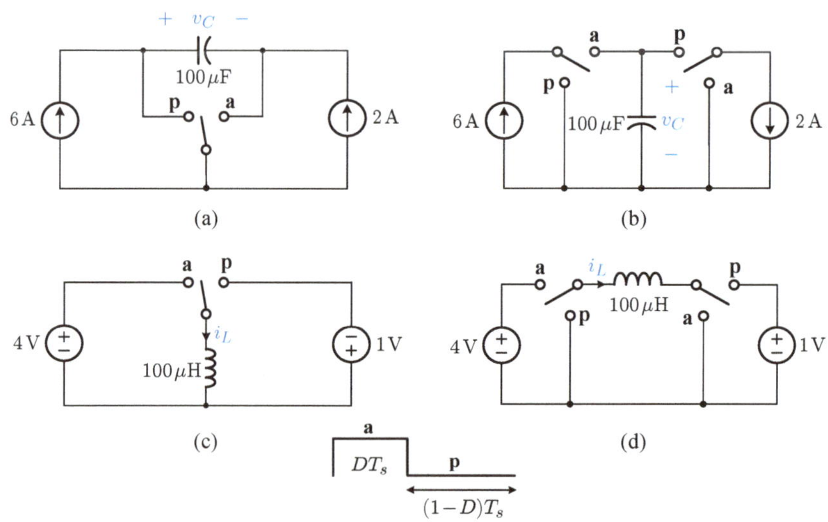

Fig. P2.7

a) For Circuits (a) and (b), determine the values for $D$ and $T_s$ so that each circuit settles into a steady-state operation with an 1.8 V peak value of the capacitor voltage, $v_{C\,peak} = 1.8$ V. Assume the capacitor is initially uncharged and the capacitor voltage does not stay at 0 V.

b) For Circuits (c) and (d), determine the values for $D$ and $T_s$ so that each circuit establishes a steady-state operation with $i_{L\,peak} = 1.2$ A. The inductor is initially unenergized and the inductor current does not stay at 0 A.

**2.8*** A switching circuit shown in Fig. P2.8 is configured using an SPDT switch and other circuit components. The SPDT switch periodically changes its position, as illustrated in Fig. P2.8. Assume the average value of $i_{L2}$ is 2 A and answer the questions.

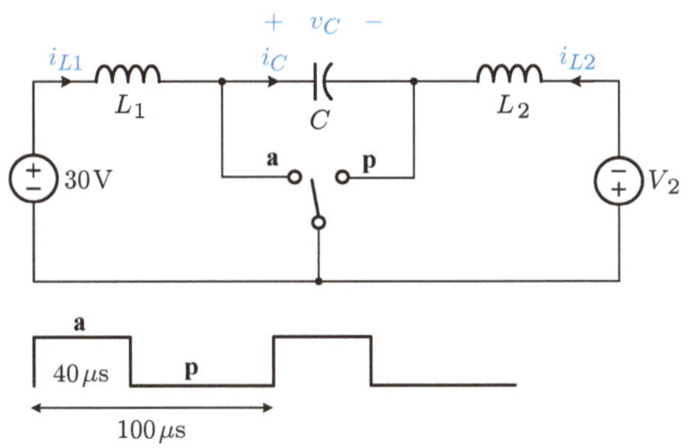


Fig. P2.8

a) Use the flux linkage balance condition on $L_1$ to evaluate the average value of $v_C$.
b) Apply the flux linkage balance condition to $L_2$ to find the value for $V_2$.
c) Use the charge balance condition on $C$ to evaluate the average value of $i_{L1}$. The average value of $i_{L2}$ is 2 A.
d) Assume $L_1 = 4.8$ mH, $L_2 = 1.2$ mH, and $C = 120$ μF. Sketch $i_{L1}$, $i_{L2}$, $i_C$, and $v_C$ for the two operational periods. Show the maximum and minimum values of each waveform.

**2.9*** Referring to Fig. P2.9, a switching circuit configured using an SPDT switch and other circuit components. The SPDT switch periodically changes its structure as illustrated in Fig. P2.9.

a) Evaluate the average value of $v_C$ and $V_2$.
b) Assume that the average value of $i_{L1}$ is 4 A. Sketch $i_{L1}$ and $i_{L2}$ for the two operational periods. Show the maximum and minimum values of the waveforms.
c) Using a) and b), sketch the $v_C$ waveform for the two operational periods. Show the maximum and minimum values.

**2.10** Figure P2.10 shows a switching circuit configured using a MOSFET and diode along with other linear circuit components. Assume the MOSFET switch is activated with a duty ratio of $D = 0.5$ at $T_s = 6$ ms.

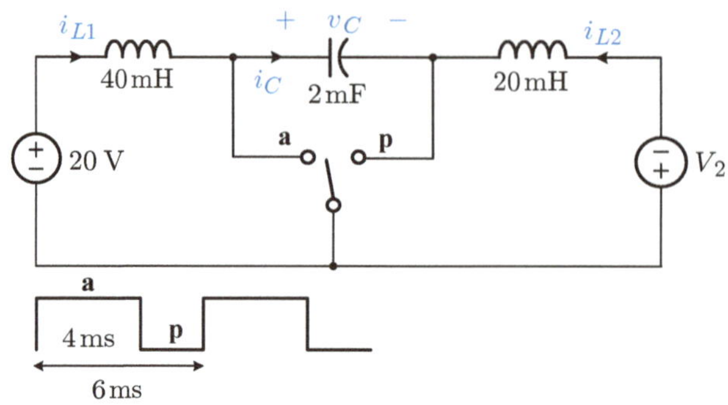


Fig. P2.9

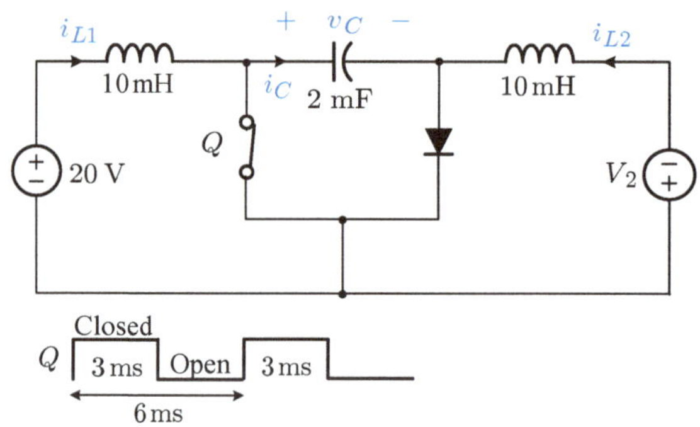


Fig. P2.10

a) Evaluate the average value of $v_C$ and $V_2$.

b) Assume that the average value of $i_{L1}$ is 0 A. Sketch $i_{L1}$, $i_{L2}$, $i_C$, and $v_C$ for the two operational periods. Show the maximum and minimum values of the waveforms.

**2.11*** The common switch drive signal is applied to four switching circuits shown in Fig. P2.11. For each circuit, sketch $i_S$, $v_T$, and $i_{D2}$ for the two operational periods. Assume the transformer is ideal.

**2.12** A transformer is built using a toroidal core with $\mu_r = 5000$, $S = 2\ \text{cm}^2$, and $l_m = 4\pi \times 10^{-1}$ cm. The turns of the primary winding are 200 and the turns of the secondary winding are 400. The permeability of the free space is $\mu_o = 4\pi \times 10^{-7}$ H/m.

a) A simple circuit shown in Fig. P2.12(a) is constructed using the transformer discussed above. Find the expressions for $i_P$ and $v_S$.

b) Now assume that the circuit is modified as shown in Fig. P2.12(b) in which the two identical transformers, each constructed as described above, are used. Find the expressions for $i_P$, $i_m$, $v_m$, and $v_O$.

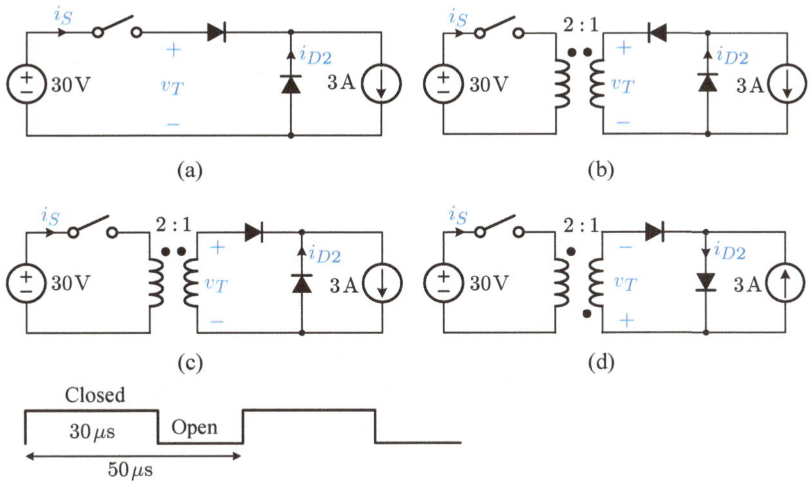


Fig. P2.11

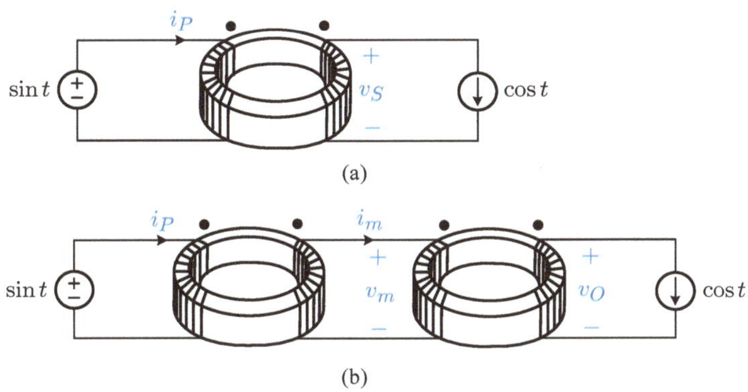


Fig. P2.12

**2.13** A couple of experiments are performed to extract the circuit parameters of a practical transformer.

- A 0.5 H inductance is measured at the primary winding with the secondary winding open-circuited.
- When $v_S(t) = 20 \sin t$ is connected to the primary winding, $v_O(t) = 5 \sin t$ is measured at the open-circuited secondary winding.

a) Draw the circuit model for the transformer. Show all the circuit parameters in your model.

b) Five different circuits in Fig. P2.13 are configured using the transformer discussed above. For each circuit, find an expression for the primary winding current, $i_P$.

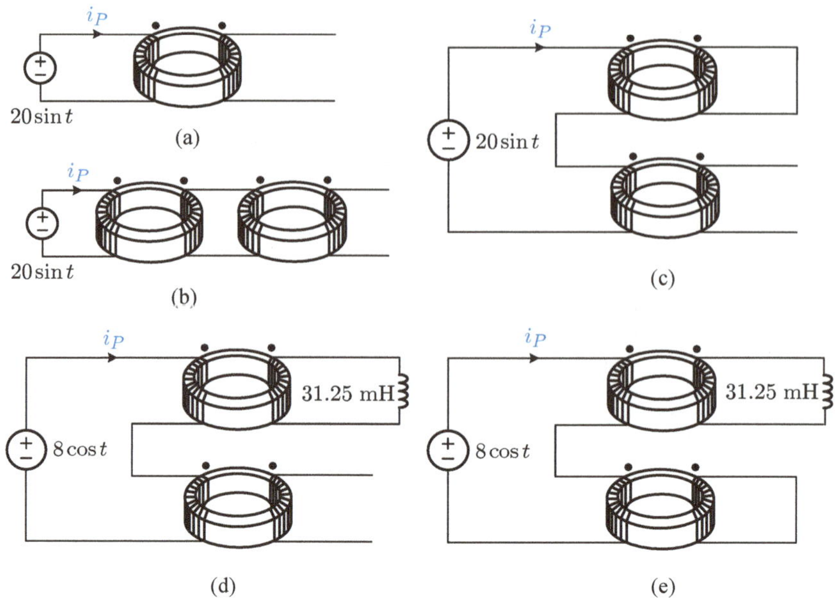


Fig. P2.13

**2.14*** An inductor is built using a toroidal core with $\mu_r = 5000$, $S = 2\ \text{cm}^2$, and $l_m = 4\pi \times 10^{-1}$ cm.

a) Figure P2.14(a) shows the $i_L$–$\lambda$ curve of the inductor.
   i) Evaluate the inductance of the inductor.
   ii) Find the number of the turns of the inductor winding.

b) A switching circuit shown in Fig. P2.14(b) is constructed using the inductor discussed in a). In the following problems, the average value of $i_P$ is zero: $\bar{i}_P(t) = 0$.
   i) Assume $T_s = 1$ ms and find the maximum value of $V_S$ for a proper operation of the circuit.
   ii) Now assume $V_S = 30$ V and find the maximum value for $T_s$.
   iii) For the case with $T_s = 1$ ms and $V_S = 30$ V, sketch $i_P$ for the two operational periods. Show the maximum and minimum values of the waveform.

c) A secondary winding is added to the inductor discussed above, thereby yielding a two-winding transformer with the turns ratio of $1 : 0.5$. As shown in Fig. P2.14(c), a 3.75 mH inductor is terminated across the secondary winding of the transformer. For the case with $T_s = 1$ ms, $V_S = 30$ V, and $\bar{i}_P(t) = 0$, sketch $i_S$ and $i_P$ for the two operational periods. Show the maximum and minimum values of the waveforms.

**2.15*** Figure P2.15 shows a conceptual circuit configured with two inductors and one ideal transformer. The following experiments are performed to determine the

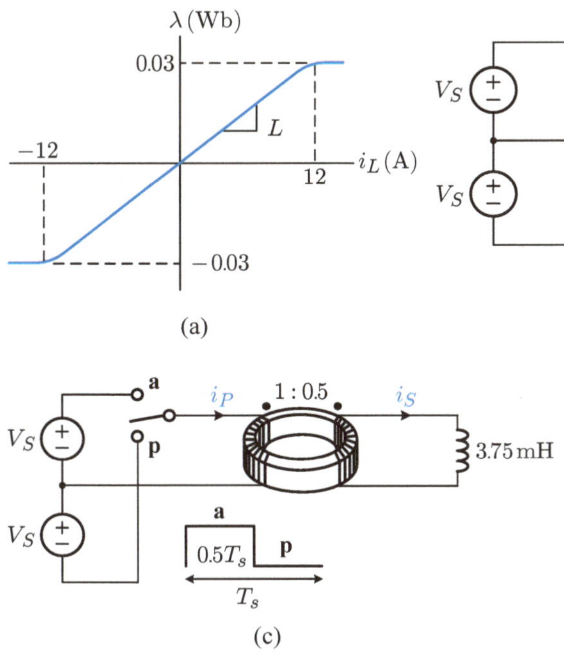


Fig. P2.14

values of the inductances and the turns ratio of the ideal transformer.

- A 2 mH inductance is measured at the terminal x–x′ with the terminal y–y′ short-circuited.
- A 7 mH inductance is measured at the terminal x–x′ with the terminal y–y′ open-circuited.
- When a voltage source $v_S(t) = \sin t$ is applied across the terminal x–x′, the same voltage $v_O(t) = \sin t$ is measured across the terminal y–y′.

a) Find the values for the inductances, $L_1$ and $L_2$, and the turns ratio $n$ of the ideal transformer.

b) Find the expression for $i_p'$ when a voltage source $v_S(t) = \sin t$ is connected across the x–x′ terminal and the y–y′ terminal is shorted.

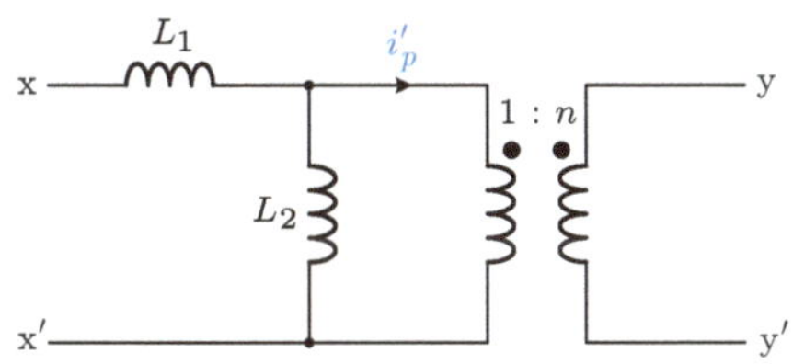


Fig. P2.15

**2.16** Referring to the two circuits shown in Fig. P2.16, answer the following questions. The transformer is built using a toroidal core with $\mu_r = 5000$, $S = 2\text{cm}^2$, and $l_m = 4\pi \times 10^{-1}$ cm. The turns of the primary winding are 10 and the turns of the secondary winding are 20.

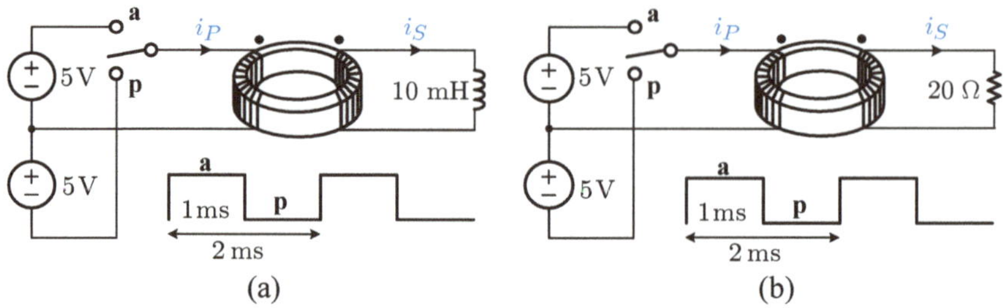


Fig. P2.16

a) For Circuit (a), sketch the steady-state waveforms of $i_P$ and $i_S$ for the two switching periods. Show the maximum and minimum values of each waveform. Assume that the average values of $i_P$ and $i_S$ are zero.

b) Repeat a) for Circuit (b).

**2.17*** In the circuit shown in Fig. P2.17, the transformer is built by winding $N_P = 2$ and $N_S = 4$ turns of copper coils around a toroidal core with $\mu_r = 5 \times 10^4$, $\mu_o = 4\pi \times 10^{-7}$ H/m, $S = 2$ cm$^2$, and $l_m = 4\pi \times 10^{-1}$ cm.

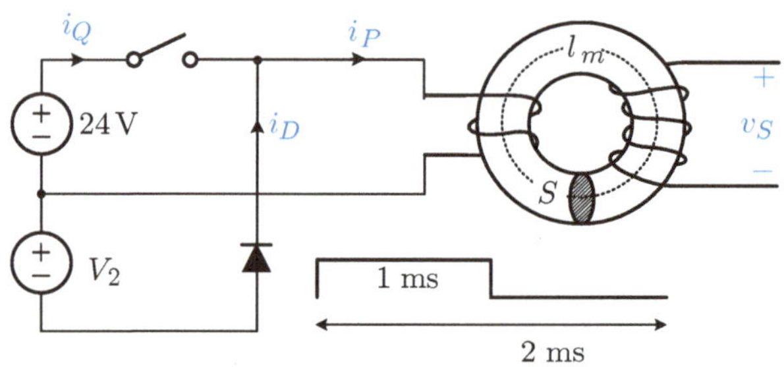


Fig. P2.17

a) Assume $V_2 = 0$ and sketch $i_P$, $i_D$, $v_S$, and $i_Q$ for the initial two operational periods. Show the maximum and minimum values of each waveform.

b) Assume $V_2 = 24$ V and repeat a).

c) Now assume $V_2 = 48$ V and repeat a).

**2.18** Referring to Fig. P2.18, the transformer is built by winding $N_P = 2$ and $N_S = 4$ turns of copper coils around a toroidal core with $\mu_r = 5 \times 10^4$, $\mu_o = 4\pi \times 10^{-7}$ H/m, $S = 2$ cm$^2$, and $l_m = 4\pi \times 10^{-1}$ cm. Sketch $i_S$, $i_P$, $i_Q$, and $i_D$ for the two initial operational periods. Assume the reactive components are initially unenergized. Show the peak value of each waveform.

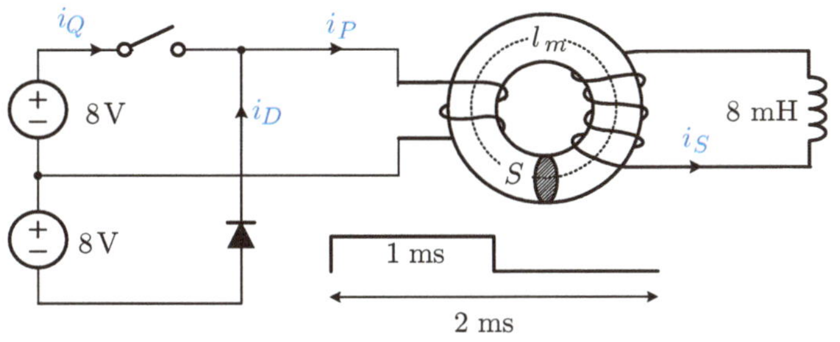


Fig. P2.18

**2.19*** Figure P2.19 shows two solenoid drive circuits, along with their switch drive signal and/or inductor current waveform.

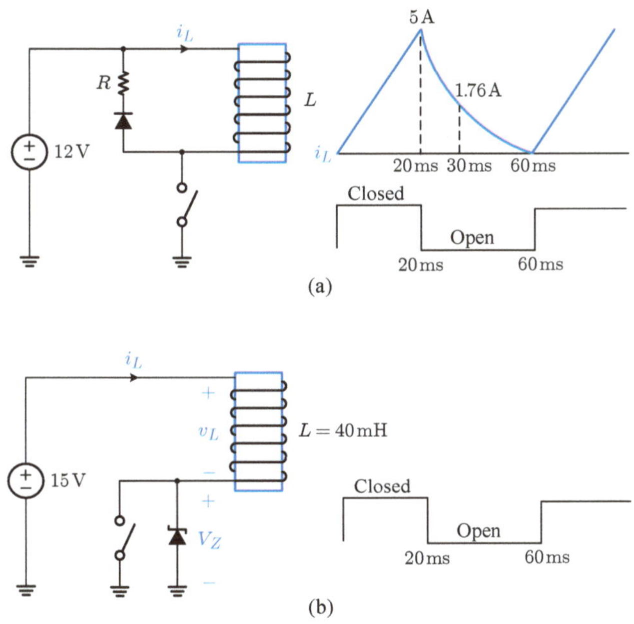


Fig. P2.19

a) Referring to Fig. P2.19(a), find the inductance $L$ of the solenoid and the resistance of $R$.

b) Use the results of a) to calculate

   i) the peak energy stored in the solenoid,

   ii) the energy consumed at the resistor in each switching period, and

   iii) the average power supplied by the source.

c) An alternative solenoid drive circuit is configured as shown in Fig. P2.19(b). Referring to the solenoid drive circuit, answer the questions.
i) Assuming $V_Z = 25$ V for the Zener diode, sketch $i_L$ and $v_L$ for the initial two operational periods. Label for the maximum and minimum values of each waveform.
ii) Now assume $V_Z = 20$ V for the Zener diode and sketch $i_L$ and $v_L$ for the initial two operational periods. Show the maximum value and the final value of $i_L$. Do you think this circuit is workable? Justify your answer.

**2.20*** Three different solenoid drive circuits are shown in Fig. P2.20. Referring to the common switch drive signal, answer the questions.

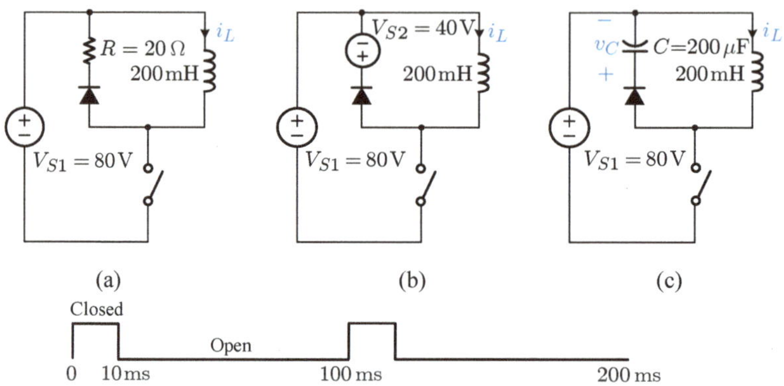


Fig. P2.20

a) Circuit (a) uses a diode-resistor branch as a freewheeling path.
i) Find the expression for $i_L(t)$ for 10 ms $< t <$ 100 ms.
ii) Find the power consumed at the resistor $R = 20\ \Omega$.

b) Circuit (b) employs an independent voltage source, $V_{S2} = 40$ V, in the freewheeling path.
i) Sketch $i_L$ for $0 < t <$ 100 ms. Label the value of $i_L$ at $t = 20$ ms on your sketch.
ii) When the switch is turned-off, what would happen to the energy stored in the inductor?

c) Circuit (c) uses a practical capacitor $C = 200\ \mu$F in place of $V_{S2}$.
i) Find the value of $v_C$ at $t = 99$ ms. Assume the capacitor is initially uncharged and the inductor current $i_L$ becomes zero at $t = 99$ ms.
ii) Sketch the general shapes of $i_L$ and $v_C$ for $0 < t <$ 200 ms.
iii) What would eventually happen to the circuit as time elapses?

**2.21*** In the circuit shown in Fig. P2.21, the transformer is built by winding $N_P = 2$ and $N_S = 4$ turns of copper coils around a toroidal core with $\mu_r = 5 \times 10^4$, $\mu_o = 4\pi \times 10^{-7}$ H/m, $S = 2$ cm$^2$, and $l_m = 4\pi \times 10^{-1}$ cm.

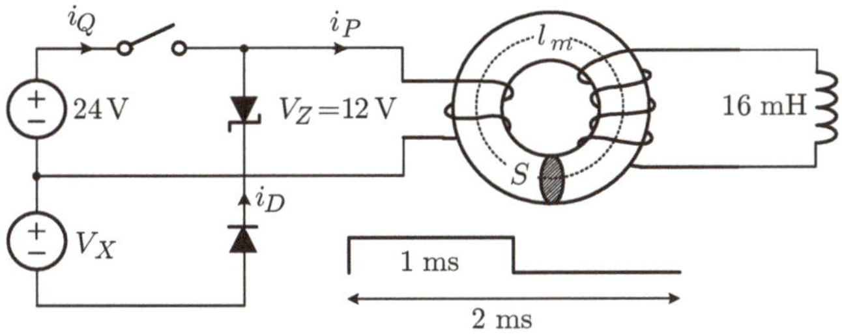


Fig. P2.21

a) Assume $V_X = 24$ V and sketch $i_P$, $i_Q$, and $i_D$ for the initial two operational periods. Show the maximum and minimum values of each waveform. Assume the reactive components in the circuit are initially unenergized.
b) Assume $V_X = 36$ V and repeat a).
c) Now assume $V_X = 48$ V and repeat a).

**2.22** Consider the circuit shown in Fig. P2.22. The transformer is built by winding $N_P = 4$ and $N_S = 4$ turns of copper coils around a toroidal core with $\mu_r = 5 \times 10^4$, $\mu_o = 4\pi \times 10^{-7}$ H/m, $S = 2$ cm$^2$, and $l_m = 4\pi \times 10^{-1}$ cm.

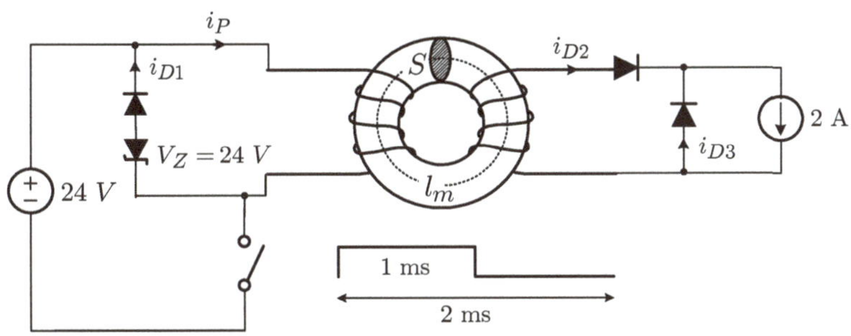


Fig. P2.22

Sketch the steady-state waveform of $i_P$, $i_{D1}$, $i_{D2}$, and $i_{D3}$ for two operational periods. Show the maximum and minimum values of the waveforms.

**2.23** A transformer is built using a toroidal core with $\mu_r = 5000$, $S = 2$ cm$^2$, and $l_m = 4\pi \times 10^{-1}$ cm. The turns of the primary winding are 10 and the turns of the secondary winding are 20. The two different switching circuits shown in Fig. P2.23 are built using the transformer. Referring to the common switch drive signal, sketch $i_S$ and $i_T$ of each circuit for the first two operational periods. Show the maximum and minimum values of the waveforms on your sketch.

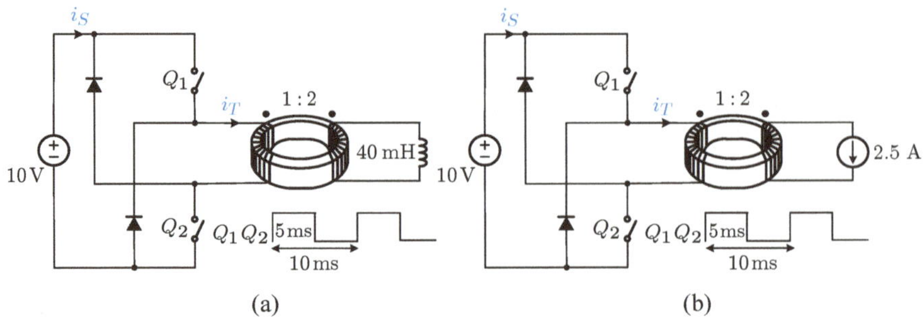


Fig. P2.23

**2.24** Consider the solenoid drive circuit shown in Fig. P2.24 and answer the questions.

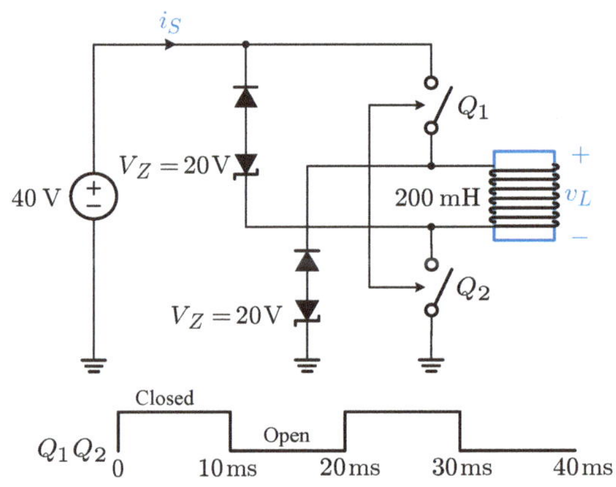


Fig. P2.24

a) Sketch the waveforms of $i_S$ and $v_L$ for $0 < t < 40$ ms. Show the maximum and minimum values on your sketch.
b) Calculate the following items evaluated during $0 < t < 20$ ms:
   i) the energy transferred from the source to the solenoid drive circuit,
   ii) the energy returned from the solenoid drive circuit to the source, and
   iii) the energy dissipated in the solenoid drive circuit.

**2.25** The non-dissipative solenoid drive circuit discussed in Section 2.3.1 is modified as shown in Fig. P2.25. Referring to the switch drive signal, sketch $i_L$, $i_{D1}$, $i_{D2}$, $i_{Q1}$, and $v_X$ for $0 < t < 6$ ms. Label the maximum and minimum values of each waveform.

**2.26** Consider the capacitor charging circuit and its major circuit waveforms shown in Fig. P2.26.

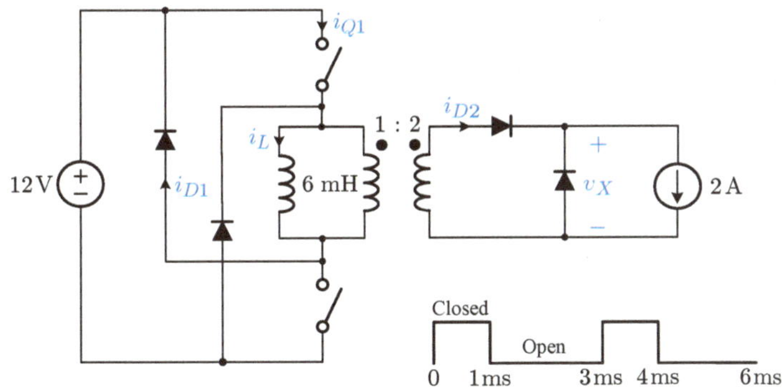


Fig. P2.25

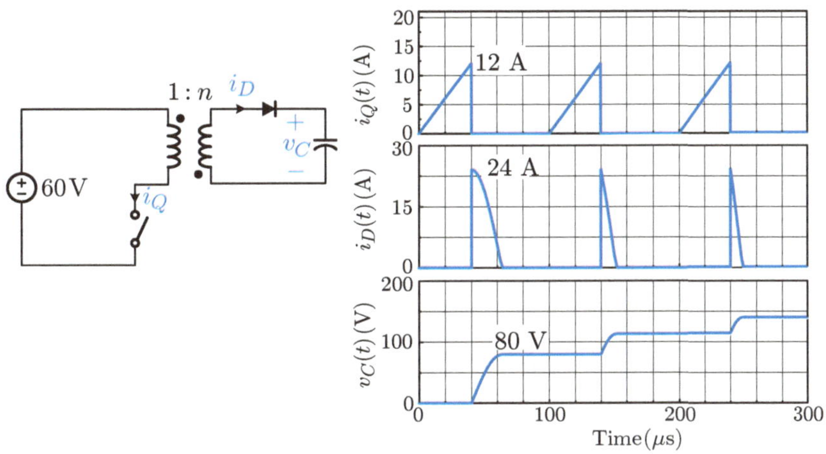


Fig. P2.26

a) Referring to the circuit waveforms, calculate the following circuit parameters and operational conditions of the solenoid driving circuit:
   i) $L_m$: magnetizing inductance of the transformer,
   ii) $n$: turns ratio of the transformer,
   iii) $D$: duty ratio of the switch, and
   iv) $C$: capacitance of the output capacitor.

b) Evaluate the following items based on the circuit waveforms:
   i) the energy transferred from the voltage source to the capacitor during the first 10 switching periods,
   ii) the capacitor voltage at the end of the $10^{\text{th}}$ switching period, and
   iii) the time instant when the capacitor voltage elevates approximately to 300 V.

c) Suggest a new switch drive scheme that would minimize the total time required to charge the capacitor to a specified voltage.

**2.27*** The operation of the circuit in Fig. P2.27 is very similar to that of the capacitor charging circuit discussed in the class. Assume the inductor and capacitor are initially unenergized.

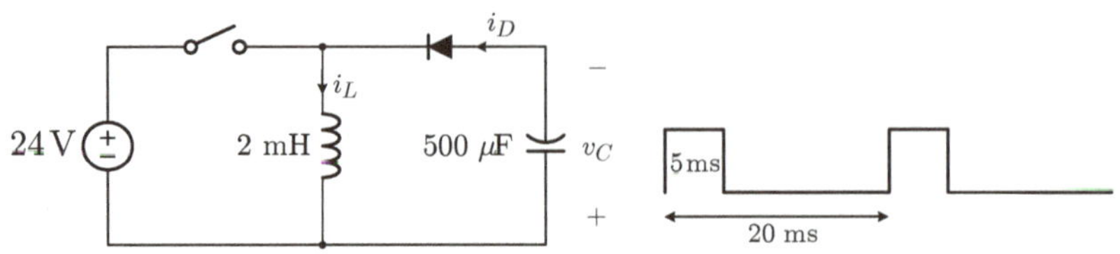


Fig. P2.27

a) Draw the circuit diagram when the switch is turned on.
b) Draw the circuit diagram when the switch is turned off.
c) Sketch $i_L$, $i_D$, and $v_C$ for the first two operational periods. Show all the characteristic features of the waveforms including their maximum and minimum values.
d) Evaluate $v_C$ at $t = 100$ ms.

**2.28*** In the circuit shown in Fig. P2.28, the transformers are built by winding $N_P = 2$ and $N_S = 4$ turns of copper coils around a toroidal core with $\mu_r = 5 \times 10^3$, $\mu_o = 4\pi \times 10^{-7}$ H/m, $S = 2$ cm$^2$, and $l_m = 4\pi \times 10^{-1}$ cm. The patterns of the transformer windings are not identical. For each circuit, sketch the primary winding current $i_P$ and the secondary winding voltage $v_S$ for the first two operational periods. Show the maximum and minimum values. Assume the reactive components in the circuits are initially unenergized.

**2.29** Figure P2.29 shows the two switching circuits are constructed using the transformers discussed in Problem 2.28.

a) For Circuit (a), evaluate the following items based on the circuit operation.
   i) the energy transferred from the 24 V voltage source to the switching circuit during one operational period.
   ii) the energy delivered to the capacitor during one operational period.
   iii) the energy dissipated in the switching circuit during one operational period.
   iv) the energy returned from the switching circuit to the 24 V voltage source during one operational period.
b) Repeat a) for Circuit (b).

**2.30*** Figure P2.30 shows the two switching circuits along with their switch drive signals.

a) For Circuit (a), evaluate the following items based on the circuit operation:
   i) the energy transferred from the 24 V voltage source to the switching

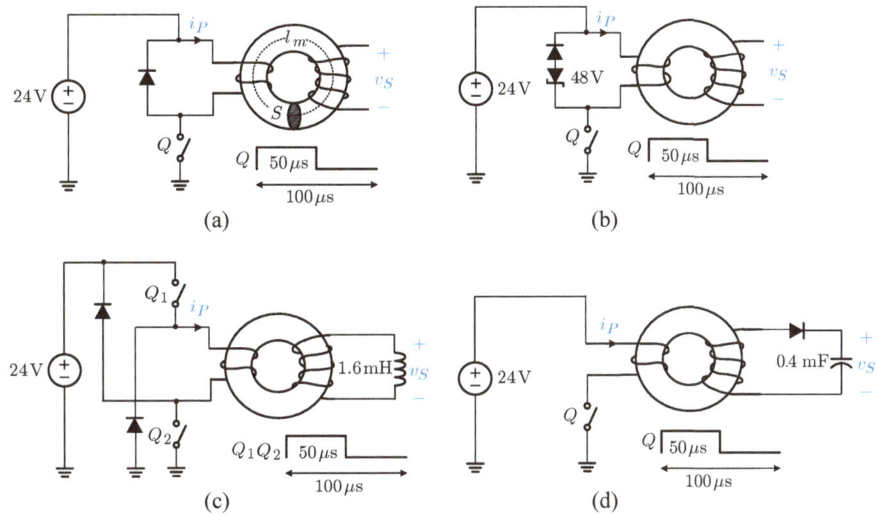


Fig. P2.28

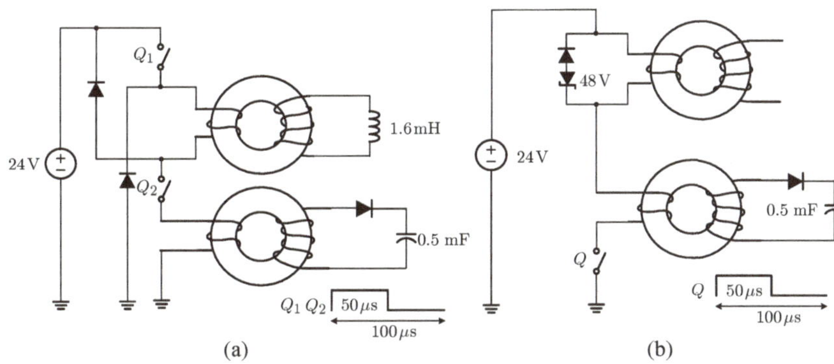


Fig. P2.29

circuit during one operational period,

ii) the energy delivered to the capacitor during one operational period,

iii) the energy dissipated in the switching circuit during one operational period, and

iv) the capacitor voltage $v_C$ at the end of the first five operational periods.

b) For Circuit (b), evaluate the following items:

i) the energy transferred from the 24 V voltage source to the switching circuit during one operational period,

ii) the energy delivered to the capacitor during one operational period,

iii) the energy returned from the switching circuit to the 24 V voltage source during one operational period, and

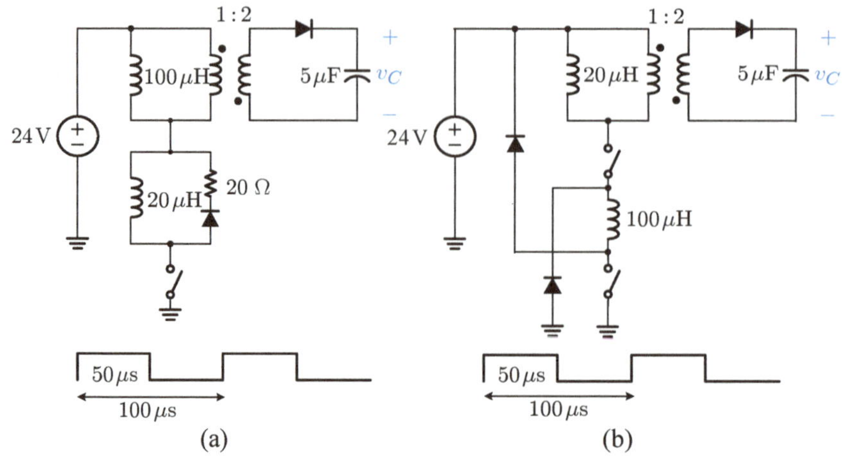


Fig. P2.30

iv) the capacitor voltage $v_C$ at the end of the first five operational periods.

**2.31*** An application example of the capacitor charging circuit is shown in Fig. P2.31(a). In the following problems, assume $L_m = 2$ mH for the magnetizing inductance of the transformer.

a) First, the application circuit is excited with a pair of switch drive signal $\{Q_1\ Q_2\}$. Sketch $v_C$ for $0 < t < 320$ ms. Show the all the characteristic features of the waveform.

b) Now, assume the application circuit is excited with the switch drive signal pair $\{Q_1\ Q_2'\}$. Repeat a).

c) The capacitor charging circuit is modified into the following three different circuits shown in Fig. P2.31(b). For each modified circuit, the circuit is derived with the switch drive signal $\{Q_1\ Q_2\}$ pair. For each circuit, describe the expected response of the circuit, in comparison with the original circuit. Do not sketch the $v_C$ waveform. Just describe the situation or operation of the circuit. State your answer in a concise and precise manner.

**2.32*** Assume all the circuit components are ideal in the circuits in Fig. P2.32.

a) Determine the value of inductance $L$ such that a 25% of the total energy delivered from the voltage source is dissipated in the circuit.

b) Assume the circuit operates with the inductance $L$ determined in a). Find the minimum value of the breakdown voltage of the Zener diode, $V_{Zmin}$, which ensures a proper operation of the circuit.

c) Assume the circuit operates with the inductance $L$ determined in a) and with the Zener diode with $V_Z = 2V_{Zmin}$ where $V_{Zmin}$ is determined in b). Sketch $i_{D1}$, $i_{D2}$, and $v_C$ for the first two operational periods. Show the

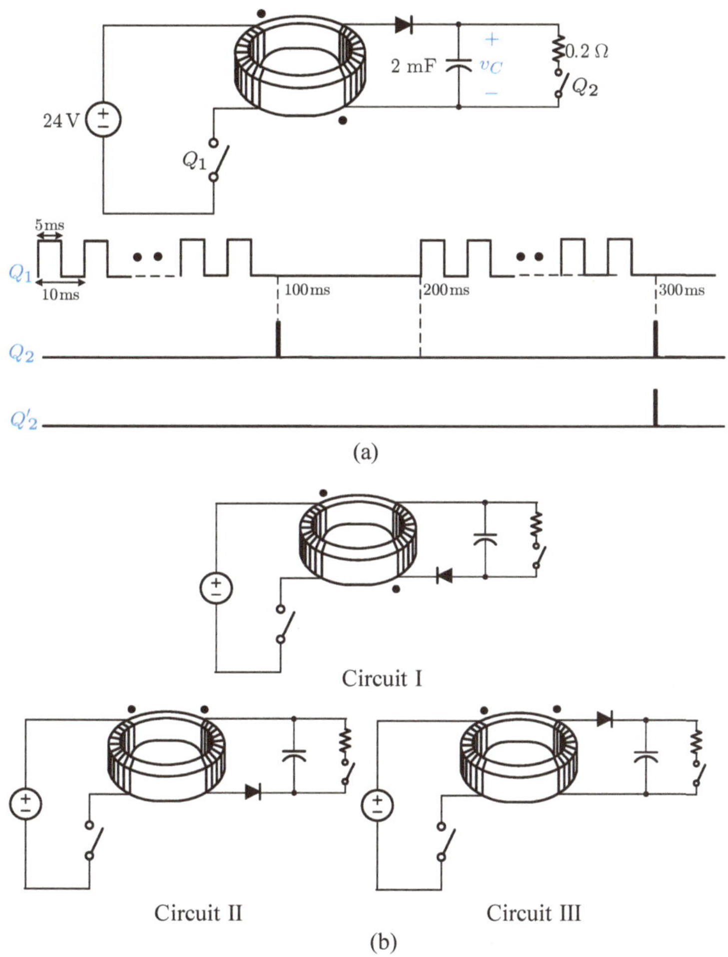


Fig. P2.31

maximum and minimum values of the waveforms. The reactive components are initially unenergized.

**2.33** Consider the circuit shown in Fig. P2.33 and answer the questions.

a) The two synchronized switches, $Q_1$ and $Q_2$, operate with the duty ratio of $D = 0.5$ at $T_s = 100\ \mu$s. Evaluate the following items based on the circuit operation.
   i) the energy transferred from the 24 V voltage source to the switching circuit during one operational period,
   ii) the energy delivered to the capacitor during one operational period,
   iii) the energy dissipated in the switching circuit during one operational period, and

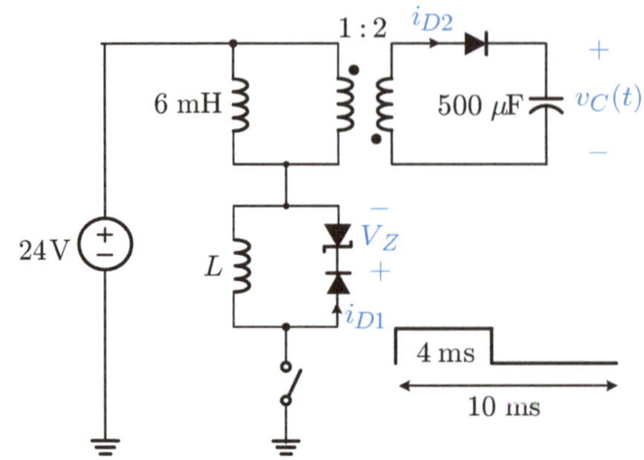


Fig. P2.32

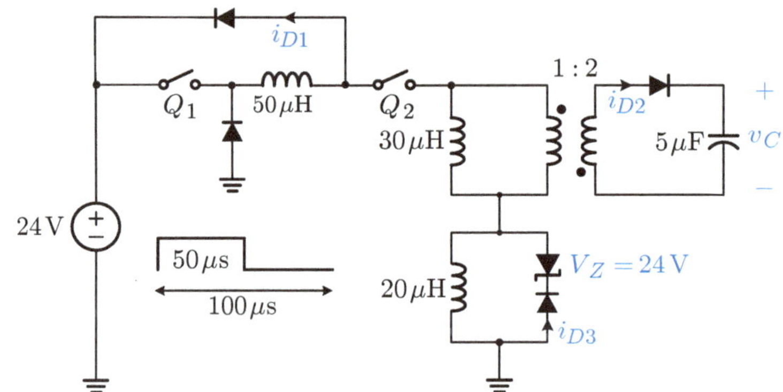


Fig. P2.33

iv) the energy returned from the switching circuit to the 24 V voltage source during one operational period.

v) the capacitor voltage $v_C$ at the end of the first five operational periods.

b) Sketch $i_{D1}$, $i_{D2}$, and $i_{D3}$ for $0 < t < 200$ $\mu$s. Specify the peak value and show all characteristic features of the current waveforms.

c) Evaluate the average value of $i_{D1}$ and $i_{D3}$.

**2.34*** In the circuit shown in Fig. P2.34, the transformer is built by winding $N_P = 2$ and $N_S = 4$ turns of copper coils around a toroidal core with $\mu_r = 5 \times 10^4$, $\mu_o = 4\pi \times 10^{-7}$ H/m, $S = 2$ cm$^2$, and $l = 4\pi \times 10^{-1}$ cm. The two synchronized switches $Q_1$ and $Q_2$ operate with a duty ratio of 0.5 at $T_s = 4$ ms.

a) Assume the switch SW is closed and sketch $i_{L1}$, $i_{D1}$, $i_T$, $i_{D2}$, and $v_C$ for the first two operational periods. Assume the reactive components are initially unenergized. Show the peak value of each waveform.

b) Now assume the switch SW is opened and answer the questions.

i) Determine the value of the inductance $L$ such that a 20% of the total

energy delivered from the voltage source is dissipated in the circuit.
ii) Assume the circuit operates with the inductance $L$ determined in i). Find the minimum of the breakdown voltage of the Zener diode, $V_{Zmin}$, which ensures a proper operation of the circuit.
iii) Assume the circuit operates with the inductance $L$ determined in i) and with a Zener diode with $V_Z = 2V_{Zmin}$ where $V_{Zmin}$ is found in ii). Sketch $i_{L2}$ and $i_{D3}$ for the initial two operational periods. Show the peak value of each waveform.

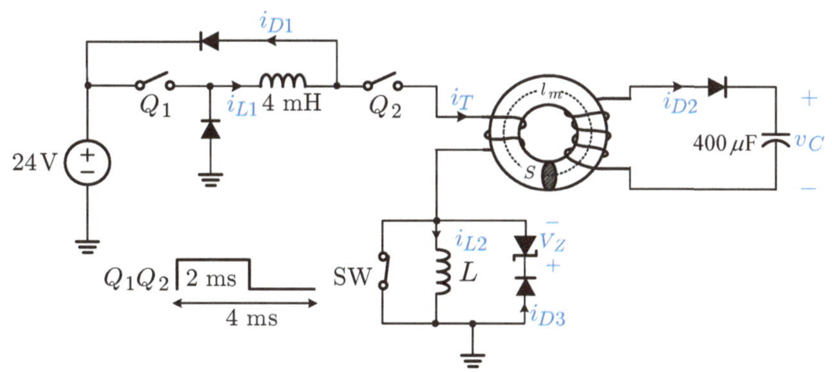


Fig. P2.34

**2.35*** Consider the composite switching circuit in Fig. 2.35 and answer the questions. The four transformers in the switching circuit are built by winding $N_P =$

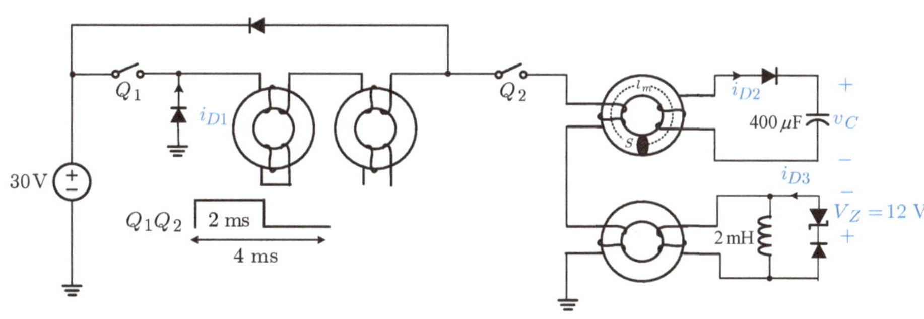


Fig. P2.35

$N_S = 2$ turns of copper coils around a toroidal core with $\mu_r = 5 \times 10^4$, $\mu_o = 4\pi \times 10^{-7}$ H/m, $S = 2$ cm$^2$, and $l_m = 4\pi \times 10^{-1}$ cm. The two synchronized switches $Q_1$ and $Q_2$ operate with a duty ratio of $D = 0.5$ with $T_s = 4$ ms.

a) Referring to the circuit schematic, evaluate or find the following items:
i) the energy delivered to the capacitor during one operational period,
ii) the energy dissipated in the switching circuit during one operational

period,
iii) the energy returned from the switching circuit to the voltage source during one operational period,
iv) the capacitor voltage $v_C$ at the end of the first three operational periods,
v) the power loss in the Zener diode, and
vi) the range for $V_Z$ of the Zener diode, which ensures a proper operation of the circuit.

b) Sketch $i_{D1}$, $i_{D2}$, and $i_{D3}$ for $0 < t < 12$ ms. Specify the peak value and all other characteristic features of the current waveforms.

c) Evaluate the average value of $\bar{i}_{D1}$ and $\bar{i}_{D3}$.

**2.36** A transformer is built using a toroidal core with $\mu_r = 5000$, $S = 2\ \text{cm}^2$, and $l_m = 4\pi \times 10^{-1}$ cm. The turns of the primary winding are 5 and that of the secondary winding is 20. The switching circuit shown in Fig. P2.36 is fabricated using the transformer described above.

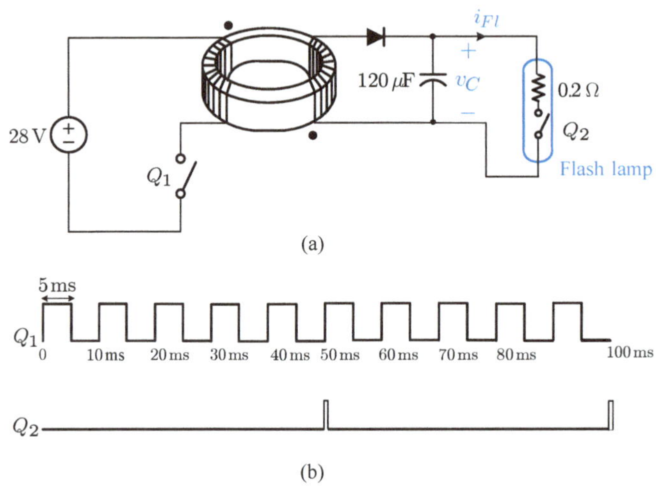


Fig. P2.36

a) Referring to the switch drive signals, sketch the waveforms $v_C$ and $i_{Fl}$ for $0 < t < 120$ ms. Show all the important information on your sketch so that your knowledge about the circuit operation is fully exposed.

b) Modify the PSIM®model **Capacitor_Charging_Circuit_02** and perform PSIM simulations to support your solution to Problem a).

c) Calculate the average power consumed in the circuit.

**2.37** Consider the circuit shown in Fig. P2.37 and answer the questions.

a) The two synchronized switches, $Q_1$ and $Q_2$, operate with the duty ratio of $D = 0.5$ at $T_s = 4$ ms. Evaluate the following items based on the circuit operation.
i) the energy transferred from the 24 V voltage source to the switching

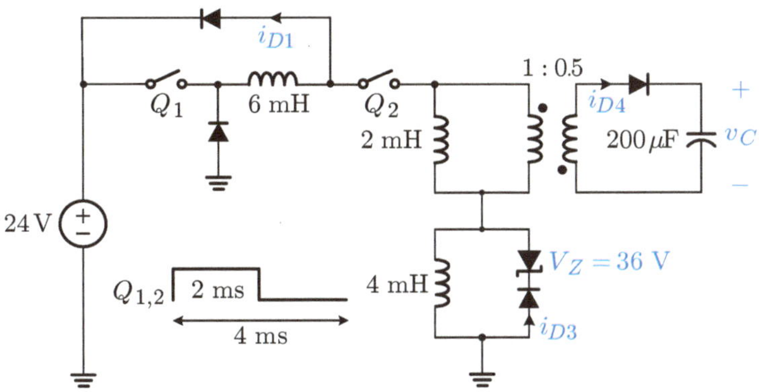


Fig. P2.37

circuit during one operational period.

ii) the energy delivered to the capacitor during one operational period.

iii) the energy dissipated in the switching circuit during one operational period.

iv) the energy returned from the switching circuit to the 24 V voltage source during one operational period.

v) the capacitor voltage $v_C$ at the end of the first three operational periods.

b) Sketch $i_{D1}$, $i_{D3}$, and $i_{D4}$ for $0 < t < 12$ ms. Specify the peak value and show all characteristic features of the current waveforms.

c) Modify the PSIM®model **Composite_Switching_Circuit_01** and perform PSIM simulations to validate your answer to Problems a)-v) and b).

# CHAPTER 3

# BUCK CONVERTER



There are numerous PWM dc-to-dc converters with different power processing functions and respective application areas. Among these, the buck converter has a very simple structure and is straightforward in operation, while also possessing all the essential features commonly found in the other PWM dc-to-dc converters. The output voltage of the buck converter is always lower than the input voltage. For this reason, the buck converter is also called the step-down converter.

There is another type of dc-to-dc converter that invariably provides a higher output voltage than the input voltage, called the boost converter or step-up converter. Furthermore, a dc-to-dc converter that either steps down or steps up the input voltage is synthesized by combining the step-down converter and step-up converter. The resulting converter is called the buck/boost converter or up-down converter.

This chapter deals with the buck converter, while the succeeding chapter covers the boost and buck/boost converters, as well as other important PWM dc-to-dc converters. Although focused on the buck converter, theoretical bases and analysis techniques of this chapter are expandable to all other PWM dc-to-dc converters including the boost and buck/boost converters.

As discussed in Chapter 1, the source of a PWM dc-to-dc converter could be any practical standalone dc source or non-ideal dc source obtained by rectifying an ac voltage source. In addition, the load might be any electronic devices, instruments, or systems operating with a constant voltage. In this chapter, however, an ideal dc source and resistive load are assumed, to concentrate on the study of the dc-to-dc power conversion itself. The issues involved with the non-ideal characteristics of the practical source and load will be studied in the third volume of the book series, which covers the analysis and design of dc power distribution systems.

This chapter discusses the functional basics and operational details of the buck converter. Several important analysis techniques are introduced and used to reveal the operation and properties of the buck converter. This chapter also illustrates the closed-loop control of the buck converter using the PWM scheme.

## 3.1 IDEAL STEP-DOWN DC-TO-DC POWER CONVERSION

The buck converter is the simplest circuit configuration that performs the step-down dc-to-dc power conversion. The principle of the step-down dc-to-dc conversion was introduced in Section 1.1.1 in conjunction with the LED string drive circuit in Fig. 1.2. The current section now presents theoretical aspects of the step-down power conversion.

The step-down dc-to-dc conversion is explained using a conceptual diagram, as shown in Fig. 3.1(a), consisting of two functional blocks: a single-pole double-throw (SPDT) switch and ideal low-pass filter. Within one switching period $T_s$, the SPDT switch holds position **a** for $T_{on}$ and position **p** for $T_{off} = T_s - T_{on}$. The time period $T_{on}$ is defined as the *on-time period*, while $T_{off}$ is denoted as the *off-time period*. The ratio $T_{on}$ to $T_s$ is defined as the *duty ratio* or *duty cycle* $D$ of the SPDT switch

$$\boxed{D \equiv \frac{T_{on}}{T_s}} \tag{3.1}$$

Similarly, the ratio $T_{off}$ to $T_s$ is defined as

$$D' \equiv \frac{T_{off}}{T_s} = \frac{T_s - T_{on}}{T_s} = 1 - D \tag{3.2}$$

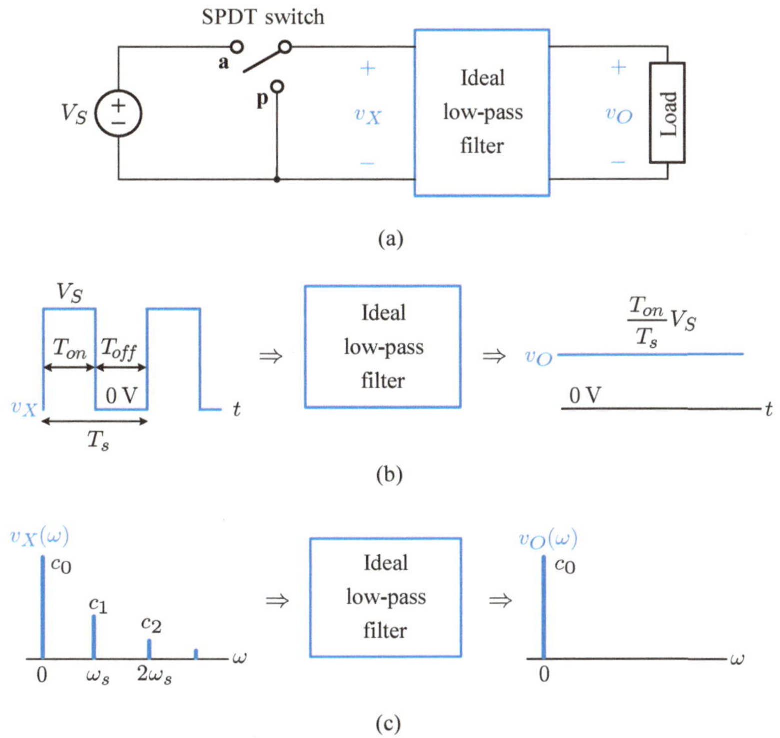


**Figure 3.1** Ideal step-down dc-to-dc power conversion. (a) Block diagram representation. (b) Time-domain description. (c) Frequency-domain interpretation.

The SPDT switch transforms the input voltage $V_S$ into a rectangular waveform $v_X$, as shown in Fig. 3.1. The rectangular waveform $v_X$ is then applied to the input port of the ideal low-pass filter.

Using Fourier series expansion, $v_X$ is expressed as the sum of a dc component and harmonic sinusoids

$$v_X(t) = c_0 + \sum_{n=1}^{\infty} c_n \sin(n\omega_s t + \theta_n) \tag{3.3}$$

where $c_0$ represents the dc component and the switching frequency $\omega_s = 2\pi/T_s$ is the fundamental frequency of $v_X$. The dc component $c_0$ is simply the average value of $v_X$

$$c_0 = \bar{v}_X(t) = \frac{T_{on}}{T_s}V_S = DV_S \tag{3.4}$$

If the cut-off frequency of the ideal low-pass filter, $\omega_c$, is lower than the fundamental frequency of $v_X$, $\omega_c < \omega_s$, all the harmonic components are completely blocked and only the dc component appears at the output of the low-pass filter

$$\boxed{v_O(t) = c_0 = DV_S} \tag{3.5}$$

Figure 3.1(b) shows the time-domain input-to-output description of the ideal low-pass filter, whereas Fig. 3.1(c) depicts the frequency-domain interpretation.

The dc-to-dc power conversion illustrated in Fig. 3.1 has the following properties:

- The circuit provides a pure dc voltage for the load, due to the ideal characteristics of the low-pass filter.
- The voltage gain of the circuit, the ratio $v_O$ to $V_S$, is simply the duty ratio $D$ of the SPDT switch. Thus, the output voltage is adjusted by controlling the duty ratio of the SPDT switch.
- Because $0 < D < 1$, the output voltage is always lower than the input voltage.

For above reasons, the power conversion illustrated in Fig. 3.1 is called the ideal step-down dc-to-dc power conversion.

## 3.2 BUCK CONVERTER: STEP-DOWN DC-TO-DC CONVERTER

Although the block diagram in Fig. 3.1 is a conceptual illustration of the ideal step-down dc-to-dc power conversion, there exists a practical converter which is very similar to Fig. 3.1 in structure and operation. This converter is called the buck converter and is one most widely-used dc-to-dc converter. This section explains the circuit configuration and theoretical basics of the buck converter, while succeeding sections describe operational details.

### 3.2.1 Evolution to Buck Converter

The functional diagram in Fig. 3.1 is transformed into the buck converter by implementing the SPDT switch with semiconductor switches and replacing the ideal low-pass filter with an LC filter. Figure 3.2 shows the circuit diagram of the buck converter along with its major waveforms. The SPDT switch is configured with the MOSFET-diode pair. The MOSFET switch is turned-on/off by the gate drive signal $v_{GS}$, meanwhile the status of the diode is determined by the condition of the MOSFET switch. When the MOSFET switch is turned-on by $v_{GS}$, the diode is turned-off because the input voltage $V_S$ reverse-biases the pn junction. Conversely, when the MOSFET switch is turned-off, the inductor current forces the diode to conduct, thereby creating a freewheeling path.

A second-order LC filter is used as a functional substitute for the ideal low-pass filter. The LC filter, despite its far from ideal characteristics, provides more than adequate filtering for most applications. It will be shown later that the consequences of the non-ideal filtering property of the LC filter are indeed negligible. A pure resistor is assumed for the load to simplify ensuing discussions.

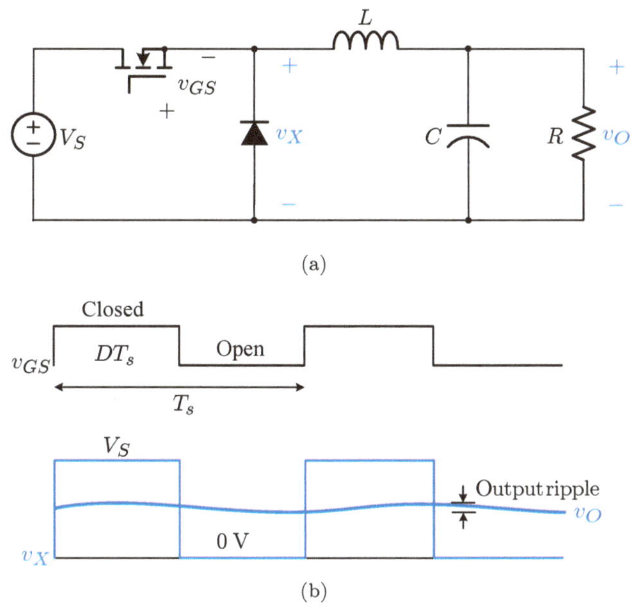


**Figure 3.2** Buck converter and major waveforms. (a) Circuit diagram. (b) Major waveforms.

### 3.2.2 Frequency-Domain Analysis

The LC filter is unable to completely remove the high-frequency harmonics from $v_X$ and the output voltage $v_O$ invariably contains an ac component, as illustrated in Fig. 3.2(b). The ac component contained in the output voltage is referred to as the *output ripple* or *switching ripple*.

A simple frequency-domain analysis is performed to determine the output ripple due to the non-ideal characteristics of the LC filter. The input-to-output transfer function of the LC filter with a load resistor $R$ is given by

$$F_f(s) = \frac{v_o(s)}{v_x(s)} = \frac{\frac{1}{sC} \parallel R}{sL + \frac{1}{sC} \parallel R} = \frac{1}{1 + \frac{s}{Q\omega_o} + \frac{s^2}{\omega_o^2}} \tag{3.6}$$

where $\omega_o$ is the pole frequency

$$\omega_o = \frac{1}{\sqrt{LC}} \tag{3.7}$$

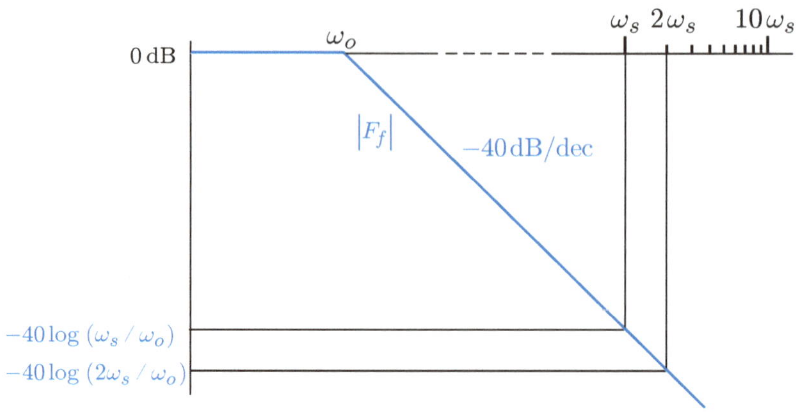


**Figure 3.3** Asymptotic plot for transfer function of second-order LC filter.

and $Q$ is the quality factor of the filter circuit

$$\boxed{Q = R\sqrt{\frac{C}{L}}} \tag{3.8}$$

Figure 3.3 shows the asymptotic magnitude plot of $|F_f(j\omega)|$,† drawn with the assumption that the switching frequency is higher than the pole frequency of the filter, $\omega_o < \omega_s$. The LC filter passes the dc component unchanged, yet attenuates the harmonic components of $v_X$. The input of the LC filter is represented by Fourier series

$$\boxed{v_X(t) = c_0 + \sum_{n=1}^{\infty} c_n \sin(n\omega_s t + \theta_n)} \tag{3.9}$$

The coefficients for the harmonic sinusoids are determined as

$$c_n = \frac{\sqrt{2}V_S}{n\pi}\sqrt{1-\cos(n2\pi D)} \tag{3.10}$$

from Fourier series expansion. Referring to Fig. 3.3, the $n^{\text{th}}$ order harmonic is attenuated by $-40\log(n\omega_s/\omega_o)$ in dB scale, while the dc component remains unaltered. Accordingly, the output of the LC filter is expressed as

$$\boxed{v_O(t) = c_0 + \sum_{n=1}^{\infty} c_n' \sin(n\omega_s t + \theta_n')} \tag{3.11}$$

†Asymptotic plots of transfer functions will be discussed in detail in the second volume of the book series.

where

$$
\begin{aligned}
c'_n &= c_n \, 10^{\dfrac{-40\log\left(\dfrac{n\omega_s}{\omega_o}\right)}{20}} \\
&= \underbrace{\frac{\sqrt{2}V_S}{n\pi}\sqrt{1-\cos(n2\pi D)}}_{c_n} \, 10^{\dfrac{-40\log\left(\dfrac{n\omega_s}{\omega_o}\right)}{20}}
\end{aligned}
\tag{3.12}
$$

The second term on the right-hand side of (3.11), representing the total sum of the harmonic sinusoids, constitutes the output ripple. Equation (3.12) shows that the output ripple is reduced to a negligible level if the condition $\omega_o \ll \omega_s$ is met. This will be achieved by either lowering the pole frequency with larger filter components or increasing the switching frequency.

## Output ripple estimation — EXAMPLE 3.1

This example demonstrates the accuracy of the previous frequency-domain ripple analysis. A buck converter operating with $V_S = 16$ V, $L = 40\ \mu$H, $C = 470\ \mu$F, $R = 1\ \Omega$, $T_s = 20\ \mu$s, and $D = 0.25$ is considered. The pole frequency of the filter is given by

$$\omega_o = \frac{1}{\sqrt{LC}} = \frac{1}{\sqrt{40\times10^{-6}\,470\times10^{-6}}} = 7.29\times10^3\ \text{rad/s}$$

and the switching frequency of the converter is

$$\omega_s = \frac{2\pi}{T_s} = \frac{2\pi}{20\times10^{-6}} = 3.14\times10^5\ \text{rad/s}$$

The dc component of the output voltage is determined as

$$c_O = DV_S = 0.25\cdot16 = 4\ \text{V}$$

The magnitude of the fundamental sinusoid in the output voltage is calculated as

$$
\begin{aligned}
c'_1 &= \underbrace{\frac{\sqrt{2}\,V_S}{1\cdot\pi}\sqrt{1-\cos(1\cdot2\pi D)}}_{c_1} \, 10^{\dfrac{-40\log\left(\dfrac{1\,\omega_s}{\omega_o}\right)}{20}} \\
&= \frac{\sqrt{2}\ 16}{1\cdot\pi}\sqrt{1-\cos(1\cdot2\pi\,0.25)} \, 10^{\dfrac{-40\log\left(\dfrac{1\cdot3.14\times10^5}{7.29\times10^3}\right)}{20}} \\
&= 3.88\times10^{-3}
\end{aligned}
$$

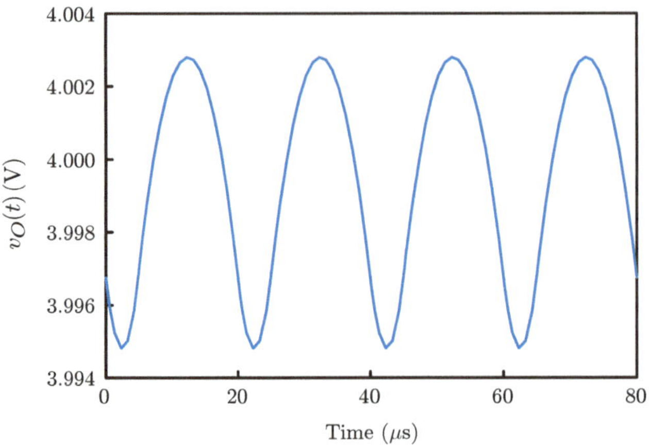


**Figure 3.4** Output voltage waveform of buck converter.

Similarly, the magnitudes of the higher-order harmonics are determined as $c_2' = 6.86 \times 10^{-4}$, $c_3' = 1.44 \times 10^{-4}$, $c_4' = 0$, and $c_5' = 3.10 \times 10^{-5} \cdots$.

Two important observations are made from the previous analysis.

1) The coefficients of the harmonic components of the output voltage are negligibly small compared to the dc component. This indicates that the LC filter provides sufficient filtering and the output of the converter is practically $V_O = c_0 = 4$ V.

2) Because the coefficient of the fundamental sinusoid, $c_1'$, is much larger than the coefficients of the higher-order harmonics, the amplitude of the output ripple is estimated as $\Delta v_O(t) \approx 2\,c_1' = 2 \cdot 3.88 \times 10^{-3} = 7.76$ mV.

Figure 3.4 shows the output voltage of the buck converter obtained from PSpice® simulation. The magnitude of the output ripple is very close to the analytical prediction of 7.76 mV.†

As demonstrated in Example 3.1, a simple second-order LC filter readily attenuates the output ripple to a negligible level. The second-order LC filter is just one specific example among many possible filter configurations. If a larger output ripple is allowed, the capacitor could be removed from the second-order filter, resulting in a first-order LR filter. Conversely, if further ripple attenuation is required, a higher-order filter or multi-stage filter could be employed.

†The average value of the output voltage is slightly less than the theoretical value of 4 V. The offset is due to the non-ideal characteristics of the diode used in the simulation.

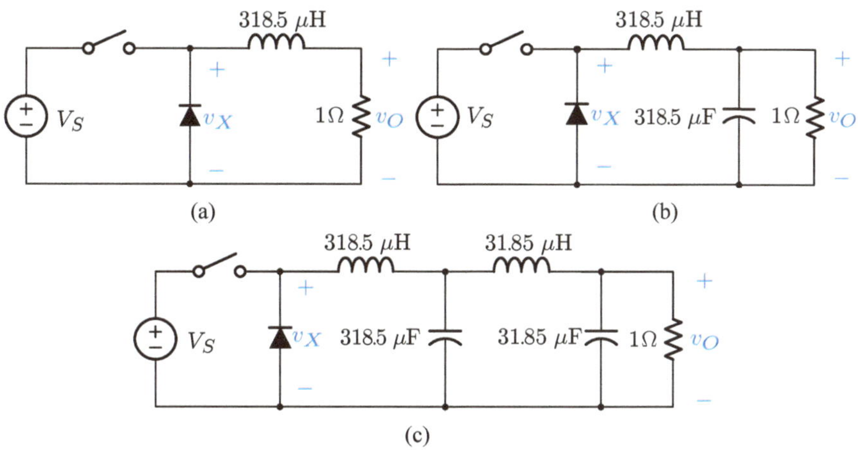


**Figure 3.5** Buck converters with different output filter configurations. (a) Buck converter with first-order output filter. (b) Buck converter with second-order output filter. (c) Buck converter with two-stage output filter.

## Output filter configuration and output ripple — EXAMPLE 3.2

This example illustrates the structure and ripple attenuation of three buck converters with different output filter configurations. Figure 3.5 shows the buck converters considered in this example: the first with a first-order LR filter, the second with a second-order LC filter, and the last with a two-stage LC filter. All the three buck converters operate with the input voltage of $V_S = 6$ V and the duty ratio of $D = 0.5$ at the switching frequency of 50 kHz. The circuit parameters of the output filters are shown in Fig. 3.5.

The three converters would identically produce the 3 V output. However, the magnitude of the output ripple component will vary widely due to the difference in the output filter configuration. PSpice® simulations are performed to investigate the output filter characteristics and output ripple attenuation.

Figure 3.6(a) shows the simulated Bode plots of the filter transfer functions, $F_f(j\omega) = v_o(j\omega)/v_x(j\omega)$, of the three buck converters. The Bode plots indicate that the first-order filter provides only a 40 dB attenuation for the fundamental component, while the second-order output filter offers a 80 dB attenuation. The two-stage output filter attenuates the fundamental component by 120 dB. Figure 3.6(b) shows the output voltage waveforms of the three buck converters. The first-order output filter generates a large triangular-shaped output ripple. The second-order filter yields a small sinusoidal output ripple. The two-stage output filter results in a nearly undetectable output ripple due to the much enhanced attenuation.

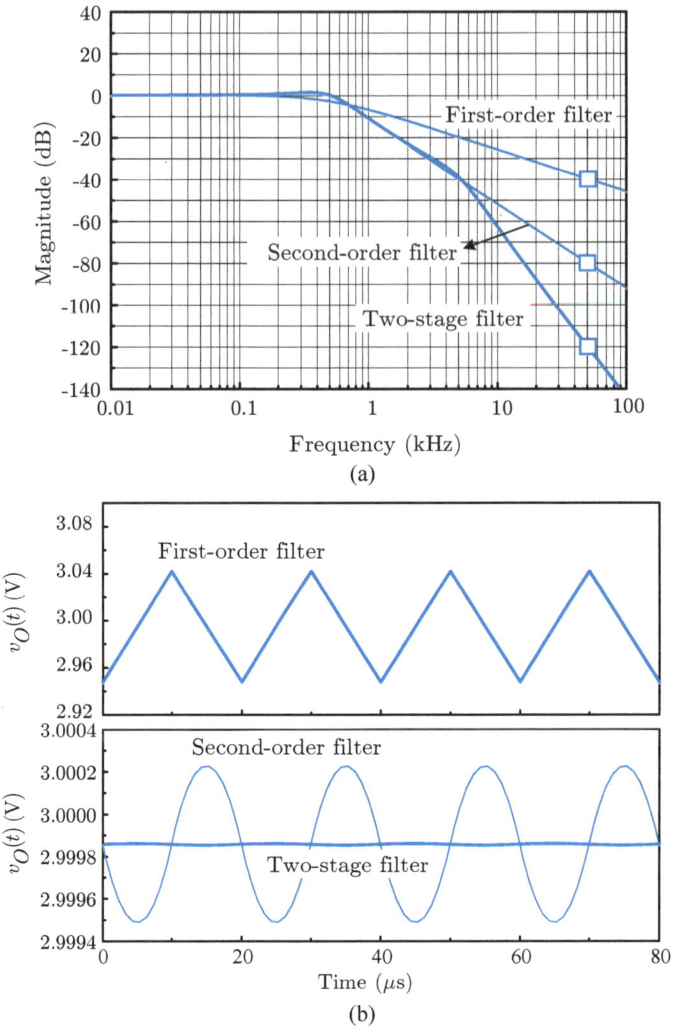


**Figure 3.6** Filter transfer functions and output voltage ripple. (a) Bode plots of filter transfer functions. (b) Output voltage of three buck converters.

## 3.3 BUCK CONVERTER IN START-UP TRANSIENT

This section explains the operation of the buck converter during the start-up process.† The concept of piecewise linear analysis is first introduced and the start-up response is then explained using the piecewise linear analysis method.

†The start up refers to starting the converter operation from the zero initial conditions up to the steady state.

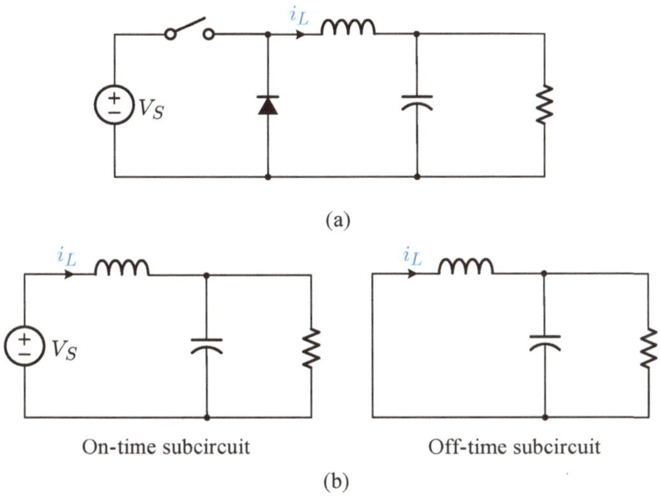


**Figure 3.7** Buck converter and on-time and off-time subcircuits. (a) Buck converter. (b) On-time and off-time subcircuits.

### 3.3.1 Piecewise Linear Analysis

The time-domain analysis of a dc-to-dc converter is not straightforward mainly because the switching action of the semiconductor devices forces the converter to periodically change its topological structure. One standard approach to analyzing the time-domain behavior of dc-to-dc converters is the piecewise linear analysis, where a dc-to-dc converter is considered as a combination of several linear circuits; each of them is valid for a specific time interval within one switching period.

The circuit that is effective when its switch drive signal is *on* (thus, the active switch is turned-on and the diode is turned-off) is defined as the *on-time subcircuit*. Likewise, the effective circuit when the switch drive signal is *off* (thus, with the active switch off and diode on) is denoted as the *off-time subcircuit*. The operation of a dc-to-dc converter is analyzed by first examining each subcircuit individually and later considering the circuit behavior of the two subcircuits collectively. Figure 3.7 shows a buck converter and its on-time and off-time subcircuits. The piecewise linear analysis using these subcircuits simplifies the analysis and, more importantly, provides significant insights on the converter operation.

The piecewise linear analysis will be used to investigate the transient and steady-state responses of the buck converter. Furthermore, this analysis method will be employed throughout the book, as an underlying principle to portray the steady-state waveforms of various PWM dc-to-dc converters.

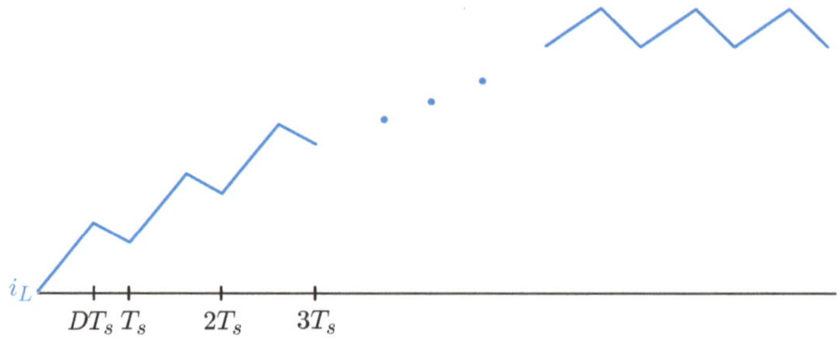


**Figure 3.8** Qualitative behavior of inductor current.

### 3.3.2 Start-up Response

The time-domain response of dc-to-dc converters could be analyzed by solving the circuit equations for the on-time and off-time subcircuits successively. For example, the response of the inductor current during the start-up process could be calculated as follows. First, the circuit equations for the on-time subcircuit are solved for $0 < t < DT_s$ with zero initial conditions. The expression for the inductor current is found from the resulting solutions. Second, the circuit equations for the off-time subcircuit are solved for $DT_s < t < T_s$ considering the circuit variables at $t = DT_s$ as the new initial conditions. By repeating this process, the transient response of the inductor current for the entire start-up period is obtained.

As an alternative to the previous iterative analysis, a qualitative method could be used to predict the behavior of the inductor current. For the time period $0 < t < DT_s$ during which the converter retains the on-time subcircuit, the voltage source transfers energy to the inductor. As a result, the inductor current is increased as illustrated in Fig. 3.8. For the time interval $DT_s < t < T_s$ in which the converter holds the off-time subcircuit, the energy stored in the inductor is released to the load and consequently the inductor current is decreased.

During the early stage of the start-up process, the energy transfer from the source is larger than the energy release to the load. Accordingly, the net change in the energy storage in the inductor for each switching period is positive. This implies that the inductor current progresses as illustrated in Fig. 3.8. As time elapses, the difference between the amounts of the transferred energy and the released energy lessens. When the on-time energy transfer becomes the same as the off-time energy release, the converter establishes steady state and the inductor current settles into a periodic triangular waveform.

**EXAMPLE 3.3 Start-up response of inductor current**

This example illustrates the start-up response of the buck converter. Figure 3.9 shows the inductor current during the start-up process of the buck converter introduced

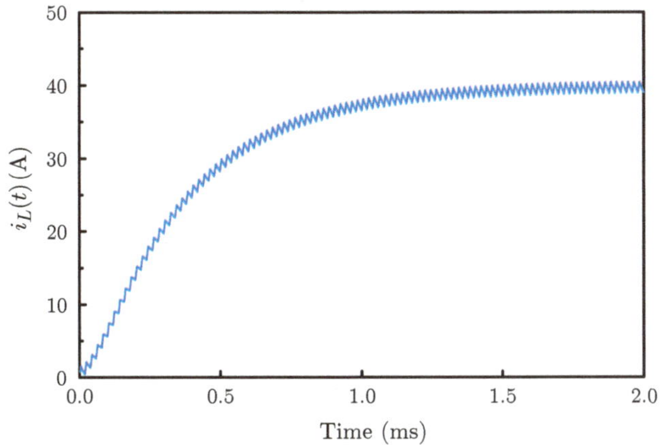


**Figure 3.9** Start-up response of inductor current.

in Example 3.1. A load resistance $R = 0.1\ \Omega$ is used in this simulation, while other parameters are the same as those of Example 3.1.

---

**PSIM® Start-up response of buck converter — EXAMPLE 3.4**

This example simulates the start-up response of the buck converter used in Example 3.3. Table 3.1 shows the PSIM model and waveforms of inductor current $i_L$ and output voltage $v_O$. The PSIM model contains Current display and Voltage display, which trace the start-up trajectories of the inductor current and output voltage. The PWM block is programmed to generate the gate drive signal specified in Example 3.1. The inductor current copies the waveform in Fig. 3.9. The output voltage gradually increases toward the steady-state value $V_O = 4$ V.

---

## 3.4 BUCK CONVERTER IN STEADY STATE

A properly designed dc-to-dc converter should reach steady state. When a dc-to-dc converter is in steady state, the inductor current settles into a periodic triangular wave and the output voltage becomes nearly constant with a small ripple component. This section first introduces several circuit analysis techniques and later analyzes the circuit waveforms of the buck converter in steady state.

### 3.4.1 Circuit Analysis Techniques

In addition to standard circuit equations, several specific analysis techniques are used for dc-to-dc power conversion circuits. These include the piecewise linear analysis,

**Table 3.1** Start-up Response of Buck Converter

PSIM model: Buck_Converter_01

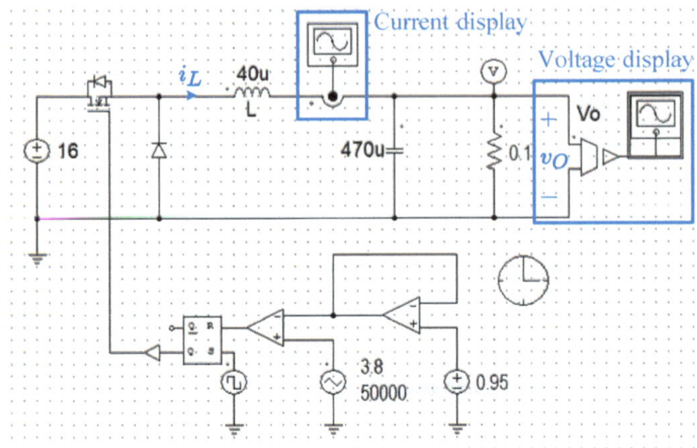


Simulation outcomes

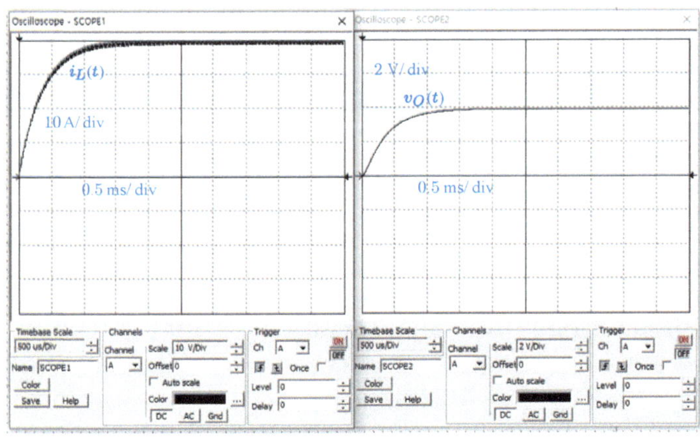


small-ripple approximation, flux linkage balance condition on inductors, and charge balance condition on capacitors.

## Piecewise Linear Analysis

The piecewise linear analysis technique discussed in Section 3.3.1 is a very effective method to analyze time-variant dc-to-dc power converter circuits. When coupled with the forthcoming small-ripple approximation, the piecewise linear analysis allows us to study dc-to-dc converters only using very simple circuit equations. In most cases, the circuit waveforms are accurately predicted from quick graphical constructions using elementary circuit equations.

### Small-Ripple Approximation

Another useful technique for the steady-state analysis of dc-to-dc converters is the *small-ripple approximation*. This technique postulates that *the ripple component contained in the output of a converter is so small that the output voltage could be considered a pure dc*. This assumption greatly simplifies the circuit analysis without compromising accuracy.

In practice, the ripple component in the output voltage is very small compared to the dc component, and its effects on the converter operation are indeed negligible. Consequently, an exact circuit analysis including the ripple component is usually unnecessary and the simplified analysis assuming a constant output voltage is fully adequate for most cases. The streamlined analysis based on the small-ripple approximation quickly yields simple circuit equations, and the resulting solutions retain sufficient accuracy for all engineering purposes.

We perform the simplified analysis assuming the output voltage is a pure dc without any ripple component. Even so, the actual output voltage ripple could always be accurately estimated from the results of the simplified analysis using the small-ripple approximation. Accordingly, the small-ripple approximation does not cause any real loss in the analysis accuracy. Details about the output ripple estimation are given in Section 3.4.3.

### Flux Linkage Balance Condition and Charge Balance Condition

Inductors used in dc-to-dc conversion circuits satisfy the flux linkage balance condition – the flux linkage increase during an on-time period should be the same as the flux linkage decrease during an off-time period. The flux linkage balance condition is transformed into the volt-sec balance condition. The volt-sec balance condition states that *the product of the voltage level and time interval, evaluated over the period when the inductor voltage is positive, should be the same as the voltage and time product calculated when the inductor voltage is negative*. Furthermore, as the generalization of the volt-sec balance condition, the average value of the inductor voltage is considered to be zero in general.

Capacitors are subject to the charge balance condition – the net change in the charge storage in a capacitor should be balanced at zero for each switching period. The charge balance condition is converted into the amp-sec balance condition. The amp-sec balance condition indicates that *the product of the positive current value (entering the capacitor) and the current-flowing period should be the same as the product of the negative current value (leaving from the capacitor) and the corresponding current-flowing time*. More generally, the average value of the capacitor current is considered to be zero, as is the case for the inductor voltage.

The volt-sec balance condition and amp-sec balance condition place constraints on the circuit variables of inductors and capacitors in dc-to-dc power converter circuits.

When dc-to-dc converters reach steady state, the circuit variables settle down to the values that satisfy the volt-sec and amp-sec balance conditions. Thus, these conditions could be used as the circuit theorems that ease the steady-state analysis of dc-to-dc converters.

### 3.4.2 Steady-State Analysis

The steady-state analysis of the buck converter is now performed on the bases of the piecewise linear analysis, small-ripple approximation, volt-sec balance condition, and amp-sec balance condition, along with other standard circuit equations. Figure 3.10 shows a buck converter, on-time and off-time subcircuits, and major circuit waveforms. According to the small-ripple approximation, the output of the converter is considered as a constant $V_O$ for both the on-time and off-time subcircuits. From Fig. 3.10(b), the voltage across the inductor is determined as

$$v_L(t) = \begin{cases} V_S - V_O & \text{for on-time subcircuit} \\ -V_O & \text{for off-time subcircuit} \end{cases} \tag{3.13}$$

By applying the volt-sec balance condition to the inductor, it follows that

$$(V_S - V_O)DT_s = V_O(1 - D)T_s \tag{3.14}$$

which is simplified to

$$\boxed{V_O = DV_S} \tag{3.15}$$

yielding the voltage gain expression. This voltage gain is identical, as it should be, to the result of the previous frequency-domain analysis.

The circuit equation for the inductor current is given by

$$i_L(t) = \frac{V_S - V_O}{L}t \tag{3.16}$$

for on-time period $DT_s$ and

$$i_L(t) = -\frac{V_O}{L}t \tag{3.17}$$

for off-time period $(1 - D)T_s$. It is evident from (3.16) and (3.17) that the inductor current ramps up during on-time period and ramps down during off-time period, resulting in a periodic triangular waveform shown in Fig. 3.10(c).

The average value of the inductor current is determined as follows. The KCL at the output port of the converter is given by

$$i_L(t) = i_C(t) + \frac{V_O}{R}$$

The averaged expression of the previous equation becomes

$$\bar{i}_L(t) = \bar{i}_C(t) + \frac{V_O}{R}$$

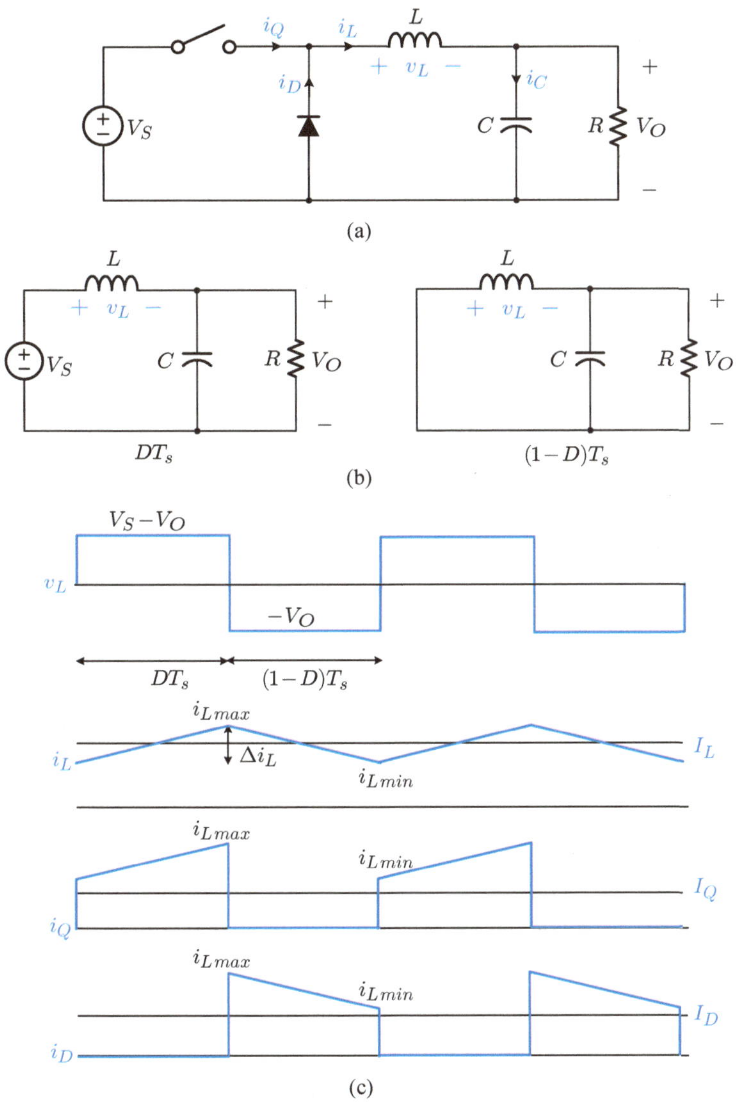


**Figure 3.10** Steady-state analysis of buck converter. (a) Buck converter. (b) On-time and off-time subcircuits. (c) Major waveforms.

The charge balance condition assures that the average value of the current flow through the capacitor $C$ is zero: $\bar{i}_C(t) = 0$. This implies that the average value of the inductor current should be identical to the dc current passing through the load resistor

$$\boxed{\bar{i}_L(t) = I_L = \frac{V_O}{R}} \tag{3.18}$$

On the other hand, it is also inferred from (3.16) and (3.17) that the difference between the maximum and minimum values of the inductor current is given by

$$\boxed{\Delta i_L = \frac{V_S - V_O}{L} DT_s = \frac{V_O}{L}(1 - D)T_s} \tag{3.19}$$

The maximum and minimum values of the inductor current then become

$$i_{L\,max} = I_L + \frac{1}{2}\Delta i_L \tag{3.20}$$

and

$$i_{L\,min} = I_L - \frac{1}{2}\Delta i_L \tag{3.21}$$

The inductor current flows through the active switch during on-time period, and then freewheels through the diode during off-time period. Figure 3.10(c) also shows the switch current $i_Q$ and diode current $i_D$, whose average values are given by

$$\bar{i}_Q(t) = I_Q = DI_L \tag{3.22}$$

and

$$\bar{i}_D(t) = I_D = (1 - D)I_L \tag{3.23}$$

**EXAMPLE 3.5 Steady-state analysis of buck converter**

This example illustrates the accuracy of the steady-state analysis based on the small-ripple approximation. For this purpose, we use the buck converter introduced in Example 3.1. The parameters of the buck converter are $V_S$ = 16 V, $L$ = 40 $\mu$H, $C$ = 470 $\mu$F, $R$ = 1 $\Omega$, $T_s$ = 20 $\mu$s, and $D$ = 0.25. The important values for the voltage and current waveforms of the buck converter are evaluated in Table 3.2. Figure 3.11 shows the simulated converter waveforms.

### 3.4.3 Evaluation of Output Voltage Ripple

The output of the buck converter contains a ripple component due to the non-ideal filtering characteristics of an LC filter. An exhaustive analysis could be performed to yield an accurate expression for the output voltage including the ripple component. In practice, however, a quick estimation of the magnitude of output ripple is more practical and useful than the exact analysis. This section presents a simple output ripple analysis.

#### Evaluation with Ideal Capacitor

Figure 3.12 shows the circuit diagram and waveforms, associated with the ripple analysis. As shown in the earlier analysis, the inductor current is a triangular waveform

**Table 3.2** Steady-State Analysis of Buck Converter

| Circuit variable | Expression |
|---|---|
| $V_O$ | $V_S D = 16 \cdot 0.25 = 4$ V |
| $v_{L\,max}$ | $V_S - V_O = 16 - 4 = 12$ V |
| $v_{L\,min}$ | $-V_O = -4$ V |
| $I_L$ | $\frac{V_O}{R} = \frac{4}{1} = 4$ A |
| $\Delta i_L$ | $\frac{V_S - V_O}{L} DT_s = \frac{16-4}{40 \times 10^{-6}} 0.25 \cdot 20 \times 10^{-6} = 1.5$ A |
| $i_{L\,max}$ | $I_L + \frac{1}{2}\Delta i_L = 4 + \frac{1.5}{2} = 4.75$ A |
| $i_{L\,min}$ | $I_L - \frac{1}{2}\Delta i_L = 4 - \frac{1.5}{2} = 3.25$ A |

consisting of dc and ac components. The dc component entirely passes through the load resistor because the capacitor presents an infinitely large impedance to the dc current.

On the other hand, the ac component, the triangular portion of the inductor current, would flow through both the capacitor and the load resistor. However, a practical dc-to-dc converter usually employs a large capacitor to provide sufficient filtering. As a result, the reactance of the capacitor evaluated at the switching frequency is much smaller than the load resistance. Thus, it is highly realistic to assume that the triangular portion of the inductor current totally passes through the capacitor. This situation is illustrated in Fig. 3.12(b).

The triangular current passing through the capacitor develops an ac voltage, thereby creating a ripple component on top of the dc output voltage. The ac voltage across the capacitor is given by

$$v_{ac}(t) = \frac{1}{C}\int i_C(t)\,\mathrm{d}t \tag{3.24}$$

where $i_C$ represents the capacitor current. It is deduced from (3.24) that $v_{ac}$ grows during the period when the capacitor current $i_C$ is positive and then decays when $i_C$ is negative, as illustrated in Fig. 3.12(b). The peak-to-peak value of $v_{ac}$ is then found

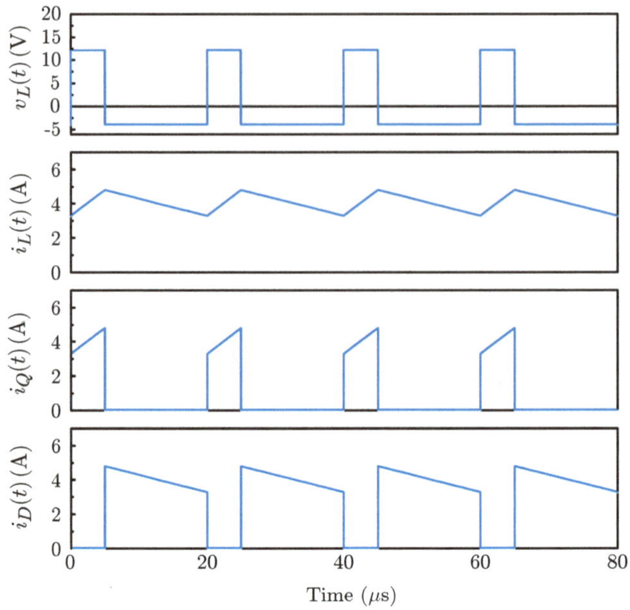


**Figure 3.11** Buck converter circuit waveforms.

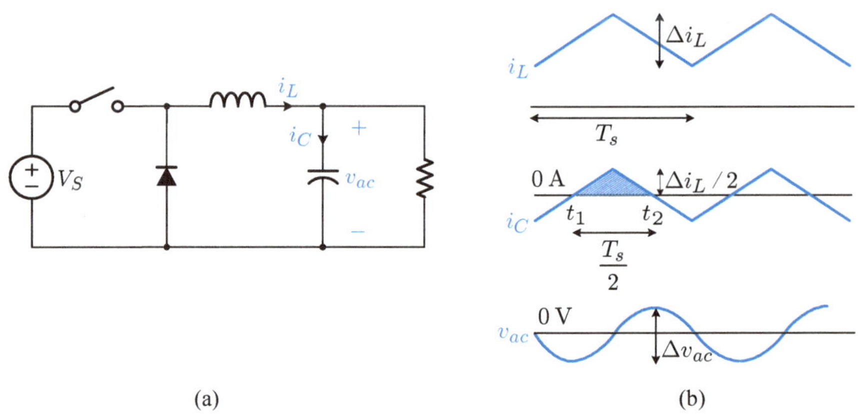


**Figure 3.12** Estimation of output ripple. (a) Circuit diagram. (b) Current waveforms and output ripple.

by integrating $i_C$ over the time period during which $i_C$ remains positive

$$\Delta v_{ac} = \frac{1}{C} \int_{t_1}^{t_2} i_C(\tau)\, \mathrm{d}\tau \tag{3.25}$$

The integration on the right-hand side of (3.25) corresponds to the area of the triangle enclosed by the positive inductor current and time axis, as highlighted in Fig. 3.12(b). By evaluating the area of the shaded triangle with incorporation of the fact $t_2 - t_1 = T_s/2$, the expression (3.25) becomes

$$\Delta v_{ac} = \frac{1}{C}\left(\frac{1}{2}\right)\left(\frac{1}{2}\Delta i_L\right)\left(\frac{1}{2}T_s\right) \tag{3.26}$$

Plugging (3.19) into (3.26) leads to the expression for the peak-to-peak variation of $v_{ac}$

$$\boxed{\Delta v_{ac} = \frac{1}{8}\frac{V_O}{LC}(1-D)T_s^2} \tag{3.27}$$

The voltage swing $\Delta v_{ac}$ given by (3.27) corresponds to the magnitude of the ripple component, $\Delta v_O$, superimposed on the dc value of the output voltage, $V_O$.

**Current waveforms and output ripple — EXAMPLE 3.6**

This example illustrates the accuracy of the preceding output ripple analysis. Figure 3.13 shows the simulated waveforms for the inductor current $i_L$, capacitor current $i_C$, and output voltage $v_O$ of the buck converter used in the previous example. Based on (3.27), the amplitude of the output ripple is estimated as

$$\begin{aligned}\Delta v_O = \Delta v_{ac} &= \frac{1}{8}\frac{V_O}{LC}(1-D)T_s^2 \\ &= \frac{1}{8}\frac{4}{40\times10^{-6}\,470\times10^{-6}}(1-0.25)\left(20\times10^{-6}\right)^2 \\ &= 7.98\ \text{mV}\end{aligned}$$

The result of this time-domain analysis is very close to the outcome of the previous frequency-domain analysis in Example 3.1. The current analysis shows $\Delta v_O =$ 7.98 mV, whereas the previous frequency-domain analysis predicted $\Delta v_O = 7.76$ mV. The exact PSpice® simulation in Fig. 3.13 shows $\Delta v_O = 7.95$ mV.

## Effects of Parasitic Resistance of Capacitor

Earlier descriptions about the steady-state operation of the buck converter assume that all the circuit components are ideal. In practice, however, the performance of the converter is influenced by non-ideal characteristics of real circuit components. The noticeable among them is the impact of the parasitic resistance of a real capacitor.

Real capacitors contain an internal parasitic resistance, referred to as the *equivalent series resistance* (*esr*), due to the non-ideal characteristics of the dielectric material inside the capacitor. Consequently, the circuit model of a real capacitor should include the esr, labeled as $R_c$ in Fig. 3.14; a real capacitor is considered as a series connection of the esr and capacitance. With the presence of the esr, the voltage ripple within a

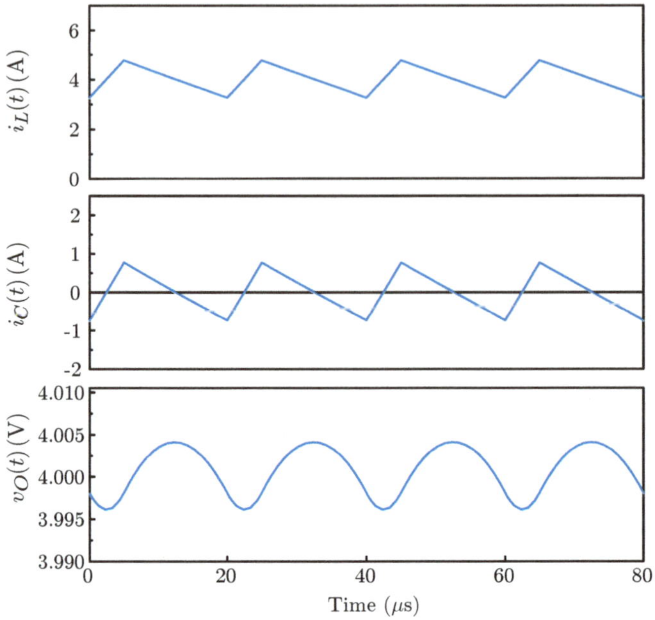


Figure 3.13 Current waveforms and output ripple of buck converter.

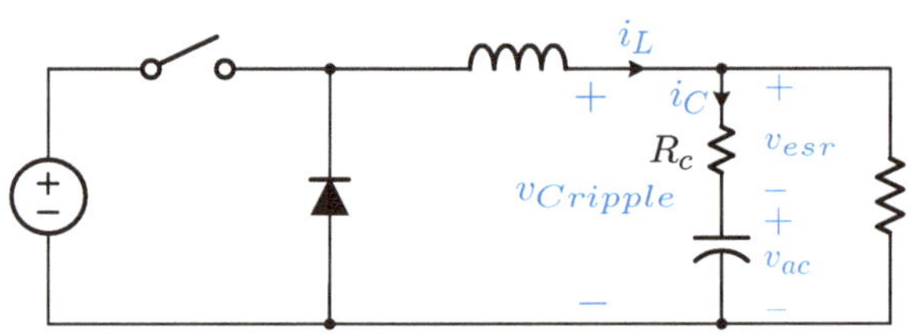


Figure 3.14 Buck converter with practical capacitor.

practical capacitor, $v_{Cripple}$, is decomposed into two components – the ripple voltage produced by the current-carrying capacitance, $v_{ac}$, and the ripple voltage due to the voltage drop at the esr, $v_{esr}$, as shown in Fig. 3.14

$$v_{Cripple}(t) = v_{ac}(t) + v_{esr}(t) = \frac{1}{C}\int i_C(t)\,\mathrm{d}t + i_C(t)R_c \qquad (3.28)$$

where $i_C$ denotes the capacitor current, which corresponds to the triangular portion of the inductor current. From (3.28), the magnitude of the output voltage ripple is expressed as†

$$\Delta v_O = \Delta v_{Cripple} \approx \frac{1}{C}\int_{t_1}^{t_2} i_C(\tau)\,\mathrm{d}\tau + \Delta i_C R_c \tag{3.29}$$

where $t_1$ and $t_2$ are defined in Fig. 3.12(b). Now, using the fact $\Delta i_C = \Delta i_L$, the magnitude of the output ripple is given by

$$\Delta v_O \approx \frac{1}{C}\int_{t_1}^{t_2} i_C(\tau)\,\mathrm{d}\tau + \Delta i_L R_c \tag{3.30}$$

The effect of the esr on the output voltage ripple is rather substantial. The magnitude of the ripple component at the esr is usually much larger than the ripple component produced by the capacitance itself. Accordingly, the magnitude of the output voltage ripple is approximated as

$$\Delta v_O \approx \Delta i_L R_c \tag{3.31}$$

considering the fact

$$\frac{1}{C}\int_{t_1}^{t_2} i_C(\tau)\,\mathrm{d}\tau \ll \Delta i_L R_c \tag{3.32}$$

**Output ripple with capacitor esr** **EXAMPLE 3.7**

This example illustrates the effects of the capacitor esr on the output voltage ripple. The esr of the 470 $\mu$F output capacitor used in the buck converter examples is $R_c = 0.05\ \Omega$. Figure 3.15 shows the simulated waveforms of the voltage across the esr, $v_{esr}(t)$, and the output voltage, $v_O(t)$, of the buck converter. As predicted, the output voltage ripple is very similar to the triangular-shaped voltage drop at the capacitor esr. The amplitude of the output voltage ripple is close to the analytical prediction of (3.31): $\Delta v_O \approx \Delta i_L R_c = 1.5 \cdot 0.05 = 0.075$ V.

The ripple component due to the current-carrying capacitance is negligible and the output voltage ripple is practically determined by the esr of the output capacitor. However, this does not imply that the previous ripple analysis with an ideal capacitor is pointless. The method of ripple calculation in (3.26) is still useful in the sense that the method itself will be applied to many cases of dc-to-dc converter circuit analyses.

## 3.5 BUCK CONVERTER IN DISCONTINUOUS CONDUCTION MODE

The operation of the buck converter is relatively simple and easy to understand. The piecewise linear analysis, along with the small-ripple approximation and flux

†Equation (3.29) is a conservative estimation because the peak values of the first term and the second term in (3.28) do not occur at the same time.

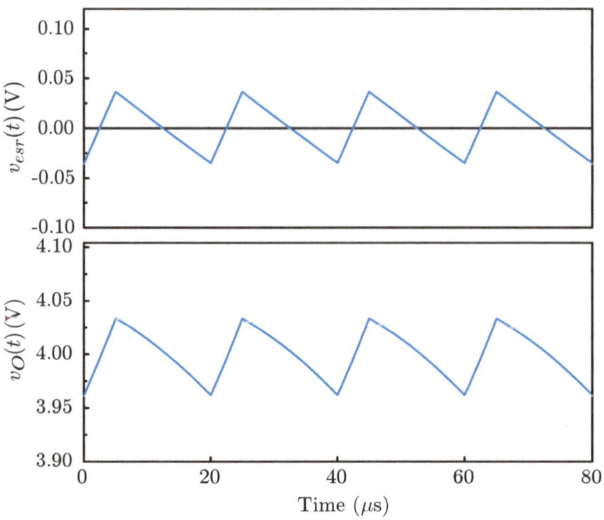


**Figure 3.15** Capacitor esr voltage $v_{esr}$ and output voltage $v_O$.

linkage/charge balance conditions, yields an accurate solution for the steady-state operation. However, there are cases where the buck converter does not follow the aforementioned operational principles and exhibits rather complex behaviors. This section deals with a new operational mode, known as the discontinuous conduction mode.

### 3.5.1 Origin of Discontinuous Conduction Mode Operation

The existence of the new operational mode is explained using Fig. 3.16 which shows a series of inductor currents for a buck converter. Each inductor current waveform is generated with a different load resistance but with the same duty ratio. As the load resistance becomes larger, the inductor current shifts downwards while retaining the same shape. This is because the average value of the inductor current is inversely proportional to the load resistance

$$I_L = \frac{V_O}{R} \tag{3.33}$$

yet the slope of the inductor current remains the same regardless of the change in the load resistance – the inductor voltage that determines the slope of the inductor current does not change as far as the duty ratio is unaltered. As the load resistance continues to increase, an instant emerges where the minimum value of the inductor current becomes zero. As shown in Fig. 3.16, this situation occurs when

$$I_L = \frac{V_O}{R} = \frac{1}{2}\Delta i_L \tag{3.34}$$

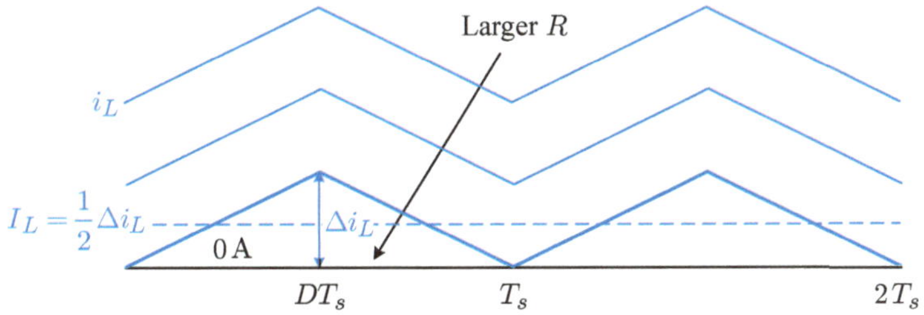


**Figure 3.16** Inductor current of buck converter as load resistance increases.

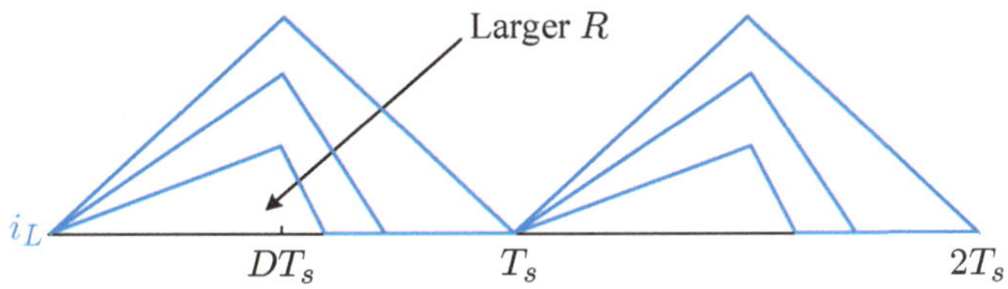


**Figure 3.17** Inductor current in DCM operation.

If the load resistance is further increased beyond the value that satisfies (3.34), while the converter obeying the same operational principles, the inductor current will shift further down, thereby becoming negative for a certain interval in the switching period. This situation implies that the inductor current has to change its direction. However, the current reversal is impossible because, during the off-time period, the inductor current flows through the diode, which is unable to deliver current in the reverse direction.

In reality, when the load resistance becomes large enough to push the average inductor current below the critical value, $I_L = \Delta i_L/2$, the converter no longer follows the operational principles described earlier and enters a new operational mode. This new operational mode is called the *discontinuous conduction mode* (*DCM*) where the inductor current remains *zero* for a certain time interval within each switching period, thereby becoming discontinuous.

The inductor current in DCM operation is illustrated in Fig. 3.17. The qualitative behavior of the inductor current is explained as follows.

1) In CCM operation, the slope of the inductor current is fixed when the duty ratio and input voltage of the converter are given, regardless of any changes in other operational conditions. This is because the output voltage stays still at the product of the fixed duty ratio and input voltage, thus retaining the inductor current slopes unchanged. In DCM operation, in contrast, the output voltage is a nonlinear function of multiple variables and could be varied even the duty

ratio is fixed. In particular, the output voltage increases in proportion to the load resistance. The consequential effects of this fact will be discussed shortly.

2) During an on-time period, the inductor voltage, $v_L = V_S - V_O$, is always positive and this forces the inductor current to grow linearly. However, as illustrated in Fig. 3.17, the slope of the on-time inductor current in DCM varies when the load resistance is changed – the larger the load resistance, the more gradual the slope. As stated in 1) and will be verified in Section 3.5.3, the DCM voltage gain is proportional to the load resistance; namely, as the load resistance becomes larger, so does the voltage gain. Accordingly, a greater load resistance produces a larger output voltage, thereby lowering the ascending slope of the inductor current, $(V_S - V_O)/L$.

3) During an off-time period, the negative inductor voltage, $v_L = -V_O$, forces the inductor current to decay linearly with the slope $-V_O/L$. The condition for DCM is $I_L < \Delta i_L/2$, and this condition implies that the inductor current reduces to zero before the next switching period begins. When the inductor current becomes zero, the diode is turned-off and remains off for the remaining part of the switching period. As addressed in 1) and 2), the output voltage in DCM increases as the load resistance becomes larger. Accordingly, as the load resistance grows, the decaying slope becomes steeper. As the decaying slope becomes steeper, the period during which the inductor current is absent widens, as illustrated in Fig. 3.17.

In contrast to the DCM operation, the case where the inductor current is present all the time is called *continuous conduction mode* (*CCM*) operation. We implicitly assumed the CCM case in earlier converter analyses in the previous sections. It should be noted that the DCM operation is as equally important and practical as the CCM operation, because all the dc-to-dc converters, even intended for CCM operation at normal conditions, will encounter DCM operation when the load current inevitably becomes lower than the critical value, $I_L = \Delta i_L/2$, during light load condition or energy-saving mode operation.

## EXAMPLE 3.8 PSIM® CCM to DCM transition

The transition from CCM operation to DCM operation is simulated in this example. Table 3.3 shows the PSIM model and inductor current $i_L$ and output voltage $v_O$ waveforms of the buck converter with successively increasing load resistances.

The converter operates with the duty ratio of $D = 1.9/3.8 = 0.5$ at the switching period $T_s = 1/50 \times 10^3 = 20\ \mu s$. As shown in the PSIM model in Table 3.3, the load resistance is increased from $R = 5\ \Omega$ to $11\ \Omega$ with a $1\ \Omega$ step using *the parameter sweep function* of PSIM®software.

The simulation results are shown in the second row of Table 3.3. The inductor current $i_L$ reveals the transition pattern illustrated in Figs. 3.16 and 3.17. With $R = 5, 6$, and $7\ \Omega$, the converter is in CCM operation. The converter is at the

**Table 3.3** CCM and DCM Operations of Buck Converter

PSIM model: Buck_Converter_02

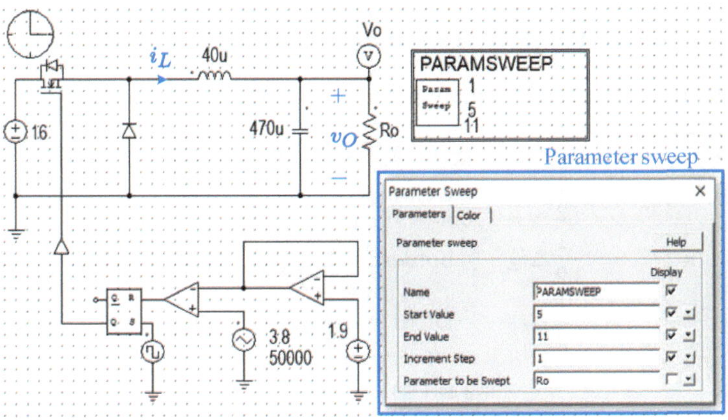


Simulation outcomes

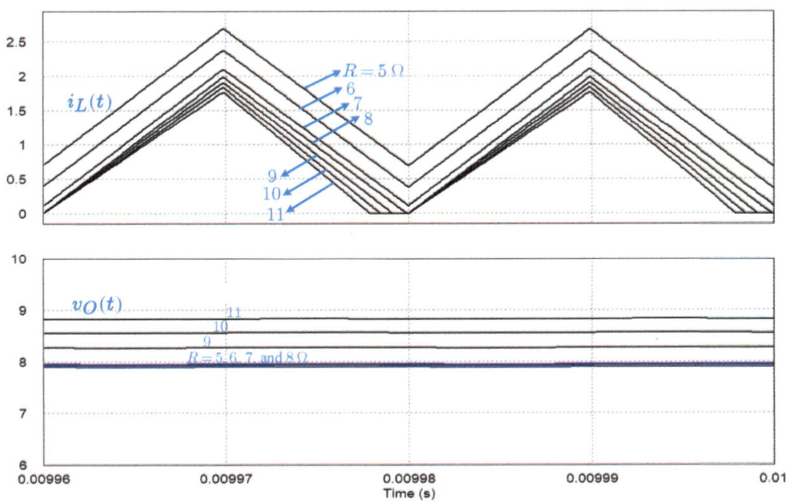


boundary of CCM and DCM with $R = 8\ \Omega$.[†] The converter is in DCM operation with $R = 9, 10,$ and $11\ \Omega$.

The output voltage $v_O$ of the converter reveals the characteristic feature of the converter in DCM operation. In the CCM and the borderline operations with $R = 5, 6, 7$ and $8\ \Omega$, the output voltage of the converter is fixed at $V_O = 0.5 \cdot 16 = 8$ V. In

[†] Section 3.5.2 will show that the converter is precisely on the CCM/DCM borderline with $R = 8\ \Omega$ for the given operational condition.

contrast, the output voltage in DCM operation with $R = 9, 10,$ and $11\ \Omega$ increases in proportion to the load resistance.

### 3.5.2 Conditions for DCM Operation

It is evident from the previous analysis that the operational mode of the converter is determined as

$$\begin{aligned} I_L &> \frac{1}{2}\Delta i_L : \quad \text{CCM} \\ I_L &= \frac{1}{2}\Delta i_L : \quad \text{borderline between CCM and DCM} \\ I_L &< \frac{1}{2}\Delta i_L : \quad \text{DCM} \end{aligned} \tag{3.35}$$

Referring to (3.18) and (3.19), the condition for the borderline between CCM and DCM operation is expressed as

$$\boxed{I_L = \frac{1}{2}\Delta i_L \quad \Rightarrow \quad \frac{V_O}{R} = \frac{1}{2}\frac{V_O}{L}(1-D)T_s} \tag{3.36}$$

Equation (3.36) is used to find the critical value for the load resistance or filter inductance that places the converter at the CCM/DCM borderline

$$\boxed{R_{crit} = \frac{2L}{(1-D)T_s}} \tag{3.37}$$

$$\boxed{L_{crit} = \frac{(1-D)RT_s}{2}} \tag{3.38}$$

Equations (3.37) and (3.38) enable us to determine the operational mode from the value of one specific circuit component. When the load resistance is larger than $R_{crit}$, the converter establishes DCM operation. Likewise, when the inductance is reduced below $L_{crit}$, the converter enters DCM operation. Readers are recommended to confirm these statements.

#### EXAMPLE 3.9 Operational mode change with resistance variation

This example illustrates the operational mode change as the load resistance is varied. The critical resistance $R_{crit}$ of the buck converter in the preceding examples is evaluated as

$$R_{crit} = \frac{2L}{(1-D)T_s} = \frac{2 \cdot 40 \times 10^{-6}}{(1-0.25)\, 20 \times 10^{-6}} = 5.33\ \Omega$$

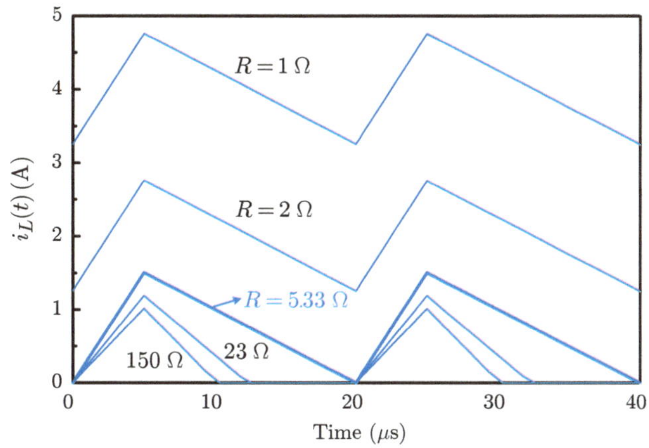


**Figure 3.18** Inductor current waveforms with different load resistances.

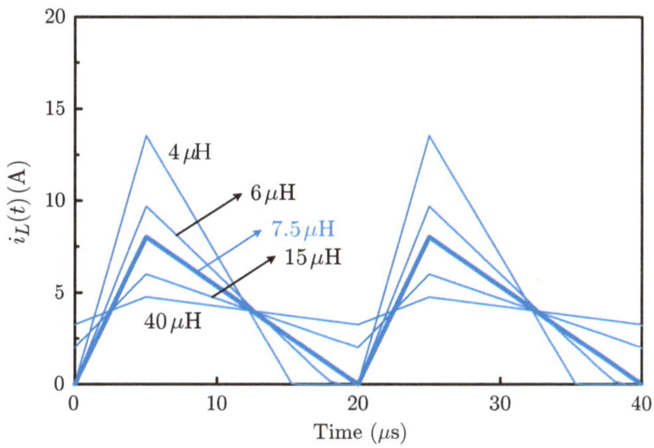


**Figure 3.19** Inductor current waveforms with different filter inductances.

The inductor current waveforms of the buck converter, whose load resistance is varied between $1\ \Omega < R < 150\ \Omega$, are shown in Fig. 3.18. The converter is on the boundary of CCM operation and DCM operation when $R = 5.33\ \Omega = R_{crit}$ and enters DCM when the load resistance is further increased.

## Operational mode change with inductance variation EXAMPLE 3.10

This example shows the operational mode change when the inductance is altered. The critical inductance $L_{crit}$ of the previous buck converter with a load resistance

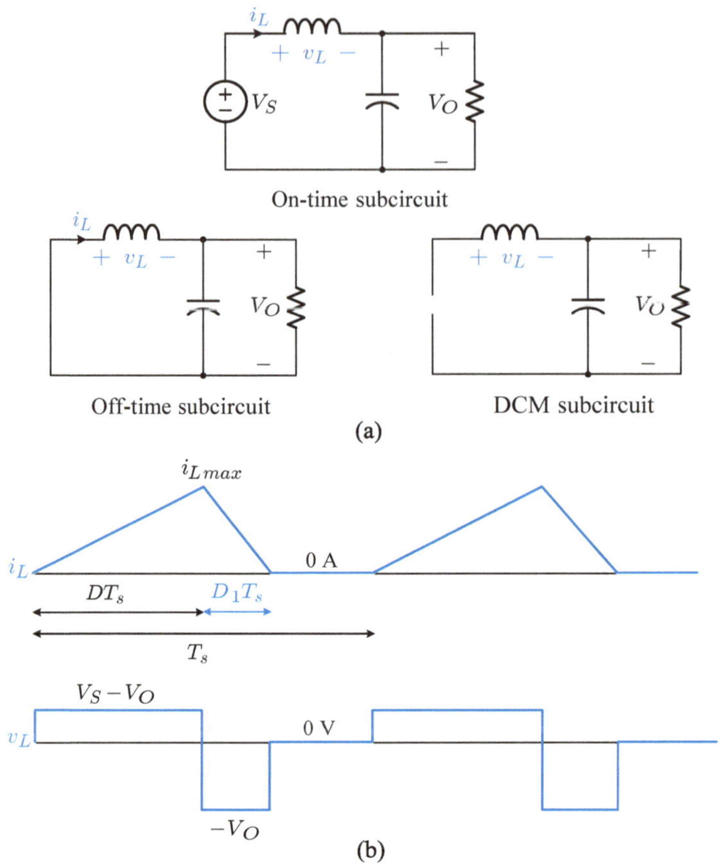


**Figure 3.20** Operation of buck converter in DCM. (a) Three subcircuits. (b) Inductor current $i_L$ and inductor voltage $v_L$.

$R = 1\ \Omega$ is given by

$$L_{crit} = \frac{(1-D)RT_s}{2} = \frac{(1-0.25)1 \cdot 20 \times 10^{-6}}{2} = 7.5\ \mu\text{H}$$

Figure 3.19 shows the inductor current waveforms of the converter where the inductance is varied between $4\ \mu\text{H} < L < 40\ \mu\text{H}$. As the inductance becomes smaller, the inductor current swing $\Delta i_L$ increases, thereby causing the converter to operate on the borderline when $L = 7.5\ \mu\text{H} = L_{crit}$, and in DCM when $L < L_{crit}$.

### 3.5.3 Steady-State Operation in DCM

In DCM operation, three topological modes exist within one switching period, as illustrated in Fig. 3.20. In addition to the on-time and off-time subcircuits found in CCM operation, a new subcircuit appears as the inductor current disappears during the off-time period. This third subcircuit is referred to as the DCM subcircuit in Fig. 3.20(a).

Figure 3.20(b) depicts typical DCM waveforms for the inductor current and inductor voltage. The notation $D_1$ in Fig. 3.20(b) is defined as

$$D_1 \equiv \frac{\text{a part of off-time period in which inductor current exists}}{\text{switching period}}$$

When the converter is reduced to the DCM subcircuit, the inductor current and inductor voltage both become zero, as shown in Fig. 3.20(b): $i_L = 0 \Rightarrow \Delta i_L = 0 \Rightarrow v_L = L(\Delta i_L/\Delta t) = 0$. By applying the volt-sec balance condition to the inductor, it follows that

$$(V_S - V_O)DT_s = V_O D_1 T_s \tag{3.39}$$

which is simplified to

$$\frac{V_O}{V_S} = \frac{D}{D + D_1} \tag{3.40}$$

The unknown variable $D_1$ should be eliminated from (3.40) to result in the complete DCM voltage gain. For this purpose, an additional equation is formulated as follows. The average value of the inductor current should be the same as the load current, because the average capacitor current is zero due to the charge balance condition. From the inductor current waveform, the average inductor current is given by

$$I_L = \frac{\frac{1}{2} i_{L\,max}(D + D_1)T_s}{T_s} \tag{3.41}$$

Using the facts that

$$I_L = I_O = \frac{V_O}{R} \tag{3.42}$$

and

$$i_{L\,max} = \frac{V_O}{L} D_1 T_s \tag{3.43}$$

the expression (3.41) is rewritten as

$$\frac{V_O}{R} = \frac{1}{2}\frac{V_O}{L} D_1 T_s (D + D_1) \tag{3.44}$$

which is simplified to

$$D_1^2 + DD_1 - \frac{2L}{RT_s} = 0 \tag{3.45}$$

By solving (3.45) for a positive $D_1$

$$D_1 = \frac{1}{2}\left(-D + \sqrt{D^2 + \frac{8L}{RT_s}}\right) \tag{3.46}$$

and finally incorporating (3.46) into (3.40), the DCM voltage gain is determined as

$$\boxed{\frac{V_O}{V_S} = \frac{2D}{D + \sqrt{D^2 + \dfrac{8L}{RT_s}}}} \tag{3.47}$$

The voltage gain is a nonlinear function of the circuit parameters and operational conditions.

A couple of informative observations are made from the previous analysis. First, it is evident from (3.47) that the voltage gain increases as the load resistance grows larger, as addressed earlier in conjunction with the behavior of the inductor current in DCM. Second, the following relationship usually holds

$$\frac{2D}{D + \sqrt{D^2 + \dfrac{8L}{RT_s}}} > D \tag{3.48}$$

for the normally selected converter parameters. Equation (3.48) indicates that the DCM voltage gain is larger than the CCM counterpart when the same duty ratio is assumed. This observation provides some insights to predict the converter behavior over CCM/DCM transitions. For example, when the converter jumps into DCM operation from CCM operation for some reasons, while the duty ratio $D$ remains the same, the output voltage becomes larger than the previous value. If we desire to keep the output voltage unchanged, we should decrease the duty ratio of the converter.

### EXAMPLE 3.11 Steady-state analysis in DCM

This example illustrates the circuit waveforms of a buck converter in DCM. The load resistance of the buck converter used in the previous examples is increased to $R = 12\ \Omega$, while the other parameters retain the original values: $V_S = 16$ V, $L = 40\ \mu$H, $C = 470\ \mu$F, $T_s = 20\ \mu$s, and $D = 0.25$. Because $R = 12\ \Omega > R_{crit} = 5.33\ \Omega$, the converter is in DCM region. Theoretical predictions for the steady-state response of the converter are listed in Table 3.4. The accuracy of the theoretical predictions is confirmed with the simulated waveforms in Fig. 3.21.

The operational mode of dc-to-dc converters also goes through the CCM/DCM transition when its duty ratio is varied. The condition for the DCM operation

$$\frac{V_O}{R} < \frac{1}{2}\frac{V_O}{L}(1 - D)T_s \tag{3.49}$$

**Table 3.4** Steady-State Analysis in DCM

| Circuit variable | Expression |
|---|---|
| $D_1$ | $\frac{1}{2}\left(-D+\sqrt{D^2+\frac{8L}{RT_s}}\right)$ <br> $=\frac{1}{2}\left(-0.25+\sqrt{0.25^2+\frac{8\cdot 40\times 10^{-6}}{12\cdot 20\times 10^{-6}}}\right)=0.47$ |
| $V_O$ | $\frac{2D}{D+\sqrt{D^2+\frac{8L}{RT_s}}}V_S$ <br> $=\frac{2\cdot 0.25}{0.25+\sqrt{0.25^2+\frac{8\cdot 40\times 10^{-6}}{12\cdot 20\times 10^{-6}}}}16=5.59\text{ V}$ |
| $i_{L\,max}$ | $\frac{V_O}{L}D_1T_s=\frac{5.59}{40\times 10^{-6}}0.47\cdot 20\times 10^{-6}=1.31\text{ A}$ |

is rearranged as

$$D<\left(1-\frac{2L}{RT_s}\right) \tag{3.50}$$

By defining the right-hand side of (3.50) as

$$1-\frac{2L}{RT_s}\equiv D_{crit} \tag{3.51}$$

it is concluded that the converter operates in DCM when $0<D<D_{crit}$ and in CCM with $D_{crit}<D<1$. Figure 3.22 depicts the voltage gain curve of a buck converter. The gain curve follows the DCM gain formula given by (3.47) until the duty ratio increases to $D_{crit}$, and tracks the CCM formula, $V_O/V_S=D$, thereafter.

It is worthwhile to note that the converter remains in CCM operation for all $0<D<1$, if the condition

$$\frac{2L}{RT_s}>1 \tag{3.52}$$

is satisfied. This condition is derived from

$$D_{crit}=\left(1-\frac{2L}{RT_s}\right)<0 \tag{3.53}$$

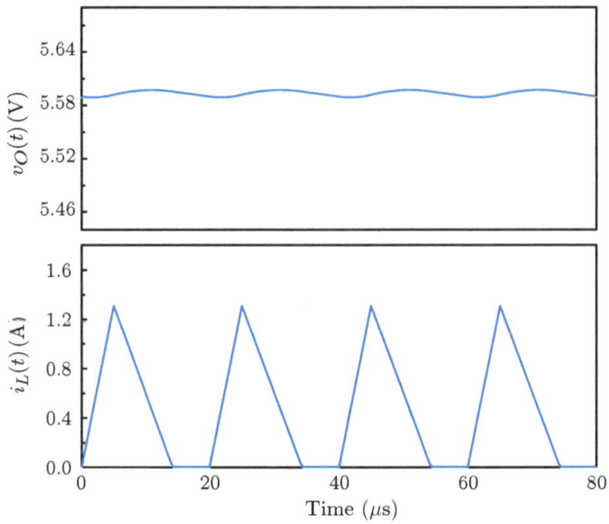


Figure 3.21 Output voltage $v_O$ and inductor current $i_L$ of buck converter in DCM.

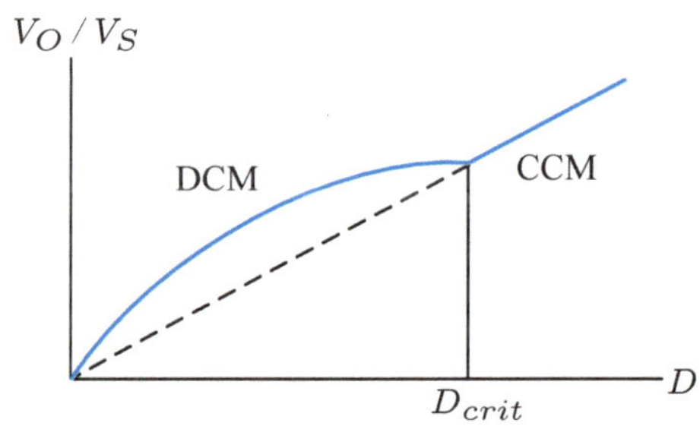


Figure 3.22 Voltage gain of buck converter as function of duty ratio.

which negates the existence of a positive $D_{crit}$.

## EXAMPLE 3.12 Buck converter example

This example shows the case where the converter retains the CCM operation for $0 < D < 1$. For the buck converter used in the previous examples with the circuit parameters of $L = 40\ \mu$H, $R = 1\ \Omega$, and $T_s = 20\ \mu$s, it follows that

$$\frac{2L}{RT_s} = \frac{2 \cdot 40 \times 10^{-6}}{1 \cdot 20 \times 10^{-6}} = 4 > 1$$

indicating the converter will remain in CCM for all $0 < D < 1$.

The DCM operation causes considerable changes in the operational principles and steady-state characteristics of the converter. Furthermore, the DCM operation also alters the dynamic characteristics of the converter. While detailed analyses of DCM dynamics will be covered in the second volume of the book series, it should be now reminded that dc-to-dc converters frequently cross the CCM/DCM boundary and therefore should be designed with both CCM operation and DCM operation in consideration.

## 3.6 CLOSED-LOOP CONTROL OF BUCK CONVERTER

In real applications, dc-to-dc converters are powered by a non-ideal voltage source rather than a constant dc source. In addition, converters are loaded with general electronic devices and instruments, not a resistor. Accordingly, dc-to-dc converters experience certain variations in the input voltage and load current.

A dc-to-dc converter is intended to be a voltage source and, as such, the converter should maintain its output voltage at the desired value, regardless of any changes in the input voltage or load current. This feature is referred to as the output voltage regulation or simply *dc regulation*. The term dc regulation implies *regulating the output voltage of a converter at a fixed dc value in steady state.*

To implement the dc regulation, a functional connection must be made between the output voltage and duty ratio of the converter. More specifically, a closed-loop feedback control should be employed around the dc-to-dc converter, which adaptively changes the duty ratio of the active switch at the cases of the input voltage and load current variations.[†] This section deals with the dc regulation of a closed-loop controlled buck converter.

Closing a feedback loop is not an easy task and requires broad knowledge about modeling, analysis, and control of dynamic systems. Yet, because the scope of the current section is limited to the issues related to the dc regulation, the dynamic modeling, analysis, and control design of PWM converters are postponed to forthcoming volumes of the book series.

### 3.6.1 Closed-Loop Feedback Controller

Figure 3.23 shows a simplified diagram of the buck converter equipped with a closed-loop feedback controller. The feedback controller consists of two functional blocks: the pulsewidth modulation (PWM) block and voltage feedback circuit. The PWM block modulates the duty ratio of the active switch using 1) the output of the voltage feedback circuit, labeled as the control voltage $v_{con}$ in Fig. 3.23, and 2) the ramp signal $V_{ramp}$ generated inside the feedback controller. The output of the PWM block is denoted as $v_q$ in Fig. 3.23.

[†]The load current change could force the converter to switch between CCM operation and DCM operation, whose voltage gain expressions are quite different.

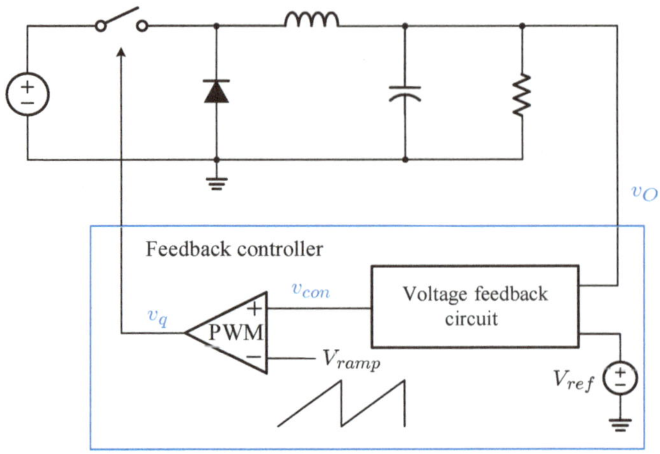


**Figure 3.23** Closed-loop controlled buck converter.

The voltage feedback circuit utilizes the converter output voltage $v_O$ and reference voltage, denoted as $V_{ref}$ in Fig. 3.23, to generate the control voltage $v_{con}$. The voltage feedback circuit and PWM block collectively enforce the output of the converter to track the reference voltage in steady state, thereby achieving dc regulation. This section describes the functional details of the closed-loop dc regulation using the PWM scheme.

## Pulsewidth Modulation

The output of the PWM block is the switch drive signal $v_q$, whose pulsewidth is modulated to yield the desired duty ratio for the active switch. Figure 3.24 depicts the PWM block and its timing diagram. As shown in Fig. 3.24(a), the PWM block employs the control voltage $v_{con}$ and ramp signal $V_{ramp}$, in order to issue the switch drive signal $v_q$.

Figure 3.24(b) shows a timing diagram for the PWM waveforms, where the control voltage is assumed to remain constant at $v_{con} = V_{con}$. First, a periodic ramp signal $V_{ramp}$ is generated inside the feedback controller. The period of the ramp signal is in fact the switching period of the converter. The switch drive signal sets *on* at the instant the ramp signal starts ramping up and resets *off* when the ramp signal intersects the control voltage. By repeating this process, the PWM block generates a periodic switch drive signal whose pulsewidth is modulated in proportion to the magnitude of the control voltage $V_{con}$.

Figure 3.25 displays the outputs of the PWM block for two different control voltage values, $V_{con}$ and $V'_{con}$. The modulation scheme generates a switch drive signal whose trailing edge is modulated to change the duty ratio within a fixed switching period. As such, the modulation scheme is called a *constant-frequency trailing edge modulation.* While alternative modulation schemes are also available

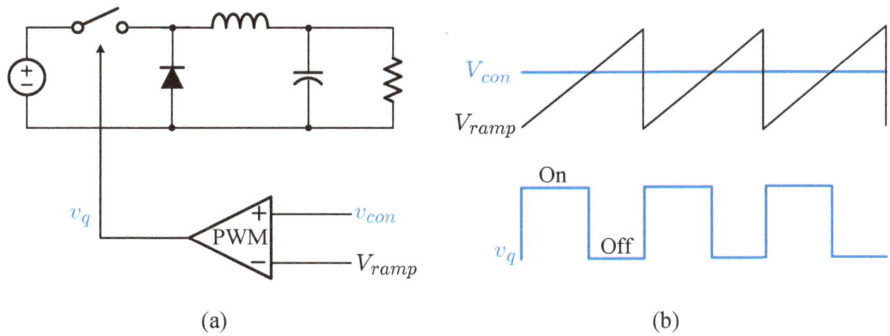


**Figure 3.24** PWM block. (a) Block diagram. (b) Timing diagram.

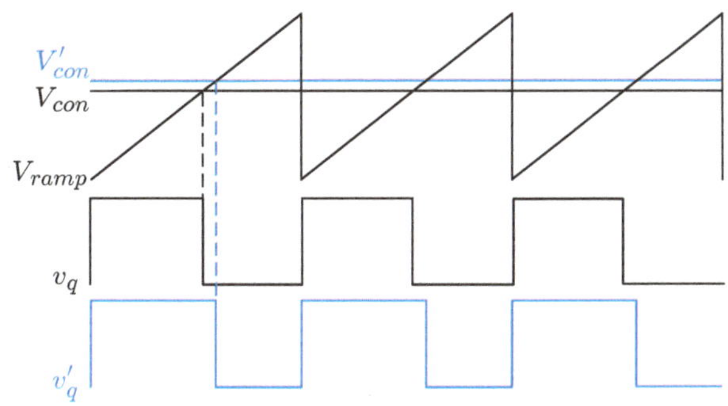


**Figure 3.25** Constant-frequency trailing edge modulation.

for PWM dc-to-dc converters, this scheme is most broadly used and thus adopted as the standard PWM in this book.

## Voltage Feedback Circuit

Figure 3.26 is a simplified diagram of a closed-loop controlled buck converter. The converter output voltage $v_O$ is fed to a voltage feedback circuit, consisting of an op amp, reference voltage $V_{ref}$, and two impedance blocks, $Z_1(s)$ and $Z_2(s)$. The output of the voltage feedback circuit is the control voltage, $v_{con}$, which is used as the input signal for the PWM block.

The voltage feedback circuit operates according to the principle of the negative feedback. When the output voltage $v_O$ grows larger than the nominal value, the output of the op amp $v_{con}$ becomes lower than the previous value, as indicated by the *minus* sign at the inverting terminal of the op amp. With the lowered $v_{con}$, the PWM block produces the switch drive signal $v_q$ whose duty ratio is reduced. The reduced duty ratio in turn brings down the output voltage towards the nominal value.

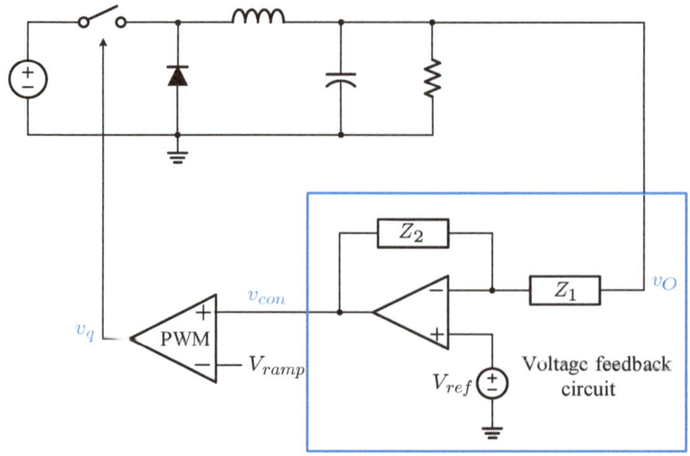


**Figure 3.26** Voltage feedback circuit and closed-loop control.

The exact mechanism of the dc regulation is further explained using the node equation at the inverting terminal of the op amp. Using the *virtual short* between the two input terminals of the op amp, the node equation is written as

$$\frac{v_O - V_{ref}}{Z_1(s)} = \frac{V_{ref} - v_{con}}{Z_2(s)} \tag{3.54}$$

which is rearranged as

$$\frac{Z_2(s)}{Z_1(s)}\left(v_O - V_{ref}\right) = V_{ref} - v_{con} \tag{3.55}$$

The nature of the dc regulation is deduced from (3.55) as follows.

1) For a proper operation of the converter, the output of the feedback circuit $v_{con}$ should have a finite value within the lower and upper bounds of the op amp output voltage.

2) Because both the two variables in the right-hand side of (3.55), $V_{ref}$ and $v_{con}$, are finite in their magnitude, the term $(|Z_2|/|Z_1|)(v_O - V_{ref})$ in the left-hand side of the equation should be also finite. Otherwise, the expression (3.55) becomes inconsistent.

3) The steady state is an equilibrium where all the time-varying components of circuit variables settle down to zero and the circuit responds only to dc components. As such, the impedance evaluated at dc, $Z(j0)$, is the factor that determines the steady-state operation of the voltage feedback circuit.

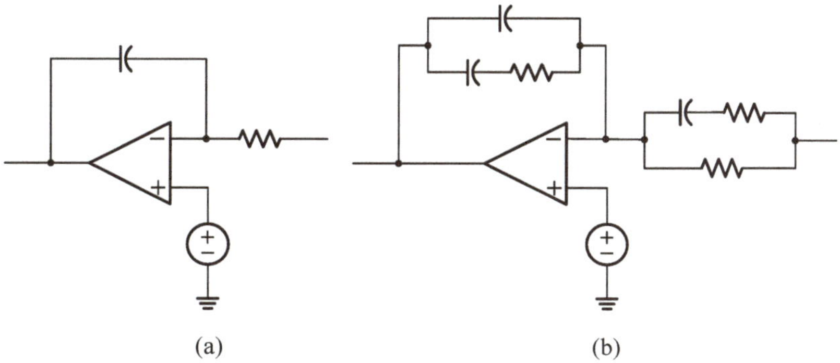

Figure 3.27 Voltage feedback circuits. (a) Miller integrator. (b) Three-pole two-zero circuit.

4) If the impedance ratio $|Z_2(j0)|/|Z_1(j0)|$ is selected to be infinite, the variable $(V_O - V_{ref})$ should converge to zero in order to make their product finite

$$\frac{|Z_2(j0)|}{|Z_1(j0)|}(V_O - V_{ref}) = \infty \cdot 0 \quad \Rightarrow \quad \text{a finite} \quad \text{constant}$$

This condition implies $V_O = V_{ref}$ in steady state, thereby achieving dc regulation.

In conclusion, a voltage feedback circuit that satisfies the condition

$$\boxed{\frac{|Z_2(j0)|}{|Z_1(j0)|} = \infty} \tag{3.56}$$

forces the output voltage to track the reference voltage in steady state. Accordingly, any circuit that meets the condition (3.56) could provide dc regulation, thus eligible for a voltage feedback circuit. The simplest among these is Miller integrator shown in Fig. 3.27(a). However, a voltage feedback circuit employing Miller integrator in its original form does not provide good dynamic performance for the reasons that will be clarified in the forthcoming volumes of the book series. Consequently, a simple Miller integrator is rarely used for high performance design.

It has been established from past researches that a three-pole two-zero circuit shown in Fig. 3.27(b) is the best feedback circuit for the buck converter. The three-pole two-zero circuit, named after its transfer function, provides excellent dynamic performance as well as tight dc regulation. It is easy to confirm that the three-pole two-zero circuit meets the requirement of (3.56). Because the design of voltage feedback circuit requires comprehensive knowledge about dynamic modeling and frequency-domain analysis, this topic will be separately treated in the second volume.

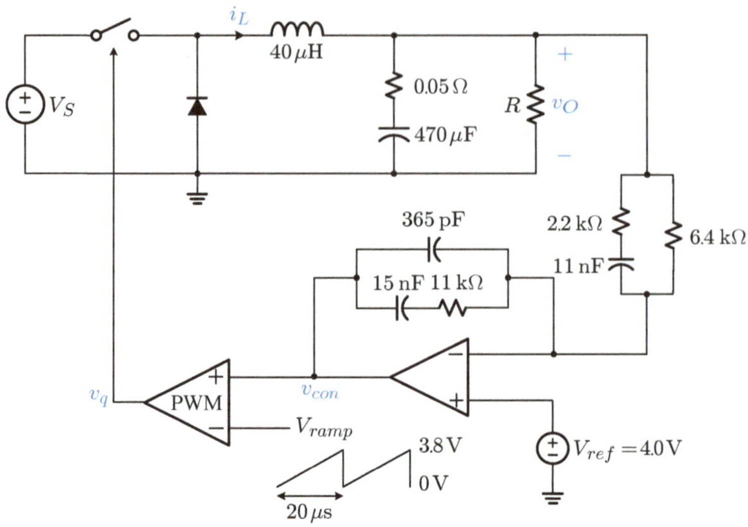


**Figure 3.28** Closed-loop controlled buck converter.

## EXAMPLE 3.13 Alternative derivation of regulation condition

This example provides an alternative derivation of the output voltage regulation condition given in (3.56). From Fig. 3.26, the control voltage $v_{con}$ is expressed as the sum of the responses of the non-inverting and inverting amplifiers

$$v_{con} = v_{ref}\left(1 + \frac{|Z_2(j\omega)|}{|Z_1(j\omega)|}\right) + v_O\left(-\frac{|Z_2(j\omega)|}{|Z_1(j\omega)|}\right)$$

The previous equation is rearranged as

$$v_{con} = \frac{|Z_2(j\omega)|}{|Z_1(j\omega)|}\left(v_{ref}\left(\frac{|Z_1(j\omega)|}{|Z_2(j\omega)|} + 1\right) - v_O\right)$$

When the equation is evaluated under the condition

$$\frac{|Z_2(j0)|}{|Z_1(j0)|} = \infty$$

it follows that

$$V_{con} = \infty\left(V_{ref} - V_O\right) \quad \Rightarrow \quad \left(V_{ref} - V_O\right) = 0 \quad \Rightarrow \quad V_O = V_{ref} \tag{3.57}$$

because $V_{con}$ is a finite constant in steady state.

### 3.6.2 Transient Responses of Closed-Loop Controlled Buck Converter

This section presents the time-domain response of a closed-loop controlled buck converter obtained from both PSpice® and PSIM® simulations. Figure 3.28 shows the schematic diagram of a closed-loop controlled buck converter. The power stage parameters are identical to those of the buck converter used in the previous examples, however, the input voltage $V_S$ and load resistance $R$ are allowed to vary during the converter operation.

The output of the converter is regulated at $V_O = V_{ref} = 4$ V and the switching frequency is $f_s = 1/(20 \times 10^{-6}) = 50$ kHz. The ramp signal ramps from 0 to 3.8 V. A three-pole two-zero circuit is employed as the voltage feedback circuit. The circuit parameters for the voltage feedback circuit are correctly selected for good closed-loop performance, as will be demonstrated in the second volume of the book series.

Before embarking on formal analyses, we glimpse the transient behavior of the converter using PSIM® simulations. An illustrative simulation is given below.

**PSIM® Transient response of buck converter EXAMPLE 3.14**

Table 3.5 shows the PSIM model of the buck converter of Fig. 3.28. Here, the load resistance of the converter periodically toggles between $R = 1\ \Omega$ and $0.5\ \Omega$. This step load change is programmed with two resistors followed by complementary-driven MOSFET switches.

Table 3.5 shows the inductor current $i_L$ and output voltage $v_O$ of the converter. The converter first goes through the start-up response and shows a couple of step load transient responses. The inductor current $i_L$ shows very stable and well-controlled behavior, in response to the step changes between $I_O = 1$ A and 2 A. The output voltage $v_O$ is tightly regulated at $V_O = 5$ V with nearly undetectable excursions in response to step load changes. Detailed discussions about the transient response are given later in this section.

#### Step Input Response

In this section, we assume some step changes in the input voltage. The buck converter undergoes a series of step changes in its input voltage, whereas the load resistance is fixed at $R = 1\ \Omega$. As shown in Fig. 3.29, the input voltage is changed as $V_S = 16\text{ V} \Rightarrow 8\text{ V} \Rightarrow 16\text{ V}$ during the converter operation.

First, the steady-state waveforms of the converter with $V_S = 16$ V and $R = 1\ \Omega$ are analyzed. The output voltage $v_O$ is regulated at 4 V with a small ripple component. The inductor current $i_L$ is a triangular waveform. The average value of the inductor current equals the load current

$$I_L = \frac{V_O}{R} = \frac{4}{1} = 4\text{ A}$$

**Table 3.5** Transient Response of Buck Converter

PSIM model: Buck_Converter_03

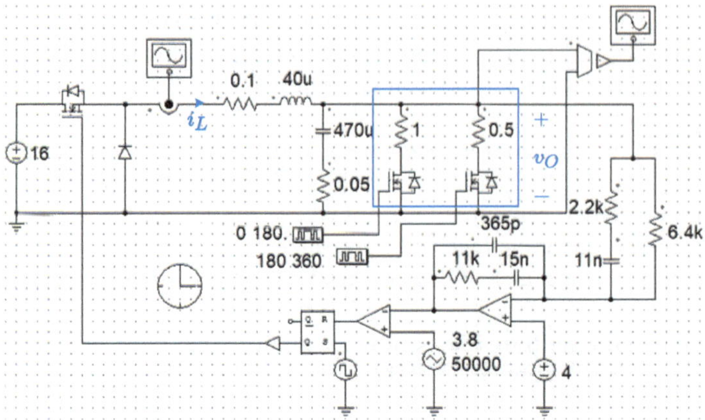


Simulation outcomes

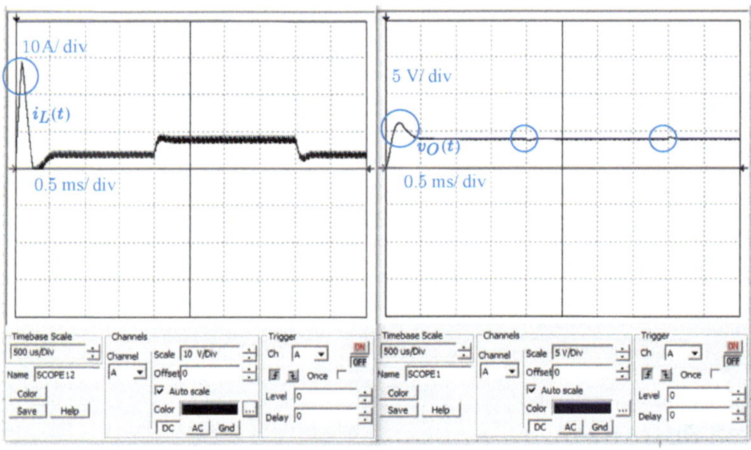


The ripple component of the inductor current is given by

$$\Delta i_L = \frac{V_S - V_O}{L} DT_s = \frac{16 - 4}{40 \times 10^{-6}} \frac{4}{16} 20 \times 10^{-6} = 1.5 \text{ A}$$

The control voltage $v_{con}$ and ramp signal $V_{ramp}$ are simultaneously shown in the third plot in Fig. 3.29, while the switch drive signal $v_q$ is displayed in the bottom plot.

The average value for the control voltage is evaluated from the PWM waveforms in Fig. 3.30. From the PWM waveforms, it is seen that

$$T_s : DT_s = V_m : V_{con} \tag{3.58}$$

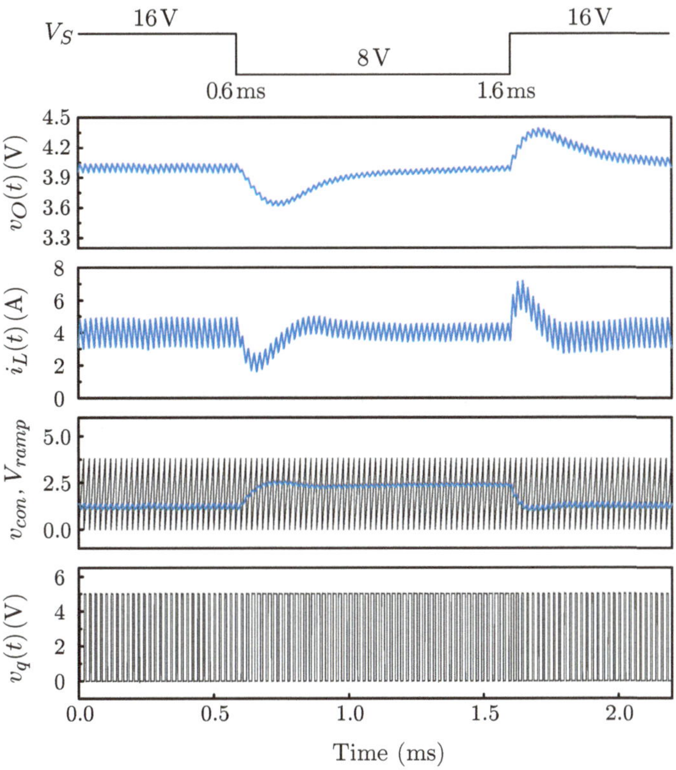


**Figure 3.29** Step input response of buck converter: output voltage $v_O$, inductor current $i_L$, control voltage $v_{con}$, ramp signal $V_{ramp}$, and switch drive signal $v_q$.

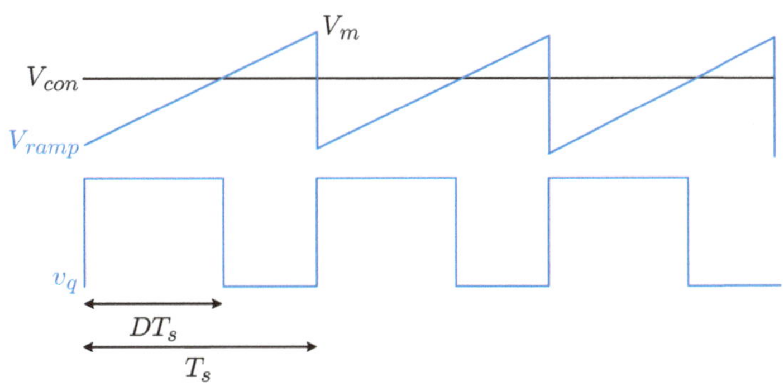


**Figure 3.30** PWM block waveforms.

where $V_m$ represents the magnitude of the ramp signal and $V_{con}$ is the steady-state control voltage. The above expression is simplified to $V_{con} = DV_m$. With the given operational conditions, the value for the control voltage is evaluated as

$$V_{con} = DV_m = \frac{V_O}{V_S}V_m = \frac{4}{16}3.8 = 0.95 \text{ V}$$

The control voltage contains a ripple component on top of the dc value evaluated above. The switch drive signal $v_q$ shows the duty ratio of the active switch. The ramp signal, control voltage, and switch drive signal altogether demonstrate the principles of the PWM control.

When the input voltage steps down from $V_S = 16$ V to 8 V at $t = 0.6$ ms, several changes occur in the converter waveforms. First, the output voltage exhibits a transitional undershoot before it returns to the nominal value $V_O = 4$ V. Upon a sudden decrease in $V_S$, the feedback controller responds to the change with a certain response time. Before the feedback controller establishes a new steady state, the change in the input voltage influences the output voltage as follows. When the input voltage is reduced, the energy delivered from the input source to the output capacitor is also reduced, incurring an energy deficit in the output capacitor. This energy deficit causes the capacitor voltage to drop, resulting in a transitional undershoot in the output voltage. As time elapses, the feedback controller establishes a new steady state and the output voltage tracks back to the nominal value. The inductor current shows the similar behavior.

With the reduced input voltage $V_S = 8$ V, the ripple component of the inductor current is decreased to

$$\Delta i_L = \frac{V_S - V_O}{L}DT_s = \frac{8-4}{40 \times 10^{-6}}\frac{4}{8}20 \times 10^{-6} = 1 \text{ A}$$

The control voltage gradually increases to produce a larger duty ratio, required to regulate the output voltage with the reduced input voltage. The new value for the control voltage is determined as

$$V_{con} = \frac{V_O}{V_S}V_m = \frac{4}{8}3.8 = 1.9 \text{ V}$$

The duty ratio of the switch drive signal also gradually grows according to the PWM principle.

The transient behavior of the converter due to a step input increase is interpreted in the contrasting way to the step-down case. The instantaneous increase of the input voltage causes the output voltage and inductor current to go through a transitional overshoot. As the control signal gradually decreases, thereby reducing the duty ratio to $D = 4/16 = 0.25$, the output voltage returns to the nominal value.

The buck converter exhibits very stable and well-controlled transient behavior. In fact, the feedback circuit parameters are carefully selected to yield the best possible transient behavior. Selections of the feedback circuit parameters will be studied in the second volume of the series.

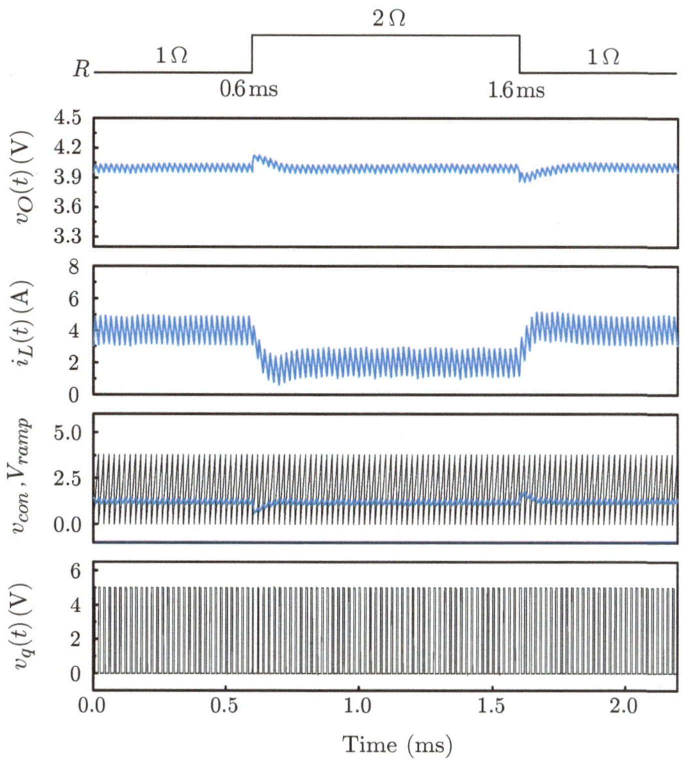


**Figure 3.31** Step load response of buck converter.

## Step Load Response

Now the converter goes through a series of step changes in the load resistance, $R = 1\ \Omega \Rightarrow 2\ \Omega \Rightarrow 1\ \Omega$, while the input voltage is fixed at $V_S = 16$ V. The pattern of the step changes and the simulated waveforms are shown in Fig. 3.31. With a step load change from $R = 1\ \Omega$ to $2\ \Omega$ at $t = 0.6$ ms, the energy release from the capacitor to the load resistor suddenly becomes smaller than the previous value, resulting in an energy surplus at the output capacitor. Accordingly, the capacitor voltage momentarily exceeds the nominal value, thus producing an output voltage overshoot. The inductor current decreases in proportion to the load current. The inductor current also shows that the converter still remains in CCM operation with $R = 2\ \Omega$. The load resistance that places the converter on the borderline between CCM operation and DCM operation is given by

$$R_{crit} = \frac{2L}{(1-D)T_s} = \frac{2 \cdot 40 \times 10^{-6}}{(1-0.25)\,20 \times 10^{-6}} = 5.33\ \Omega$$

Figure 3.31 also shows the transient waveforms of the control voltage and switch drive signal. After passing through a short transition period, the control voltage and switch drive signal return to their original waveforms – the duty ratio is invariant to the load resistance, as far as the converter operates in CCM while maintaining the output voltage in regulation.

The transient behavior in response to the step load change from $R = 2\ \Omega$ to $1\ \Omega$ could be understood in the contrasting way to the previous analysis. In this case, the output voltage exhibits an undershoot due to the transitional energy deficit.

## EXAMPLE 3.15 PSIM® Effects of voltage feedback circuit

Transient responses of the closed-loop converters are critically affected by the voltage feedback circuit. Accordingly, the circuit components of the voltage feedback circuit should be carefully chosen to obtain the best possible transient responses. In particular, the transient response of the output voltage due to step load changes is a prime concern in most practical applications for the reasons discussed in Section 1.3.

The design of voltage feedback circuit is very important in the power conversion technology and deserves an extensive study. The second volume of this book series is dedicated to this important topic.

In this example, we just observe the impacts of the voltage feedback circuit on the step load response of the buck converter. For this purpose, three different voltage feedback circuits are considered, as shown in Table 3.6. The three PSIM models in Table 3.6 have different circuit components in their voltage feedback circuit, while the other parameters are the same.

**Buck_Converter_04** adopts the voltage feedback circuit shown in Fig. 3.28. The voltage feedback circuit is optimally designed to provide the best possible step load responses. This design achieves the phase margin of 65° for converter loop gain,† which guarantees stability with a superior step load response

The voltage feedback compensation is now modified to yield the phase margin of 16° that produces the worse step load response. The converter with this inferior design is referred to as **Buck_Converter_05.**

The voltage feedback circuit is further altered to produce **Buck_Converter_06**. Here, the phase margin is reduced to −5°, thus failing to meet the requirement of the converter stability. Thus, the converter shall be unstable and will produce unstable oscillations in the output voltage.

Table 3.7 shows the step load responses of the three converters with different voltage feedback circuits. The step changes of 4 A are introduced and the output voltage waveforms are recorded. **Buck_Converter_04** with the 65° phase margin produces an excellent response with small overshoot/undershoot and fast response time. On the other hand, **Buck_Converter_05** with the 16° phase margin reveals a stable but oscillatory behavior with slower response time. Finally, **Buck_Converter_06** with the −5° phase margin never reaches the steady-state and continues to exhibit unstable oscillations.

†Details about the phase margin and stability of the closed-loop controlled systems will be given in the second volume of the series.

**Table 3.6** PSIM Models with Different Voltage Feedback Circuit

PSIM model: Buck_Converter_04

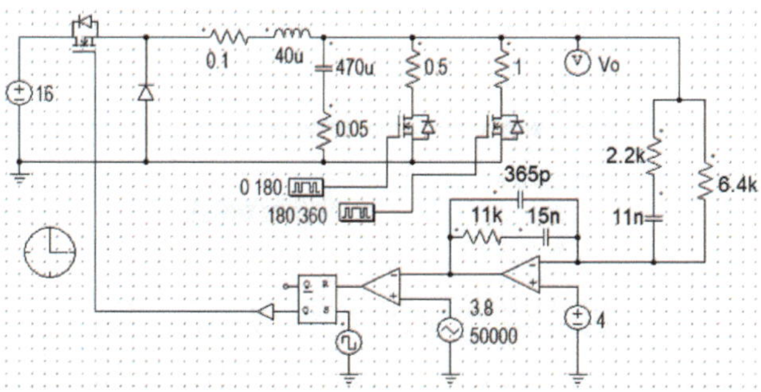


PSIM model: Buck_Converter_05

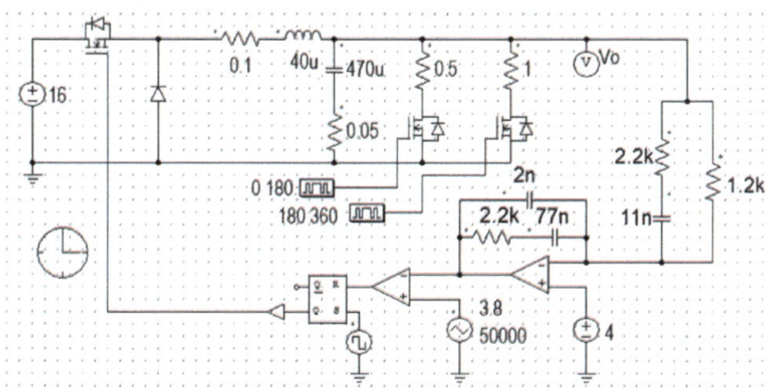


PSIM model: Buck_Converter_06

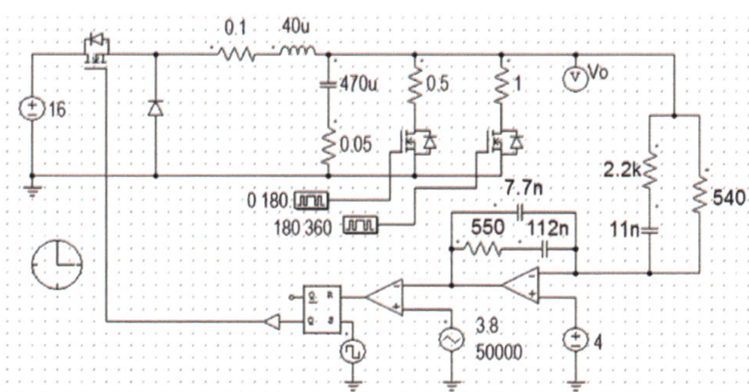

**Table 3.7** PSIM Simulations with Different Voltage Feedback Circuits

Simulation with $65°$ phase margin: Buck_Converter_04

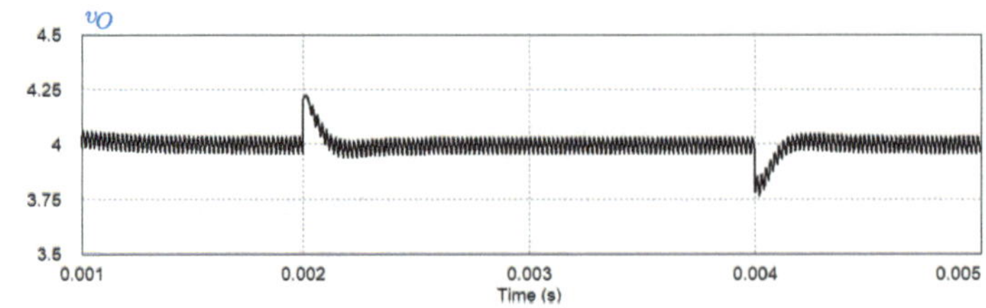


Simulation with $16°$ phase margin: Buck_Converter_05

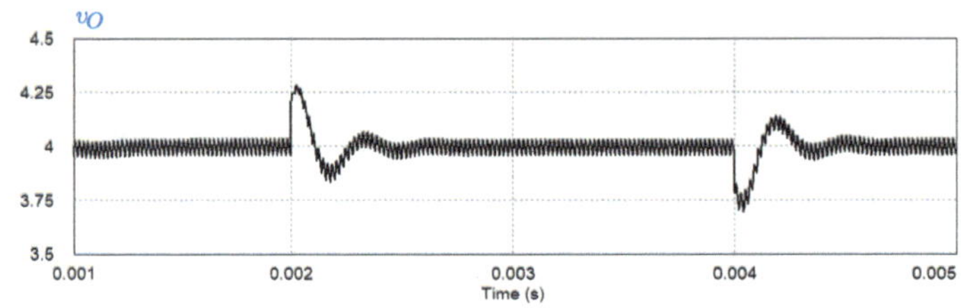


Simulation with $-5°$ phase margin: Buck_Converter_06

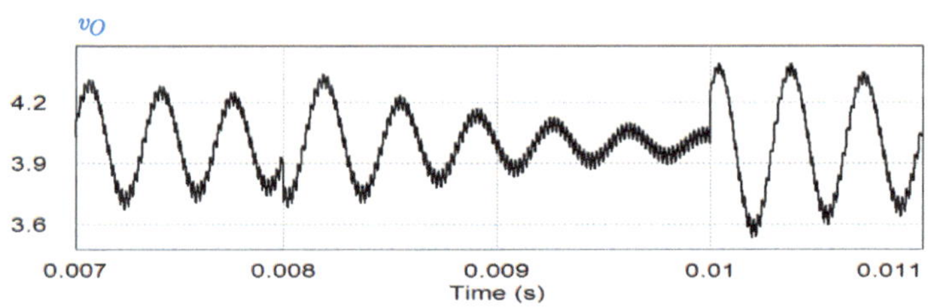


## Operational Mode Change Response

For this case, the load resistance temporarily steps up to a higher value, $R = 8\ \Omega$, well above the critical value of $R_{crit} = 5.3\ \Omega$ that places the converter on the CCM/DCM boundary. The input voltage remains constant at $V_S = 16$ V. The converter is expected to undergo transitions in the operational mode because it crosses the CCM/DCM borderline. The simulated transient responses of the converter are shown in Fig. 3.32.

With the step increase from $R = 1\ \Omega$ to $8\ \Omega$ at $t = 0.6$ ms, the output voltage exhibits an overshoot and at the same time the converter enters DCM operation within a few switching cycles, as demonstrated by the inductor current waveform. The output voltage then gradually settles to the target value $V_O = 4$ V, because *the feedback controller guarantees dc regulation regardless of the change in operational mode.*

One important comment is given regarding the transient responses in Fig. 3.32. It is important to notice that, unlike the previous step load response case where the converter remains in CCM operation all the time and the control voltage retains the same value in steady state, the control voltage and duty ratio both decrease when the converter enters DCM operation. This is because the gain formula in DCM operation differs from the CCM case. As such, the converter must settle in a new duty ratio pertinent to the DCM operation. The DCM gain is larger than the CCM gain for the same duty ratio. Thus, when the converter enters DCM operation, the duty ratio should be lessened to produce the same output voltage, as shown in Fig. 3.32. As a matter of fact, Section 3.5.3 provided some discussions about the duty ratio changes upon the CCM/DCM switchover in the closed-loop controlled converter. With the step decrease in the load resistance from 8 Ω to 1 Ω at $t = 1.6$ ms, the converter returns to CCM operation after a short transition period.

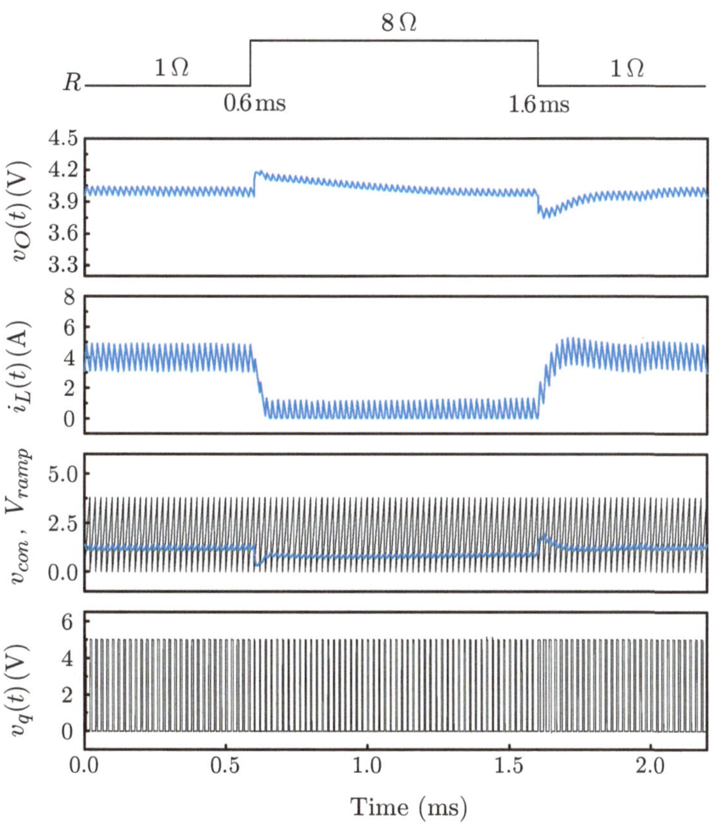


Figure 3.32 Operational mode change response of buck converter.

## EXAMPLE 3.16 PSIM® Step load responses in CCM and DCM

This example demonstrates that the operational mode of PWM converters, CCM or DCM, exerts direct and explicit impacts on the transient response of the converter. Table 3.8 illuminates this fact by contrasting the step load responses of the buck converter in DCM and that of the converter in CCM.

The load resistance of the buck converter in Fig. 3.28 is initially set at $R = 6\ \Omega$ to ensure DCM operation. The transient response of the output voltage with the step changes between $R = 6\ \Omega$ and $12\ \Omega$ is shown in the upper plot of Table 3.8. The lower plot is the step load response of the converter in CCM operation. The converter was in CCM with $R = 1\ \Omega$. The output voltage waveform in response to the step changes between $R = 1\ \Omega$ and $0.5\ \Omega$ is displayed.

As elucidated in Table 3.8, DCM operation produces much slower step load response than CCM operation. This is due to the differences in the converter dynamics in the two different operational modes. In CCM, the converter exhibits the fast second-order dynamics and generates a fast step load response. When the converter enters DCM operation, the second-order dynamics are altered to the slow first-order dynamics. As a consequence, the transient response becomes sluggish, as shown in the upper plot in Table 3.8. Detailed discussions about this topic will be provided in the second volume of the series.

## 3.7 CHAPTER SUMMARY

A step-down dc-to-dc power conversion is achieved by altering a dc input into a rectangular waveform and filtering the resulting rectangular waveform into another dc value. The buck converter is the simplest circuit that performs the step-down dc-to-dc power conversion.

The buck converter employs an active-passive switch pair and LC low-pass filter. The output of a buck converter invariably contains a ripple component. However, the magnitude of the ripple component is so small that the output voltage could be considered as a pure dc, given by the product of the input voltage and the duty ratio of the active switch – this presumption is called the small-ripple approximation. The small-ripple approximation is proved to be very useful for steady-state analysis.

The piecewise linear analysis is adapted to describe the behavior of the buck converter during the start-up period and in steady state. In the piecewise linear analysis, the converter is decomposed into the on-time and off-time subcircuits. The operation of the converter is understood by studying each subcircuit individually and later considering the behavior of the two subcircuits collectively. The piecewise linear analysis, when applied along with the small-ripple approximation, offers accurate predictions for steady-state converter waveforms. Readers will further appreciate the usefulness of this analysis in the next chapter which deals with various dc-to-dc converter circuits.

The performance of a dc-to-dc converter is influenced by the non-ideal characteristics of the circuit components. In particular, the equivalent series resistance (esr) of the output capacitor determines the magnitude of the output ripple. Furthermore, it

**Table 3.8** Step Load Change Response in CCM and DCM

PSIM model: Buck_Converter_07

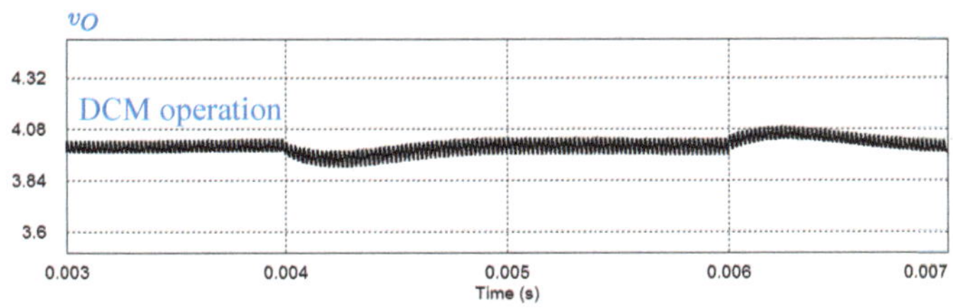


PSIM model: Buck_Converter_04

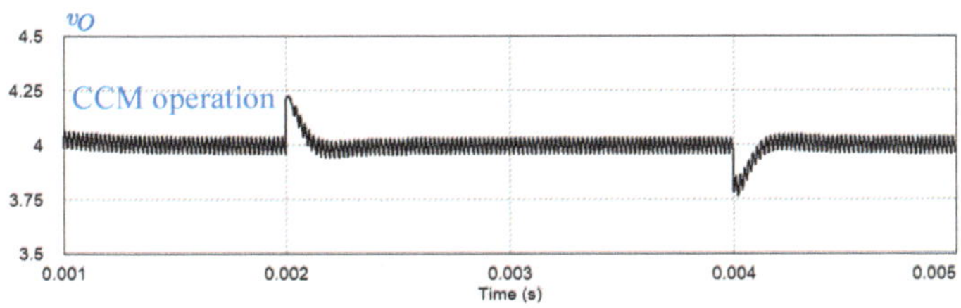


will be shown in later volumes of the book series that the esr of the output capacitor has significant impacts on the dynamic characteristics of the converter.

Most dc-to-dc converters employ unidirectional switches which only deliver the current in one direction. Accordingly, the switch current cannot alter its direction regardless of any changes in operational conditions. This constraint forces dc-to-dc converters to move into a new operational mode, referred to as the discontinuous conduction mode (DCM), when the operational conditions cross certain boundaries. In DCM operation, there exist time intervals in which the inductor current stays at zero value. The converter operation in DCM considerably differs from the continuous conduction mode (CCM) operation in which the inductor current is present all the time. This operational difference causes notable changes in the steady-state characteristics and transient response of the converter. Details about the DCM operation are addressed in Section 3.5.

The DCM operation solely stems from the use of the unidirectional switches, particularly the diode as the passive switch. When the *passive* diode switch is replaced with the *active* MOSFET switch and the MOSFET switches are complementary derived, the buck converter always maintains CCM operation regardless of the operational conditions. Such a converter is referred to as the synchronous buck converter, which finds an increasing use in high-performance step-down power conversion. Problems 3.12 and 3.13 cover the operational details of the synchronous buck converter. Further discussions about this type of converters are given in Section 4.6.

A dc-to-dc converter should provide a fixed output voltage for all the operational conditions. For this purpose, the feedback control is employed to adaptively adjust the duty ratio so that the converter produces a desired output voltage regardless of the changes in the input voltage, load current, or operational mode – CCM or DCM. The feedback controller consists of the PWM block and voltage feedback circuit. The PWM block generates a pulsewidth-modulated switch drive signal, using the control voltage provided by the voltage feedback circuit. To regulate the output voltage at a fixed value, the voltage feedback circuit should satisfy the specific condition given by (3.56), $|Z_2(j0)|/|Z_1(j0)| = \infty$, as is the case with the three-pole two-zero circuit shown in Fig. 3.27(b).

PSpice®and PSIM®simulations are very useful in visualizing converter operations and substantiating theoretical predictions. The simulations shown in Section 3.6.2 provide insights on both steady-state and transient responses of a closed-loop controlled PWM dc-to-dc converter.

## REFERENCES


1. Byungcho Choi, *Pulsewidth Modulated Dc-to-Dc Power Conversion*, 2nd Ed., Wiley, Hoboken, NJ, 2021.
2. R. W. Erickson and D. Maksimović, *Fundamentals of Power Electronics*, 3rd Ed., Springer Nature, 2020.
3. L. Dixon, "Spice simulation of switching power supply performance," Unitrode Power Supply Design Seminar Manual: SEM-800, pp.1-1–1-14, 1991.

## PROBLEMS

**3.1*** Listed below are useful circuit theorems or analysis techniques that are frequently applied to dc-to-dc power converter circuits. For each item, give a brief description, mathematical expression, or illustrative example.

i) volt-sec balance condition ii) amp-sec balance condition
iii) small-ripple approximation iv) piecewise linear analysis
v) flux linkage balance condition vi) charge balance condition

**3.2*** A secondary filter stage is added to a conventional buck converter, yielding a buck converter with a two-stage output filter. The resulting buck converter is shown in Fig. P3.2. Circuit analyses reveal that the transfer function of the two-stage filter could be approximated as

$$\frac{v_o(s)}{v_x(s)} \approx \frac{1}{\left(1+\frac{s}{Q_1\omega_{o1}}+\frac{s^2}{\omega^2_{o1}}\right)\left(1+\frac{s}{Q_2\omega_{o2}}+\frac{s^2}{\omega^2_{o2}}\right)}$$

$$Q_1 = R\sqrt{\frac{C_1}{L_1}} \qquad \omega_{o1} = \frac{1}{\sqrt{L_1C_1}}$$

$$Q_2 = R\sqrt{\frac{C_2}{L_2}} \qquad \omega_{o2} = \frac{1}{\sqrt{L_2C_2}}$$

when the conditions $L_2 \ll L_1$ and $C_2 \ll C_1$ are met.

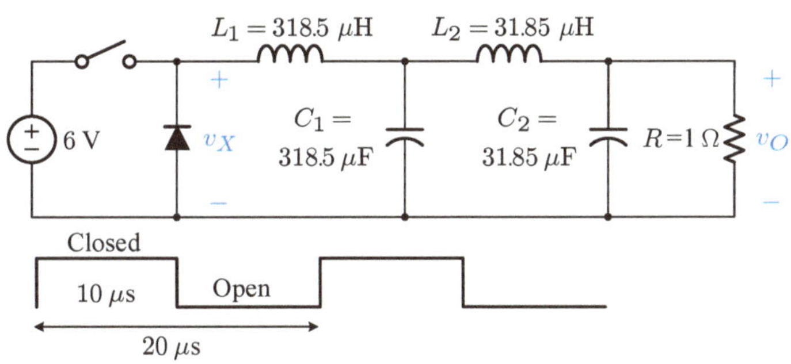


Fig. P3.2

a) Perform the frequency-domain analysis to predict the magnitude of the output voltage ripple. Consider only the first harmonic component of the output voltage.

b) Now assume that the secondary filter stage is removed, $L_2 = C_2 = 0$. For this case, find new values for $L_1$ and $C_1$ so that the converter produces the same output voltage ripple as that of a). Assume $Q_1 = R\sqrt{C_1/L_1} = 1$ for this problem.

c) Referring to the results of a) and b), state the advantage of using the two-stage output filter rather than the single-stage output filter.

**3.3** A buck converter and its inductor current waveform are shown in Fig. P3.3.

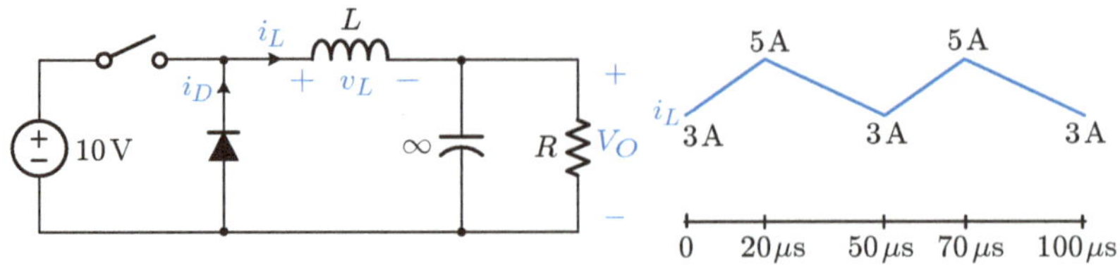


Fig. P3.3

a) Evaluate the output voltage $V_O$.
b) Sketch the steady-state waveforms of the inductor voltage $v_L$ and diode current $i_D$ for the two operational periods. Label the maximum and minimum values on your sketch.
c) Evaluate the inductance $L$ and the load resistance $R$.

**3.4*** Consider the dc-to-dc conversion system shown in Fig. P3.4. The system is configured with two buck converters powered from a common voltage source. Referring to the switch drive signals for $Q_1$ and $Q_2$, answer the questions.

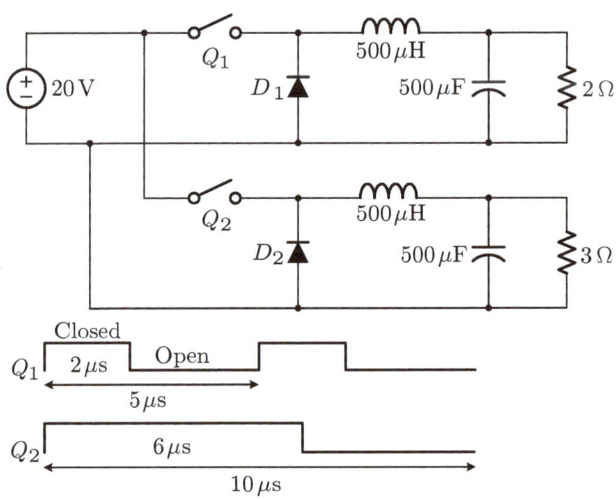


Fig. P3.4

a) First assume that all the circuit components are ideal and the power conversion is performed losslessly. Evaluate the average current flowing through each semiconductor switch, $\bar{i}_{Q1}$, $\bar{i}_{D1}$, $\bar{i}_{Q2}$, and $\bar{i}_{D2}$.
b) Now, assume the following practical devices for the semiconductor switches:
$Q_1$ and $Q_2$: MOSFET with the channel resistance of $R_{DS} = 0.5\,\Omega$

$D_1$ and $D_2$: Schottky diode with the turn-on voltage of $V_D = 0.5$ V
Calculate the average power dissipated at each practical switch.

c) Lastly, calculate the overall efficiency of the system using the following formula

$$\eta = \frac{P_{out}}{P_{out} + P_{loss}}$$

where $P_{out}$ is the power delivered to the load resistors and $P_{loss}$ is the sum of the power losses, each evaluated in b) for the respective practical switch.

3.5 Shown in Fig. 3.5 is a buck converter along with its switch drive signal.

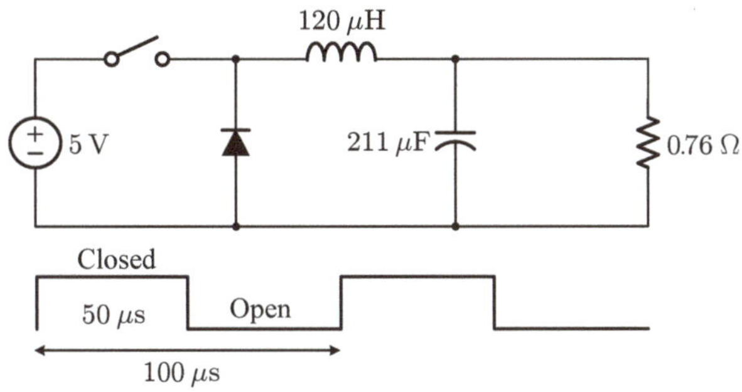


Fig. P3.5

a) Perform the frequency-domain analysis illustrated in Example 3.1, in order to predict the magnitude of the output voltage ripple.
b) Use the equation (3.27) in the text to estimate the magnitude of the output voltage ripple. Compare the result with the outcome of a).

3.6* Consider the three converter circuits shown in Fig. P3.6 and answer the questions. The duty ratio of the active switch is $D = 0.4$ for all the converter circuits.

a) Converter (a) is a simple variation of the buck converter where a redundant diode is added to the regular buck converter. *Do you think the converter will operate properly? If you claim a proper operation, find the steady-state value of the output voltage $v_O$. If you claim a fatal failure, explain the reason of the failure.*
b) Converter (b) is created by inserting a $1 : 2$ transformer to Converter (a). Assume that the relative permeability of the core material of the transformer is infinite, $\mu_r = \infty$. Answer the questions shown in *italic font* in a).
c) Now assume the relative permeability of the core material is $\mu_r = 5 \times 10^4$ for Converter (b). Answer the questions shown in *italic font* in a).

Converter (c) is created by inserting a diode and Zener diode pair to Converter (b). Now assume the relative permeability of the core material is $\mu_r = 5 \times 10^4$ for Converter (c) for the Problems d) through g).

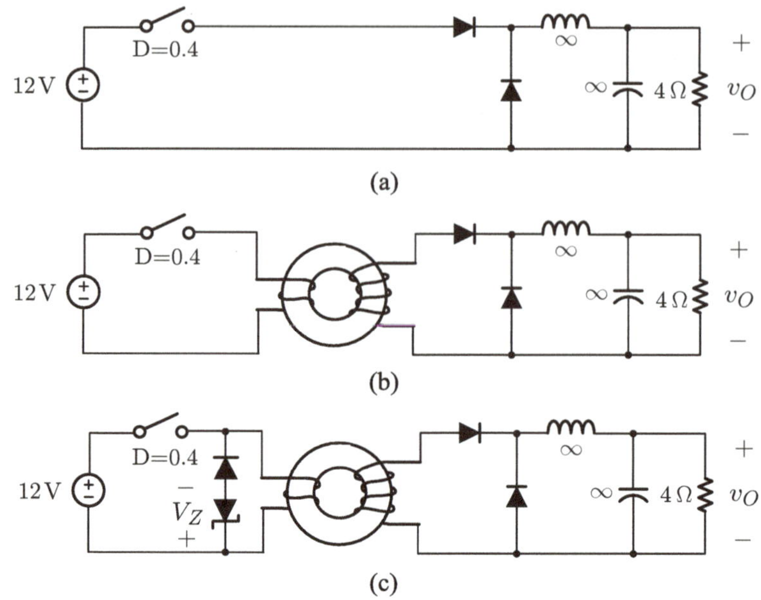


Fig. P3.6

d) Assume that the breakdown voltage of the Zener diode is $V_Z = 10$ V for Converter (c). Answer the questions shown in *italic font* in a).
e) Now assume that the breakdown voltage of the Zener diode is $V_Z = 6$ V for Converter (c). Answer the questions shown in *italic font* in a).
f) Now assume that the breakdown voltage of the Zener diode is $V_Z = 8$ V for Converter (c). Answer the questions shown in *italic font* in a).
g) Assume that Converter (c) operates properly with an appropriate value of $V_Z$. State the merit and demerit of the operational Converter (c).

**3.7** Figure P3.7 shows two buck converters with different output filter configurations. Referring to the circuit diagrams, answer the questions.

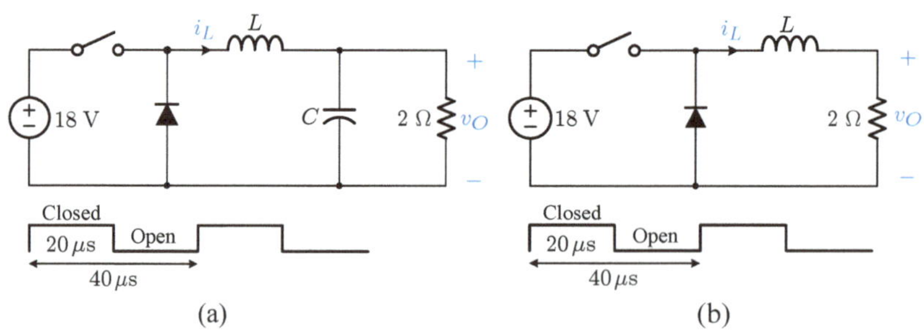


Fig. P3.7

a) For Converter (a), determine the values for $L$ and $C$ to meet the design specifications of $\Delta i_L/I_L = 0.2$ and $\Delta v_O/V_O = 0.02$.
b) For Converter (b), determine the value for $L$ to comply with the design specification of $\Delta v_O/V_O = 0.02$.

**3.8** Shown in Fig. P3.8(a) are a buck converter and the Bode plot of its output filter transfer function, $|F_f| = |v_o|/|v_x|$.

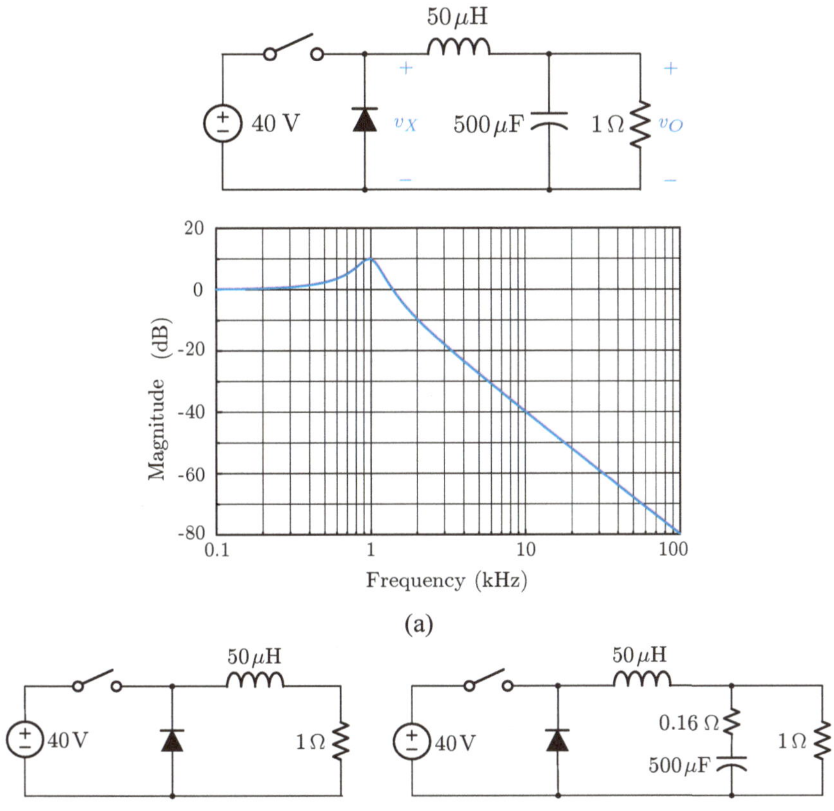


Fig. P3.8

a) Referring to Fig. P3.8(a), estimate the magnitude of the output voltage ripple when the buck converter operates at 30 kHz with $D = 0.5$. Use the information given in the Bode plot.
b) Now assume the buck converter is modified into two other converter circuits shown in Fig. P3.8(b). For each of the two converter circuits, repeat a). *Modify the Bode plot in Fig. P3.8(a) to evaluate the output voltage ripple with the respective filter configuration.*

**3.9*** Referring to Fig. P3.9, answer the following questions.

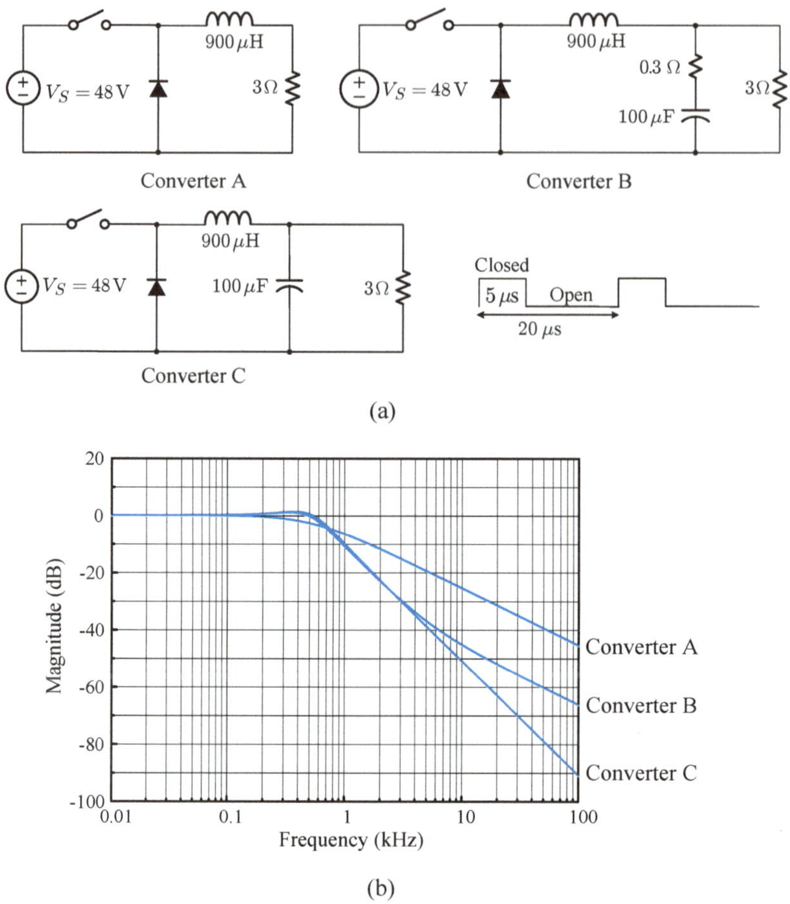


Fig. P3.9

a) Figure P3.9(a) shows the three buck converters. The common switch drive signal is applied to the three converters. Perform the *time-domain analysis* to estimate the output voltage ripple of each converter.

b) Shown in Fig. P3.9(b) are the Bode plots of the output filter transfer functions of the three converters. Referring to the Bode plots, perform the frequency-domain analysis to predict the magnitude of the output voltage ripple.

**3.10*** Shown in Fig. P3.10 are the four different buck converters, each with a dissimilar low-pass filter configuration. For all the four converters, assume $D = 0.5$ and $\omega_s = 100 \times 10^3$ r/s. You are required to use (in part) the graphical method based on the asymptotic Bode plot technique. Use the semi-log paper shown

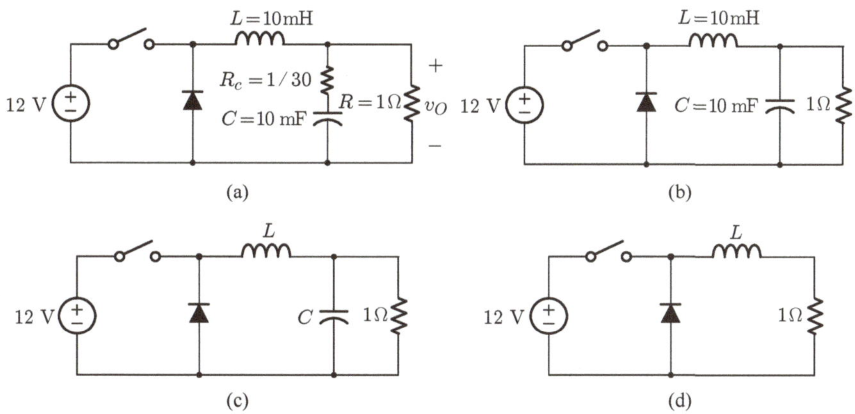


Fig. P3.10

in Fig. P.3.10P and clearly illustrate the graphical procedures of your solutions to the following questions.

a) For Converter (a), prove that the magnitude of the output ripple can be estimated as $\Delta v_O(t) \approx 4.83 \times 10^{-4}$ V. You may use the following information regarding Fourier coefficient and the transfer function of the low-pass filter in Converter (a).

$$C_n = \frac{\sqrt{2}V_S}{n\pi}\sqrt{1 - \cos n2\pi D}$$

$$F_f(s) = \frac{1 + s/\omega_z}{1 + s/(Q\omega_o) + s^2/{\omega_o}^2}$$

$$\text{with} \quad \omega_z = \frac{1}{CR_c} \quad Q \approx R\sqrt{\frac{C}{L}} \quad \omega_o \approx \frac{1}{\sqrt{LC}}$$

b) For Converter (b), provide an appropriate estimate for $\Delta v_O(t)$.

c) For Converter (c), determine the values for $L$ and $C$ such that the converter produces the same $\Delta v_O(t) \approx 4.83 \times 10^{-4}$ V as the case of Converter (a). Use $Q = 1$ for this problem.

d) For Converter (d), determine the value of $L$ such that the converter produces the same $\Delta v_O(t) \approx 4.83 \times 10^{-4}$ V.

**3.11*** Figure P3.11 shows the inductor current of a buck converter that operates at the boundary between the continuous conduction mode (CCM) and discontinuous conduction mode (DCM).

a) Assume that the converter operates in an open-loop condition with a fixed duty ratio. The load resistance is increased beyond the critical value that places the converter at the CCM/DCM borderline, while the input voltage remains the same. Sketch a family of inductor currents, each with a different

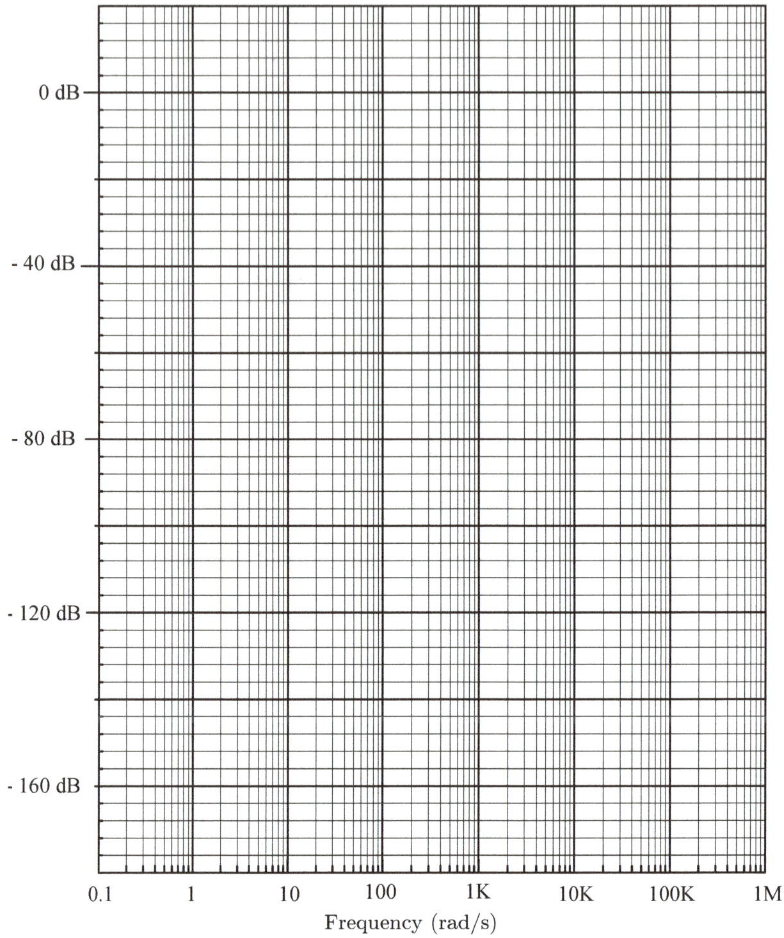


Fig. P3.10P

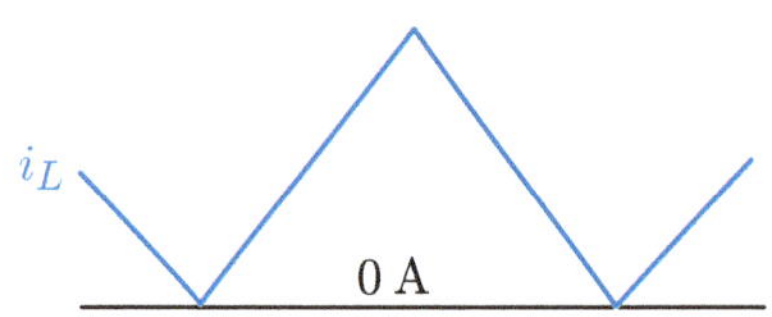


Fig. P3.11

load resistance, in order to illustrate the transition pattern of the inductor current as the load resistance is increased.

b) The converter is now closed-loop controlled, *thereby achieving the output voltage regulation*, and again operates at the CCM/DCM boundary. Assume that the load resistance is increased beyond the critical value, while the input voltage remains the same, and sketch a family of inductor currents to show the transition pattern of the inductor current.

c) Again assume that the *closed-loop control* and the borderline operation for the converter. For this case, the input voltage is increased beyond the critical value for the CCM/DCM boundary, while the load resistance remains the same. Sketch a family of the inductor currents that exhibit the transition pattern of the inductor current as the input voltage is increased.

**3.12*** Consider the converters shown in Fig. P3.12 and answer the questions.

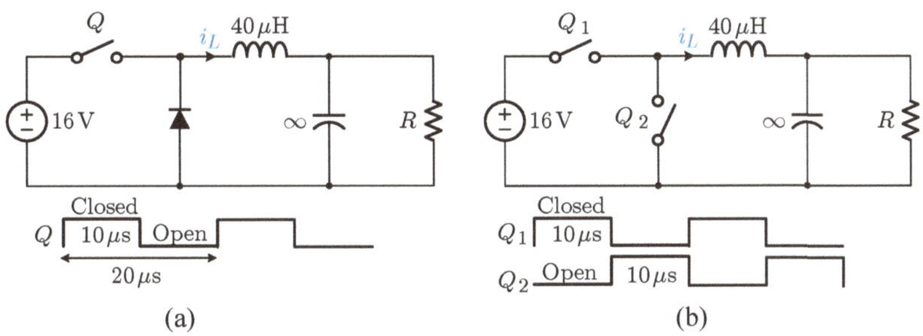


Fig. P3.12

a) In Converter (a), a unidirectional switch $Q$ is driven with $D = 0.5$ for $T_s = 20$ $\mu$s. Answer the following questions.
i) Sketch the steady-state waveforms of $i_L$ for $R = 2$ $\Omega$, $R = 8$ $\Omega$, and $R = \infty$ $\Omega$. Show the three $i_L$ waveforms simultaneously on the same plot for the two operational periods. Show the maximum and minimum values of the waveforms.
ii) On the plot obtained in i), sketch $i_L$ waveforms for the case of $2\ \Omega < R < 8\ \Omega$ and for the case of $8\ \Omega < R < \infty$, in order to show the evolution pattern of $i_L$ as $R$ varies between $2\ \Omega < R < \infty$.

b) For Converter (b), assume that $Q_1$ and $Q_2$ are ideal bidirectional switches which deliver the current in both directions. The two bidirectional switches are complementary driven with $D = 0.5$ for $T_s = 20$ $\mu$s. Repeat i) and ii) in a) for Converter (b).

c) For Converter (b), sketch $i_L$ for the case of $R = 16\,\Omega$ for the two operational periods. Show the maximum and minimum values.

**3.13*** Consider the two buck converters shown in Fig. P3.13 and answer the following questions. Converter (a) employs the unidirectional switches. In contract, Converter (b) uses the bidirectional switches, which could deliver the current in both directions. The bidirectional switches in Converter (b) are driven in complementary manner. While the inductance L is fixed, a variable resistance

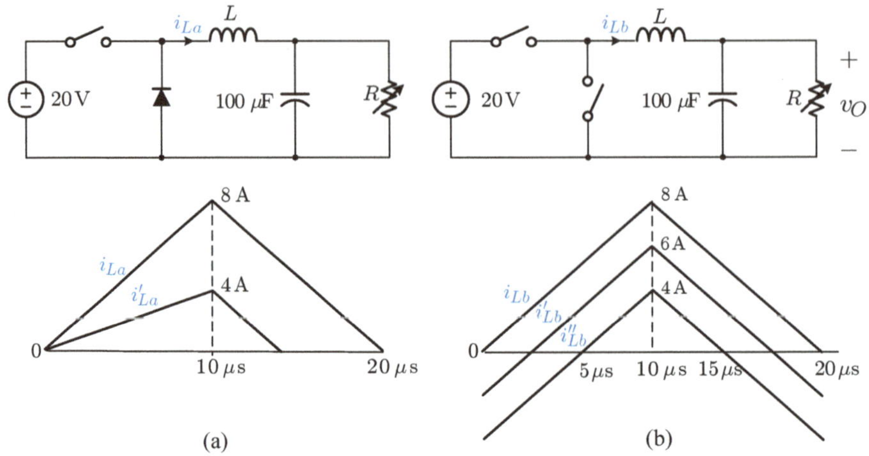


Fig. P3.13

$R$ is used in the two converters. For each converter, a family of inductor currents produced with different $R$ values are shown for one operational period.

a) Determine the inductance $L$ in the two converters.

b) Find the $R$ value that is used to generate $i_{La}$ for Converter (a).

c) For Converter (a), state the condition needed to claim the transient behavior illustrated with $i'_{La}$. Justify the correctness of the $i'_{La}$ waveform based on the afore-stated condition.

d) Find the $R$ value that is used to generate $i'_{Lb}$ for Converter (b).

e) Find the $R$ value that is used to generate $i''_{Lb}$ for Converter (b).

f) For the case of e), sketch $v_O(t)$ waveform for the two steady-state operational periods. Label the maximum and minimum values of $v_O(t)$.

**3.14** Figure P3.14 shows a buck converter and its inductor current waveform. Based on the given information, evaluate the output voltage $V_O$, load current $I_O$, inductance $L$, and the load resistance $R$. You are not allowed to use the DCM gain formula in this problem.

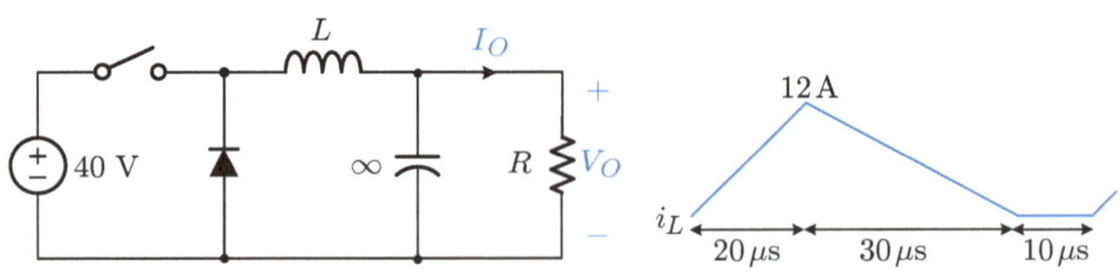


Fig. P3.14

**3.15*** Shown in Fig. P3.15 are the circuit diagram of a buck converter and its inductor current waveform.

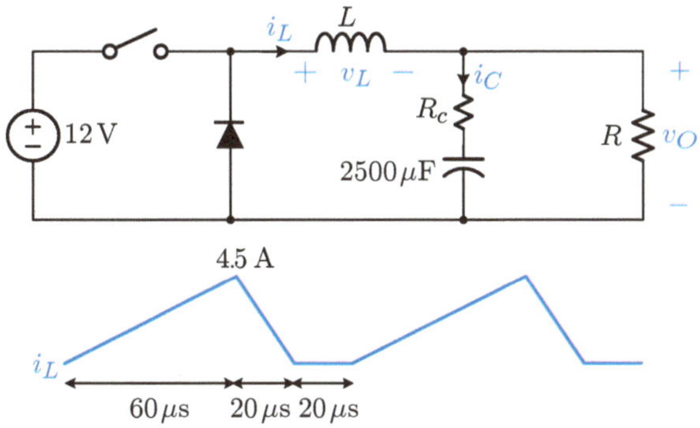


Fig. P3.15

a) Referring to the inductor current $i_L$, sketch the inductor voltage $v_L$ for the two switching periods. Show the maximum and minimum values on your sketch.
b) Find values for the inductance $L$ and resistance $R$.
c) Sketch the capacitor current $i_C$ for the two switching periods. Show the maximum and minimum values.
d) Evaluate the output voltage ripple $\Delta v_O$ when $R_c = 0.12\ \Omega$. Assume $C = 2500\ \mu$F is sufficiently large so that the ripple component across the capacitance is negligible, compared with the ripple component across $R_c = 0.12\ \Omega$.
e) Evaluate $\Delta v_O$ when $R_c = 0$.

**3.16** Figure P3.16 shows a family of voltage conversion curves of a buck converter, as a function of the duty ratio $D$ and dimensionless parameter $\tau = 2L/(RT_s)$. The converter operates with the period $T_s = 20\ \mu$s.

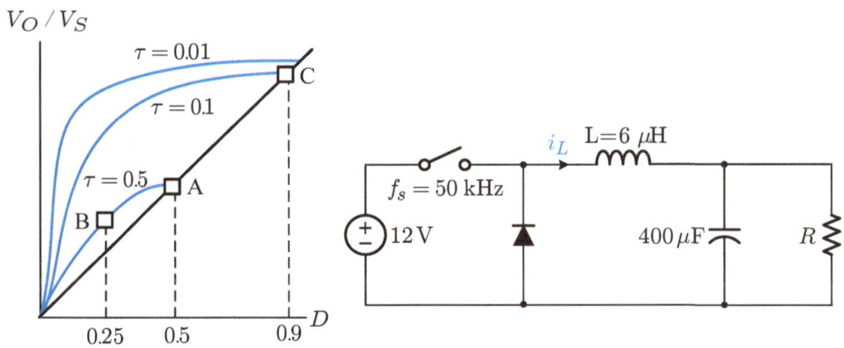


Fig. P3.16

a) Sketch the steady-state waveform of $i_L$ when the operational condition of the converter is located at Point A. Show the maximum, minimum, and average values of the waveform on your sketch.

b) Assume that the operational condition is located at Point B and repeat a).

c) Repeat a) for Point C.

**3.17** Referring to the buck converter in Fig. P3.17, answer the following questions.

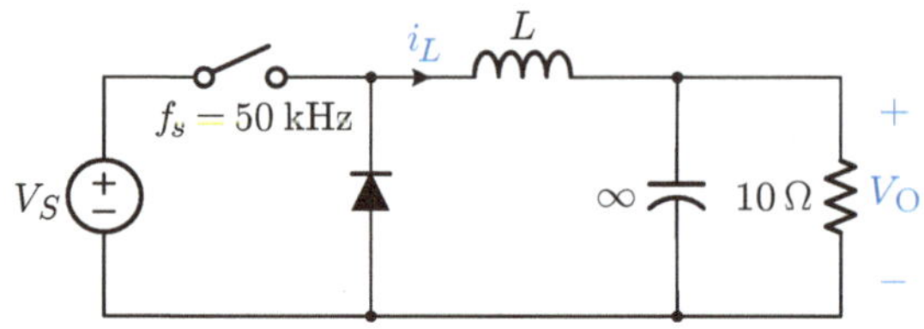


Fig. P3.17

a) Determine the inductance $L$ so that the converter operates at the CCM/DCM borderline with $D = 0.6$.

b) Find the minimum inductance $L$ that ensures CCM operation of the converter for all $0 < D < 1$.

c) Now assume $L = 60\ \mu$H and sketch the general shape of the dc voltage gain curve, $V_O/V_S$ vs $D$, for all $0 < D < 1$.

**3.18** Figure P3.18 shows two different buck converters. Converter (a) operates in an open-loop fashion, while Converter (b) operates with a closed-loop control.

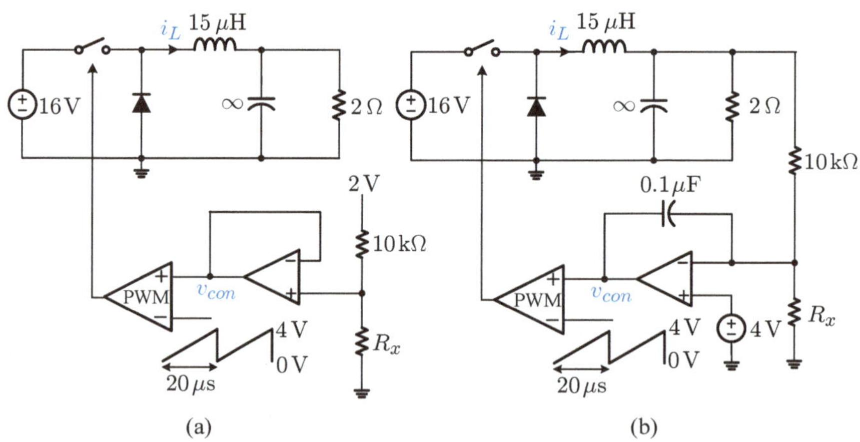


Fig. P3.18

a) For Converter (a), answer the following questions.

i) Assume $R_x = \infty$ and sketch the steady-state waveforms of $i_L$ and $v_{con}$

for the two operational periods.

ii) Now assume $R_x = 10$ kΩ and repeat i).

b) Repeat a) for Converter (b).

**3.19*** In the four buck converters shown in Fig. P3.19, assume the circuit components are all ideal. For each converter, calculate the average values of the control voltage $\bar{v}_{con}$ and output voltage $\bar{v}_O$ in CCM operation.

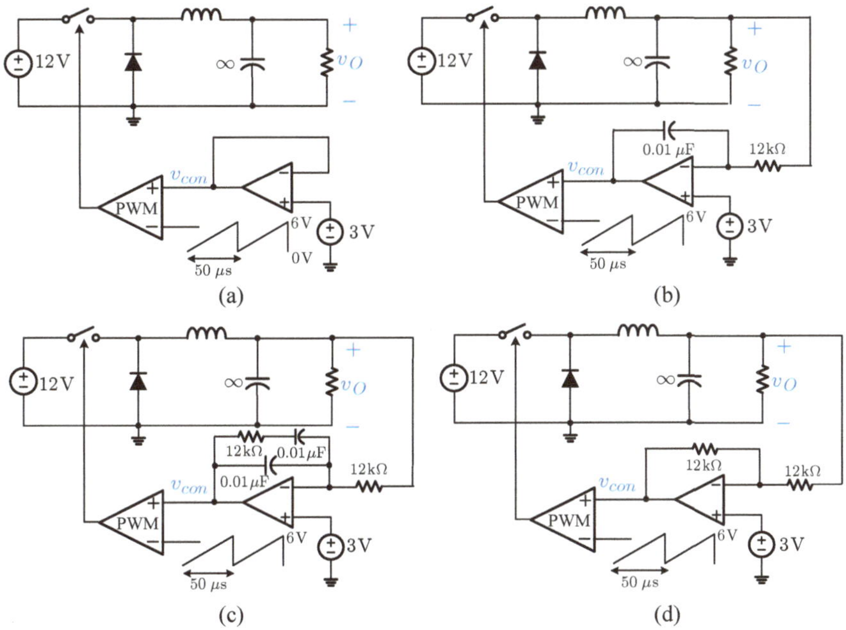


Fig. P3.19

**3.20*** Assume all the circuit components are ideal in the two converter circuits shown in Fig. P3.20.

a) For Converter (a), draw $i_L$ waveforms for the two operational periods when $R = 5$ Ω, $R = 8$ Ω, and $R = 12$ Ω. Show the average value, ascending slope, and transition pattern of $i_L$ in a comparative manner. Show the peak value of $i_L$ when $R = 5$ Ω.

b) Repeat a) for Converter (b).

**3.21** Consider the closed-loop controlled buck converter shown in Fig. P3.21 and answer the questions.

a) With $V_S = 20$V, $R = 1$Ω, and $R_1 = \infty$, sketch the steady-state waveforms of the inductor current $i_L$ and control voltage $v_{con}$ for the two switching periods. Label the maximum and minimum values on your sketch.

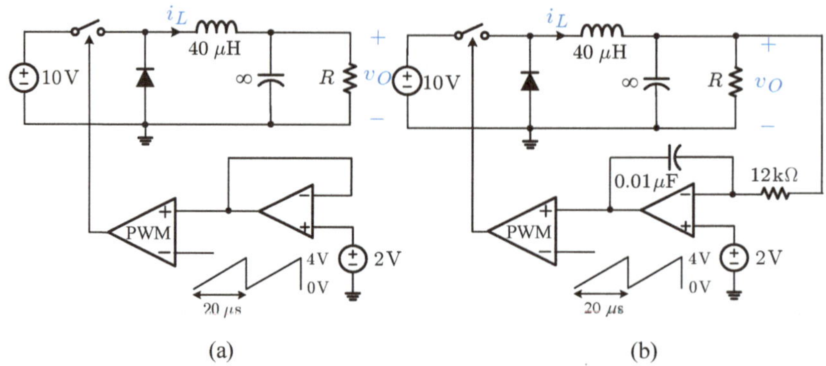


Fig. P3.20

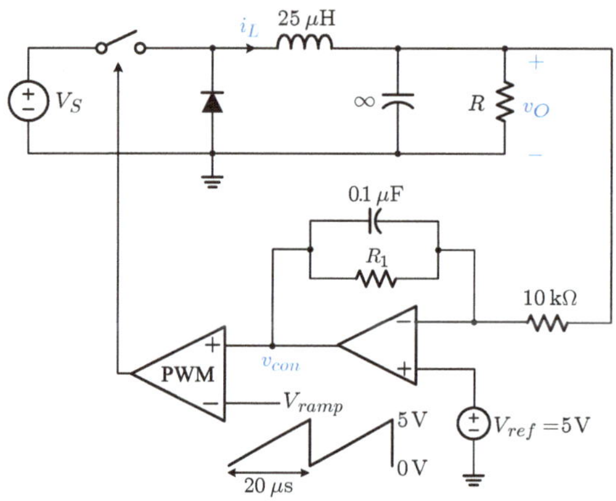


Fig. P3.21

b) When the operational conditions of the converter vary as $15\text{ V} < V_S < 25\text{ V}$ and $0.5\ \Omega < R < 2\ \Omega$ with $R_1 = \infty$, find the range of the control voltage $v_{con}$ for the entire operating range: $(\quad) < v_{con} < (\quad)$.

c) For $R_1 = 10\text{ k}\Omega$, the converter fails to achieve dc regulation. Explain the reason for this.

d) When $V_S = 20$ V and $R = 1\ \Omega$ with $R_1 = 10\text{ k}\Omega$, find the steady-state values of the output voltage $v_O$ and control voltage $v_{con}$.

**3.22** Figure P3.22 shows a closed-loop controlled buck converter and its voltage gain curve. Answer the following questions.

a) Find the ranges for the duty ratio $D$ and control voltage $v_{con}$, when the input voltage is varied between $20\text{ V} < V_S < 60$ V.

b) Find the inductance of the filter inductor $L$.

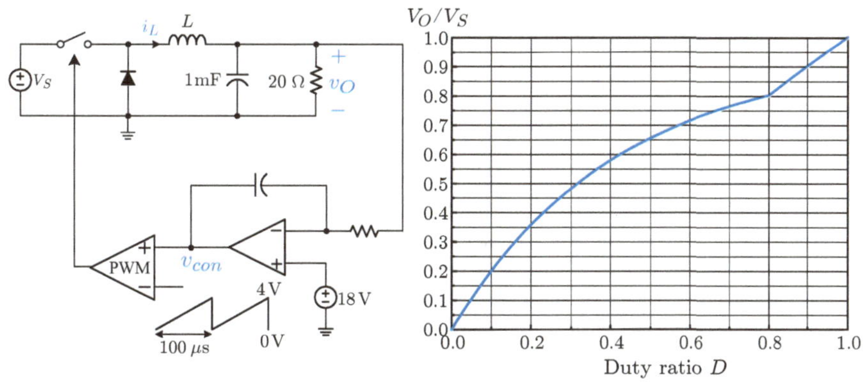


Fig. P3.22

c) Sketch the inductor current $i_L$ for the two switching periods for the three cases of $V_S = 20$ V, $V_S = 22.5$ V, and $V_S = 60$ V. Show the maximum and minimum values of $i_L$.

**3.23** Consider the buck converter shown in Fig. P3.23 and answer the following questions.

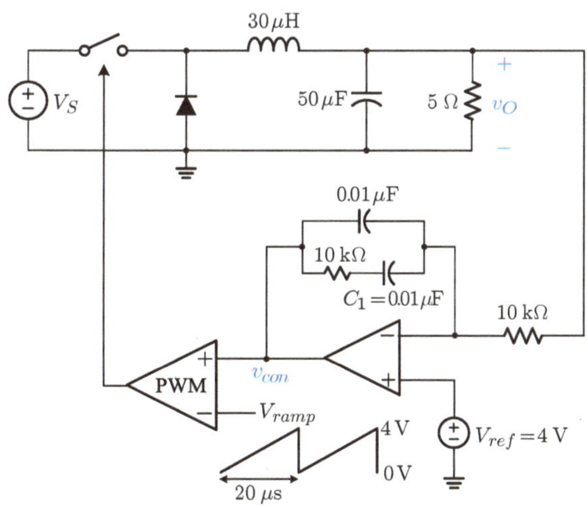


Fig. P3.23

a) Assume that the input voltage is varied between 8 V < $V_S$ < 20 V. For this condition, the converter could operate in either CCM or DCM, depending on the input voltage range. Fill in the blanks in the following descriptions about operation of the buck converter:
i) The converter operates in CCM for (   ) < $V_S$ < (   ) and the control voltage varies between (   ) < $v_{con}$ < (   ) in this CCM operation.

ii) The converter operates in DCM for ( ) $< V_S <$ ( ) and the control voltage varies between ( ) $< v_{con} <$ ( ) in this DCM operation.

b) Now assume $V_S = 10$ V and answer the questions.

i) Find the steady-state values of the output voltage $v_O$ and control voltage $v_{con}$.

ii) Assume that the capacitor $C_1$ is accidentally open-circuited. Do you expect any change in the steady-state waveform of $v_O$ or $v_{con}$? If you do not expect any change, give the reason for your claim. If you do predict changes in $v_O$ or $v_{con}$, evaluate the average value of the corresponding waveform in the new steady state.

iii) Assume the compensation capacitor $C_1$ is accidentally short-circuited and repeat ii).

**3.24** Figure P3.24 shows a closed-loop controlled buck converter. Answer the questions.

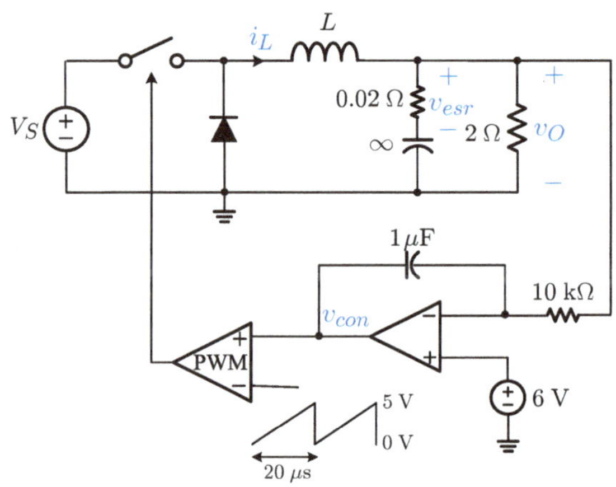


Fig. P3.24

a) Determine the inductance $L$ such that the converter operates at the borderline between CCM and DCM with the input voltage of $V_S = 24$ V. For this case, sketch the inductor current $i_L$ and the ripple component of the output voltage, $\tilde{v}_O = v_O - V_O$, for the two operational periods. Show the maximum and minimum values.

b) Assume the input voltage is varied between 12 V $< V_S <$ 24 V, while the inductance found in a) is used in the output filter. Under these conditions, find the range of the following operational condition and circuit variables of the converter.

i) ( ) $< D <$ ( ) ii) ( ) $< v_{con} <$ ( )

iii) ( ) $< \Delta i_L <$ ( ) iv) ( ) $< \Delta v_O <$ ( )

**3.25** In the closed-loop controlled buck converter shown in Fig. P3.25, assume the three-pole two-zero feedback circuit is well designed for stability and good transient performance.

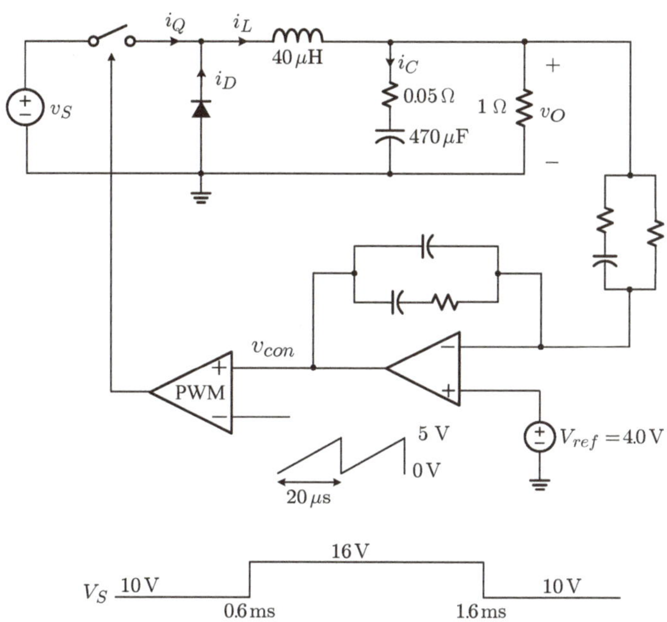


Fig. P3.25

a) Assume $v_S = 10$ V and sketch the steady-state waveform of $i_L, i_Q, i_D, i_C, \bar{v}_{con}$, and $\tilde{v}_O = v_O - \bar{v}_O$ for two operational periods. Show the maximum, minimum, and average value of the waveforms.

b) Now assume the step changes are occurred in the input voltage of the converter as illustrated in the circuit diagram. Sketch the transient waveform of $v_O$, $i_L$ and $v_{con}$. Show and explain (for the reasons) all the characteristic features of the transient responses.

**3.26** Shown in Fig. P3.26 are the transient responses of the major circuit waveforms of a closed-loop controlled buck converter. Case (a) is the transient response of the inductor current $i_L$ and control voltage $v_{con}$ in response to the step changes in the input voltage, $V_{S1} \Rightarrow V_{S2} \Rightarrow V_{S1}$ with $V_{S2} < V_{S1}$. Case (b) is the transient responses due to the step changes in the load resistor, $R_1 \Rightarrow R_2 \Rightarrow R_1$ with $R_1 < R_2$.
Table P3.26 summarizes 1) the prominent behavior of the inductor current $i_L$ and control voltage $v_{con}$, observed in the period circled on the corresponding circuit waveform, and 2) a brief explanation for the cause/origin of the respective circuit behavior. Based on the given information and operational principles of the converter, fill in the blanks in Table P3.26 with an appropriate statement.

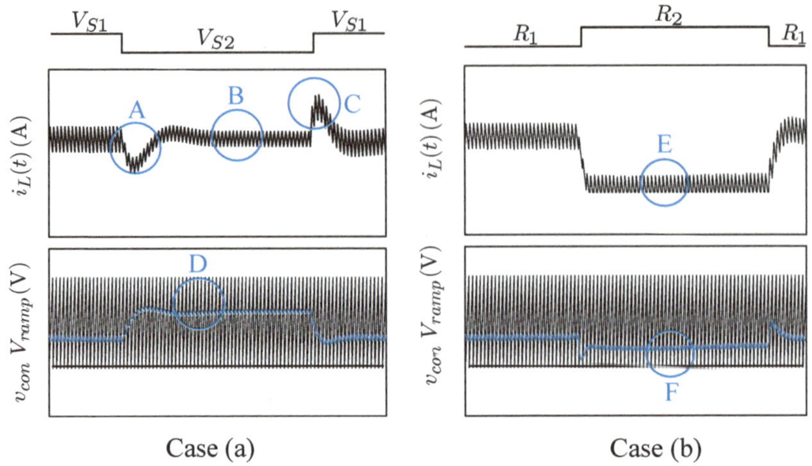


Fig. P3.26

Table P3.26

| Circled period | Observed behavior | Cause/origin of behavior |
|---|---|---|
| A | undershoot in $i_L$ | ( ) |
| B | decrease in the ripple of $i_L$ | ( ) |
| C | overshoot in $i_L$ | ( ) |
| D | increase in $v_{con}$ | ( ) |
| E | discontinuity in $i_L$ | DCM operation with $R > R_{crit}$ |
| F | decrease in $v_{con}$ | ( ) |

**3.27*** Consider the following buck converter shown in Fig. P3.27 and answer the questions.

a) Determine the inductance $L$ such that the buck converter operate in a deep continuous conduction mode (CCM) with the condition $\Delta i_L / I_L = 0.2$.

b) Assume the value of inductance $L$ found in 1) and sketch the steady-state waveforms of $v_O$ for the two switching periods. Show the maximum/minimum value of $v_O$.

c) Assume the input voltage is altered from $V_S = 24$ V to $V_S = 12$ V, while the other parts of the converter remain the same as the case of 2). Under this situation, repeat b).

d) Now assume $V_S = 24$ V again. Modify Block A such that the converter regulates the output voltage at $V_O = 12$ V. You are allowed to change or add some passive components but you have to use the reference voltage $V_{ref} = 6$ V. Minimize the number of the component changes and reduce the component count as much as possible. Draw the circuit diagram of the closed-loop controlled converter.

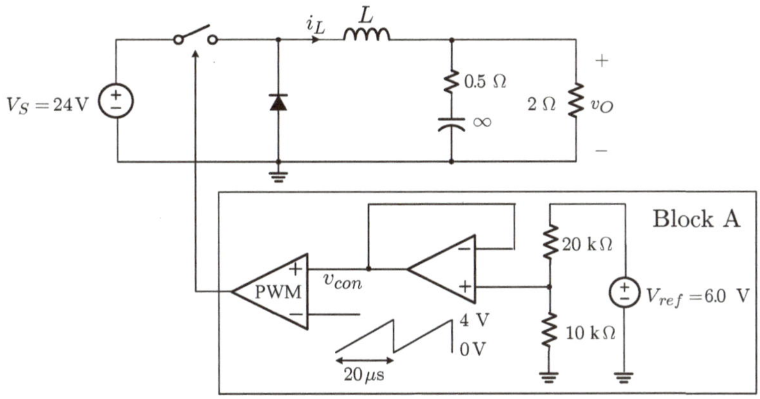


Fig. P3.27

e) With the closed-loop converter you designed in d), sketch $v_O$ and $v_{con}$ for the two operational periods. Show the maximum/minimum values of the wave forms.

CHAPTER 4

# DC-TO-DC POWER CONVERTER CIRCUITS

Numerous dc-to-dc converters are currently being employed in various consumer and industrial electronics. These converters outwardly look so varied in topology and operation that the diversity of dc-to-dc converters is seemingly amazing.

Despite the variety and dissimilarity in topological structures, most dc-to-dc converters have evolved from the *three basic converters* known as the buck converter, boost converter, and buck/boost converter. Furthermore, among these three converters, the buck converter is the forerunner of the other two converters – specifically, the boost converter is derived, at least in its functional abstraction, from the buck converter and the buck/boost converter is created by combining the buck converter and boost converter. The buck converter was studied in Chapter 3.

In this chapter, we study an important class of dc-to-dc converters including the boost converter, buck/boost converter, and other converters derived from the three basic converters. For each converter, the origin of the circuit topology is first illustrated by showing how the converter circuit has evolved from its respective forerunner. This chapter then investigates the operational details of the converter and presents steady-state waveforms for both continuous conduction mode (CCM) operation and discontinuous conduction mode (DCM) operation.

In Chapter 3, we studied the closed-loop control of the buck converter. The outcomes of this study are generally applicable to all PWM converters. Thus, this chapter mainly deals with the circuit configuration and power stage operation of various PWM converters. Even so, we will investigate the transient behavior of PWM converters operating under a closed-loop feedback control. This will provide more pragmatic and revealing information about the converter operation. This chapter presents the final outcome of the feedback control, while omitting theoretical and technical details. Comprehensive treatments of the controller design will be covered in the second book of the series.

There are many ways to group dc-to-dc converters, each with a different criterion for grouping. One way is to divide PWM dc-to-dc converters into non-isolated converters and isolated converters. In a non-isolated converter, the input port of the converter shares a common ground with its output port. The buck converter is an example of the non-isolated converter. On the other hand, in an isolated converter, the input and output ports of the converter are electrically isolated. A transformer is commonly used to provide galvanic isolation. In this chapter, we study important classes of non-isolated and isolated dc-to-dc converters. Discussions start with the non-isolated boost and buck/boost converters and later extend to isolated converters.

## 4.1 BOOST CONVERTER

The boost converter is a step-up dc-to-dc power conversion circuit that always produces a higher output voltage than the input voltage. In this section, the topological structure of the boost converter is first discussed. The steady-state operations in both CCM and DCM are then investigated using the analysis techniques established in Chapter 3.

### 4.1.1 Evolution to Boost Converter

The boost converter is viewed as an evolution of the buck converter, specially configured to offer a larger output voltage than the input voltage. Figure 4.1 illustrates the procedure of transforming the buck converter into the boost converter through steps of circuit manipulations. The buck converter is shown in Fig. 4.1(a), where **a** represents the node to which the active switch is connected, while **p** is the node where the passive switch is present. Figure 4.1(a) is transformed to Fig. 4.1(b), by replacing the active-passive switch pair with a single-pole double-throw (SPDT) switch. The SPDT switch is connected to **a** for $DT_s$ and **p** for $(1 - D)T_s$. In Fig. 4.1(b), the circuit is partitioned into the source section, middle section, and load section.

The middle section includes the SPDT switch and inductor. Referring to the voltage across the left-hand side of the middle section as $v_1$ and that of the right-hand side as $v_2$, the volt-sec balance condition is applied to the inductor

$$(v_1 - v_2)DT_s = v_2(1 - D)T_s \tag{4.1}$$

resulting in

$$\boxed{v_2 = v_1 D} \tag{4.2}$$

As the first-step circuit manipulation, we interchange the location of the source section and load section, while keeping the middle section unchanged, as shown in Fig. 4.1(c). With this modification, $v_2$ is viewed as the input voltage associated with the source section, while $v_1$ becomes the output voltage of the load section. Now, the following two facts are reminded before continuing the circuit manipulation.

1) The relation (4.2), derived from the volt-sec balance condition on the inductor, is also valid in Fig. 4.1(c), regardless of the position change between the source and load sections.

2) The parameter $D$ in (4.2) denotes the fraction of the switching period in which the SPDT switch is connected to the positive side of $v_1$.

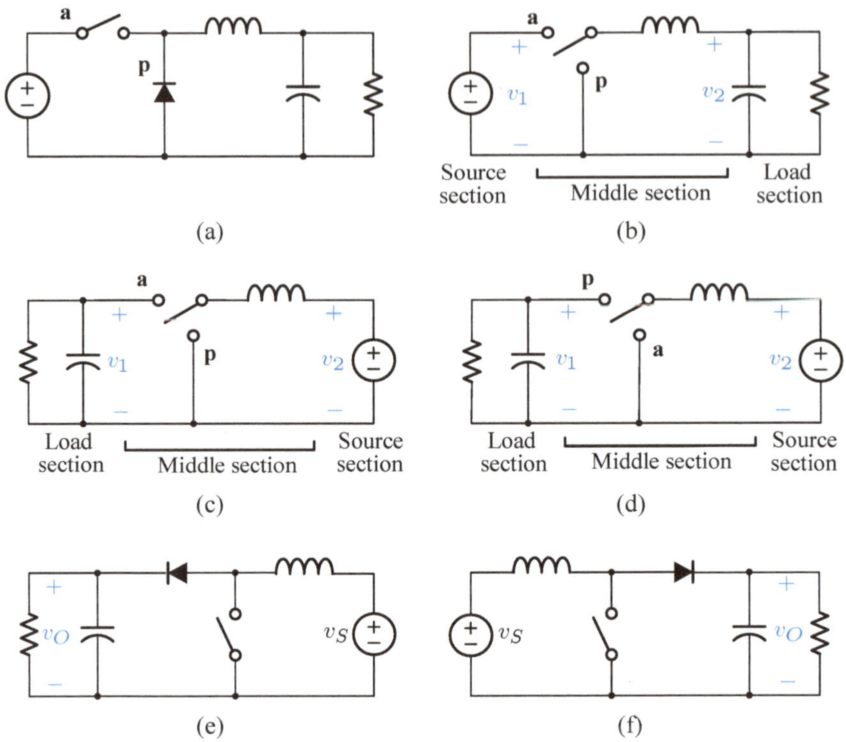


**Figure 4.1** Derivation of boost converter. (a) Buck converter. (b) Alternative representation. (c) Interchange of source section and load section. (d) Interchange of passive switch and active switch. (e) Realization of SPDT switch. (f) Boost converter.

As the second step, for the purpose of allowing energy flow from the source section to the load section, the locations of the active and passive switches are interchanged,† resulting in the circuit shown in Fig. 4.1(d). Here, the fraction of the switching period in which the SPDT switch resides on the positive terminal of $v_1$ is $(1-D)$ rather than $D$, because the $D$ is the duty ratio of the active switch, not the passive switch. Accordingly, the parameter $D$ in (4.2) should be replaced with $(1-D)$, producing in a new relationship

$$\boxed{v_2 = v_1(1-D)} \tag{4.3}$$

†For proper operation as a dc-to-dc converter, the circuit should transfer energy from the source section to the inductor when the active switch is turned-on, and release energy from the inductor to the load section when the active switch is turned-off. Although the on-time energy transfer is achieved in Fig. 4.1(c), the off-time energy release is not feasible. The on-time energy transfer and off-time energy release become both possible when the locations of the active and passive switches are interchanged, as shown in Fig. 4.1(d). Additional comments about this feature are given in Section 4.3.

Figure 4.1(d) is then transformed to Fig. 4.1(e), where the SPDT switch is represented by the active-passive switch pair, $v_1$ is renamed as $v_O$ for the output voltage, and $v_2$ is renamed as $v_S$ for the input voltage. Finally, Fig. 4.1(e) is rearranged into the standard form in Fig. 4.1(f). This new converter circuit is called the boost converter because its voltage gain

$$\boxed{\frac{v_O}{v_S} = \frac{1}{1-D}} \tag{4.4}$$

is always larger than unity for all $0 < D < 1$, thereby boosting the input voltage to a higher value.

The boost converter is viewed as an evolution from the buck converter, created by interchanging the source and load connections and rearranging the switches in such a manner that allows energy flow from the source to the load section. During this modification, the step-up feature of the boost converter has naturally arisen.

**PSIM® Start-up response of boost converter EXAMPLE 4.1**

Table 4.1 shows the start-up response of a boost converter. The power stage parameters and operational conditions are shown in the PSIM model in the table. The duty ratio of the converter is set at

$$D = \frac{1.6}{4} = 0.4$$

where a 1.6 V control voltage, $V_{con} = 1.6$ V, and a 4 V ramp signal, $V_{ramp}$.

The start-up responses of the output voltage $v_O(t)$ and inductor current $i_L(t)$ are shown in the lower row of the table. After initial transient responses, the output voltage settles in the steady-state value

$$V_O = \frac{V_S}{1-D} = \frac{12}{1-0.4} = 20 \text{ V}$$

As will be shown later, the steady-state inductor current of the converter is given by

$$I_L = \frac{1}{1-D}\frac{V_O}{R} = \frac{1}{1-0.4}\frac{20}{1} = 33.3 \text{ A}$$

### 4.1.2 Steady-State Analysis in CCM

The previous section described the evolution of the boost converter from the buck converter. The steady-state analysis of the boost converter in CCM operation is now performed using the analysis techniques established in Chapter 3.

#### Steady-State Operation in CCM

Figure 4.2 shows the boost converter, on-time and off-time subcircuits, and major circuit waveforms in CCM operation. The standard techniques used in Chapter 3, such

**Table 4.1** Start-up Response of Boost Converter

| PSIM model: Boost_Converter_01 |
| --- |
| 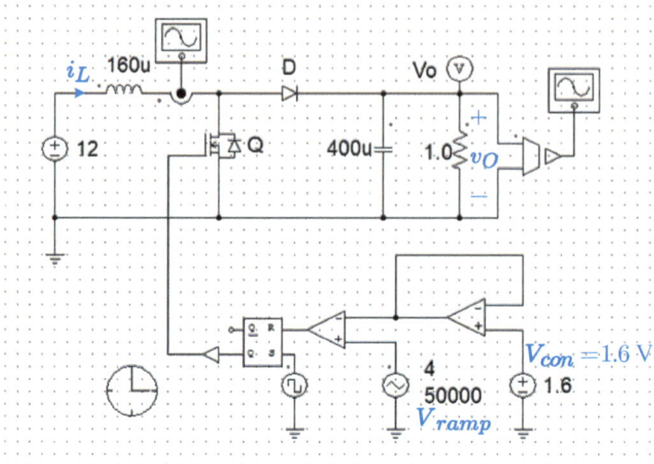 |
| Simulation outcomes |
| 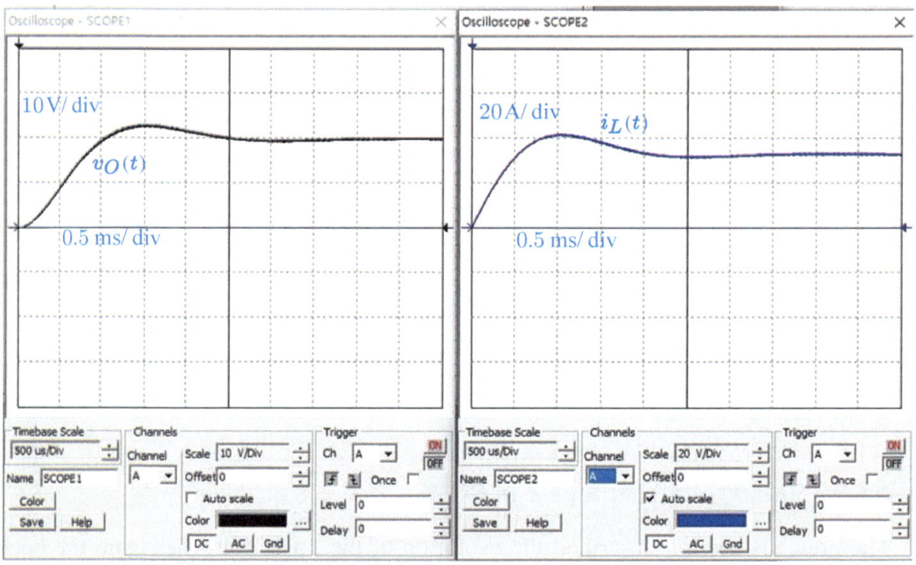 |

as the small-ripple approximation, volt-sec balance condition, and amp-sec balance condition, are employed in Fig. 4.2. This figure also incorporates that the output voltage is always larger than the input voltage, $V_O > V_S$.

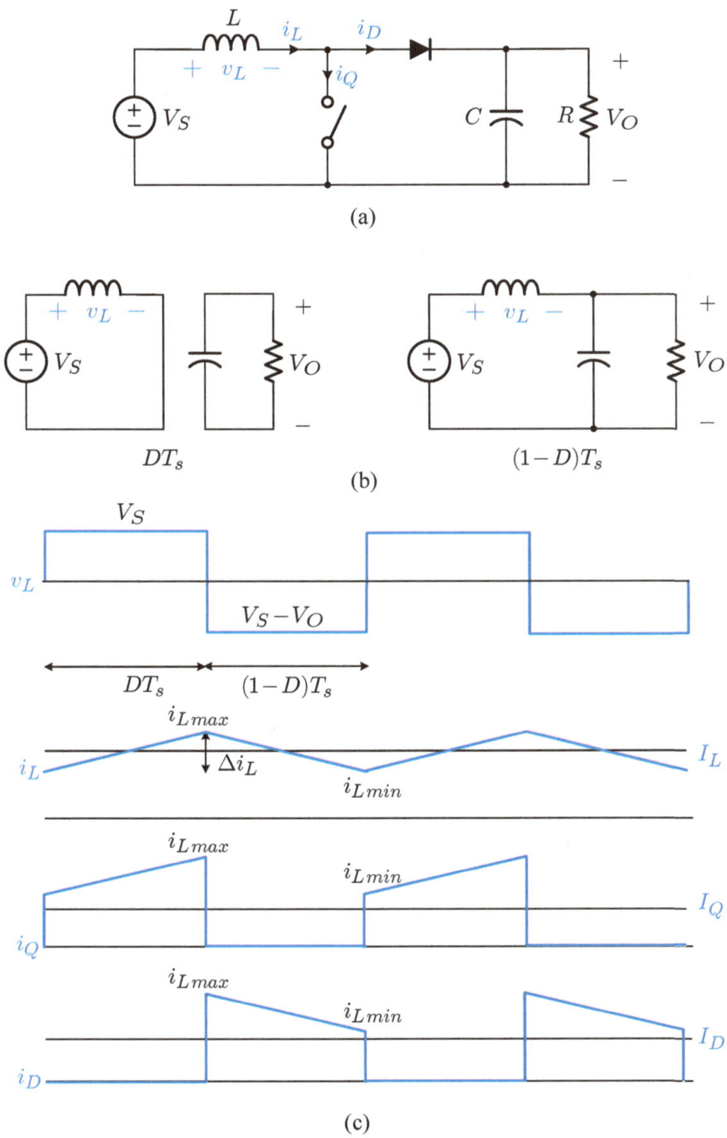


**Figure 4.2** Steady-state analysis of boost converter in CCM. (a) Boost converter. (b) On-time and off-time subcircuits. (c) Major waveforms.

The volt-sec balance condition is now applied to the inductor

$$V_S D T_s = (V_O - V_S)(1 - D)T_s \tag{4.5}$$

resulting in the voltage gain expression

$$\boxed{\frac{V_O}{V_S} = \frac{1}{1-D}} \tag{4.6}$$

which is identical to (4.4). The gain expression indicates that the voltage gain increases from unity to infinity when the duty ratio is varied between $0 < D < 1$.

During an on-time period, the inductor current ramps up with a slope $V_S/L$. Meanwhile, during an off-time period, the inductor current ramps down with a slope $(V_S - V_O)/L$. The excursion of the inductor current is determined as

$$\boxed{\Delta i_L = \frac{V_S}{L} DT_s = \frac{V_O - V_S}{L}(1-D)T_s} \tag{4.7}$$

As illustrated in Fig. 4.2(c), the inductor current flows through the active switch during the on-time period and freewheels through the diode during the off-time period. Accordingly, the following relations hold among the average value of the switch current, $I_Q$, the average diode current, $I_D$, and average inductor current, $I_L$

$$I_Q = DI_L \tag{4.8}$$

$$I_D = (1-D)I_L \tag{4.9}$$

In the boost converter, the diode is connected to the load, whereas the inductor is located at the source side. For this case, the average value of the inductor current is determined as follows. Due to the charge balance condition, the average value of the capacitor current is zero. Accordingly, the average diode current is to be equal to the load current, which is given by the ratio of the output voltage to load resistance

$$I_D = I_O = \frac{V_O}{R} \tag{4.10}$$

Using (4.9) and (4.10), the average inductor current is found as

$$\boxed{I_L = \frac{1}{1-D} I_D = \frac{1}{1-D}\frac{V_O}{R}} \tag{4.11}$$

The maximum and minimum values of the inductor current are expressed as

$$i_{L\,max} = I_L + \frac{1}{2}\Delta i_L \tag{4.12}$$

and

$$i_{L\,min} = I_L - \frac{1}{2}\Delta i_L \tag{4.13}$$

where $I_L$ is given by (4.11) and $\Delta i_L$ is determined by (4.7).

Table 4.2 Steady-State Analysis of Boost Converter

| Circuit variable | Expression |
|---|---|
| $V_O$ | $\frac{1}{1-D}V_S = \frac{1}{1-0.4}12 = 20$ V |
| $v_{L\,max}$ | $V_S = 12$ V |
| $v_{L\,min}$ | $V_S - V_O = 12 - 20 = -8$ V |
| $I_L$ | $\frac{1}{1-D}\frac{V_O}{R} = \frac{1}{1-0.4}\frac{20}{5} = 6.67$ A |
| $\Delta i_L$ | $\frac{V_S}{L}DT_s = \frac{12}{160\times10^{-6}}0.4\cdot 20\times10^{-6} = 0.6$ A |
| $i_{L\,max}$ | $I_L + \frac{1}{2}\Delta i_L = 6.67 + \frac{0.6}{2} = 6.97$ A |
| $i_{L\,min}$ | $I_L - \frac{1}{2}\Delta i_L = 6.67 - \frac{0.6}{2} = 6.37$ A |

**Steady-state operation of boost converter EXAMPLE 4.2**

This example shows the steady-state analysis and circuit waveforms of a boost converter. The operational conditions and circuit parameters are $V_S = 12$ V, $L = 160\ \mu$H, $C = 400\ \mu$F, $R = 5\ \Omega$, $T_s = 20\ \mu$s, and $D = 0.4$. The steady-state values of the major voltage and current waveforms are evaluated in Table 4.2. The simulated steady-state circuit waveforms are given in Fig. 4.3.

## Estimation of Output Voltage Ripple

The magnitude of the output voltage ripple is determined from the waveforms in Fig. 4.4. The diode current $i_D$ is split into the capacitor current $i_C$ and load current $I_O$. As discussed in Section 3.4.3, the load current $I_O$ carries the dc component of $i_D$ while the capacitor current $i_C$ absorbs the ac component of $i_D$: $I_O = I_D$ and $i_C = i_D - I_D$. This situation is illustrated in Fig. 4.4(b).

The peak-to-peak value of the capacitor voltage $\Delta v_{ac}$ or the magnitude of the output voltage ripple $\Delta v_O$ is found by integrating $i_C$ over the time period in which

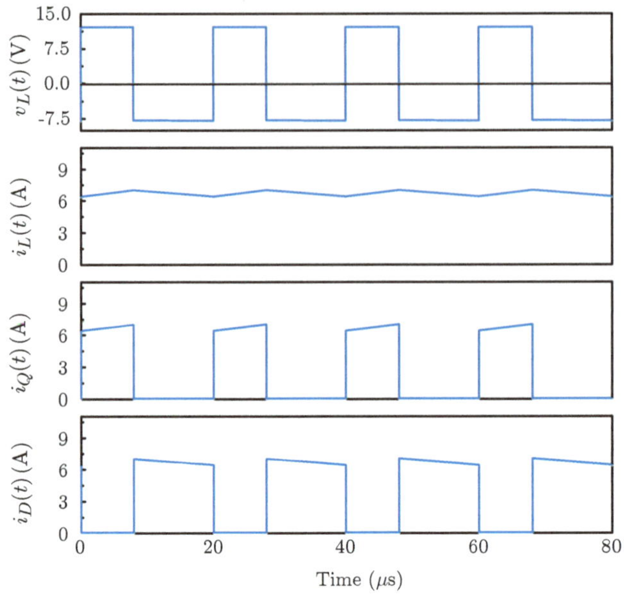


**Figure 4.3** Circuit waveforms of boost converter.

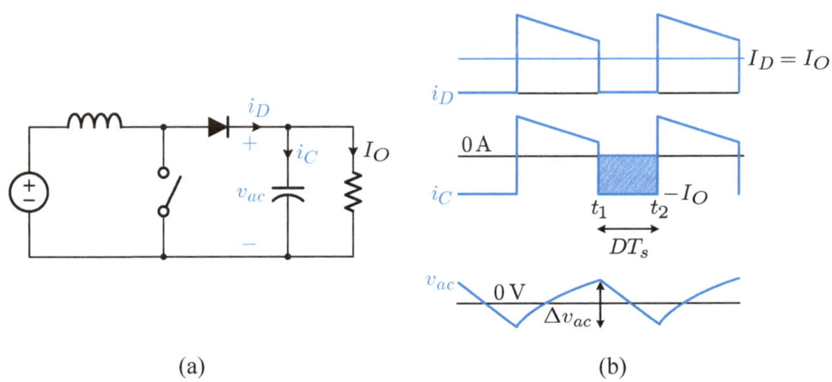


**Figure 4.4** Estimation of output ripple. (a) Circuit diagram. (b) Current waveforms and output ripple.

$i_C$ remains *negative* and by dividing the result by the capacitance

$$\Delta v_{ac} = \Delta v_O = \frac{1}{C} \int_{t_1}^{t_2} i_C(\tau)\, d\tau \tag{4.14}$$

The operation of (4.14) is equivalent to evaluating the area of the highlighted rectangle in Fig. 4.4(b) and dividing the area by the capacitance. By noting that $t_2 - t_1 = DT_s$, the magnitude of the output voltage ripple is given as

$$\Delta v_O = \frac{1}{C} I_O D T_s = \frac{1}{C} \frac{V_O}{R} D T_s \tag{4.15}$$

It should be noted that (4.15) is only valid when the boost converter is in CCM and the output capacitor does not contain an equivalent series resistance (esr). When the esr $R_c$ is present, the output ripple is given by

$$\Delta v_O \approx \Delta i_D \, R_c = i_{D\,max} \, R_c = i_{L\,max} \, R_c \tag{4.16}$$

for the boost converter operating in either CCM or DCM. The output ripple of the boost converter, given by $\Delta v_O \approx i_{Lmax} \, R_c$ in (4.16), is significantly larger than that of the buck converter, given by $\Delta v_O \approx \Delta i_L R_c$ in (3.31).

### 4.1.3 Steady-State Analysis in DCM

The boost converter operates at the borderline between CCM and DCM when the condition $I_L = \Delta i_L / 2$ is satisfied. The CCM/DCM boundary condition is expressed in terms of circuit parameters and operating condition

$$\frac{1}{1-D} \frac{V_O}{R} = \frac{1}{2} \frac{V_S}{L} D T_S \tag{4.17}$$

based on (4.11) and (4.7). The expressions (4.17) and (4.6) are used to find the critical value for the load resistance

$$R_{crit} = \frac{2L}{D(1-D)^2 T_s} \tag{4.18}$$

that determines the operational mode. When the load resistance is larger than $R_{crit}$, the converter enters DCM operation. Otherwise, the converter is in CCM operation.

Figure 4.5 shows the major circuit waveforms of a boost converter in DCM operation. Application of the volt-sec balance condition to the inductor yields

$$V_S D T_s = (V_O - V_S) D_1 T_s \tag{4.19}$$

which indicates that

$$\frac{V_O}{V_S} = \frac{D + D_1}{D_1} \tag{4.20}$$

The unknown variable $D_1$ should be removed from (4.20) to yield a complete voltage gain expression.

For the boost converter, the average value of the diode current should be the same as the load current, $I_D = I_O$, because the diode is connected to the output of the converter. From the waveform of the diode current $i_D$, the following equation is set forth

$$\frac{\frac{1}{2} i_{L\,max} D_1 T_s}{T_s} = I_O \tag{4.21}$$

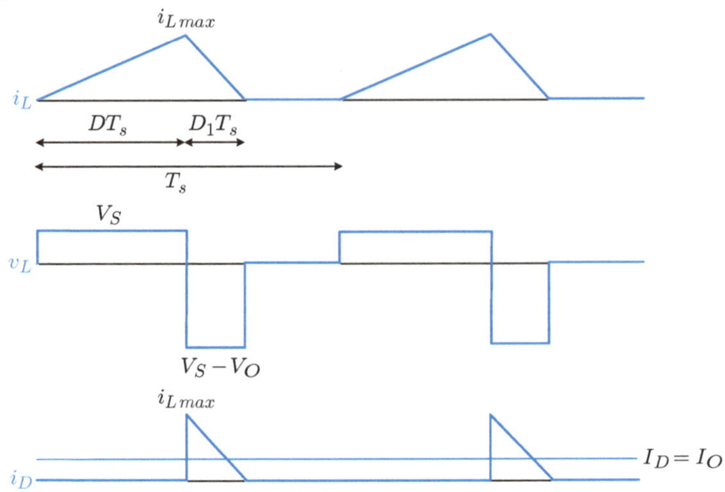


**Figure 4.5** Circuit waveforms of boost converter in DCM operation.

By incorporating the fact

$$i_{L\,max} = \frac{V_S}{L} DT_s \tag{4.22}$$

and

$$I_O = \frac{V_O}{R} \tag{4.23}$$

the equation (4.21) is rearranged as

$$\frac{1}{2}\frac{V_S}{L} DT_s D_1 = \frac{V_O}{R} \tag{4.24}$$

to yield the expression for $D_1$

$$D_1 = \frac{V_O}{V_S}\frac{2L}{RDT_s} \tag{4.25}$$

Now, the expressions (4.20) and (4.25) are combined, resulting in a quadratic equation for $V_O/V_S$

$$\left(\frac{V_O}{V_S}\right)^2 - \left(\frac{V_O}{V_S}\right) - \frac{D^2RT_s}{2L} = 0 \tag{4.26}$$

The non-negative solution to (4.26) becomes the DCM voltage gain of the boost converter

$$\boxed{\frac{V_O}{V_S} = \frac{1}{2}\left(1 + \sqrt{1 + \frac{2D^2RT_s}{L}}\right)} \tag{4.27}$$

The DCM voltage gain is a nonlinear function of circuit parameters and operational conditions. The following two observations, previously found in the buck converter, also hold true for the boost converter.

1) The voltage gain grows larger as the load resistance is increased.
2) For most practical converter parameters, the DCM voltage gain is larger than the CCM counterpart when the same duty ratio is assumed.

### 4.1.4 Effects of Parasitic Resistance on Voltage Gain

The voltage gain of the ideal boost converter in CCM operation, $V_O/V_S = 1/(1-D)$, predicts that the output voltage increases without limit as the duty ratio $D$ approaches unity, thereby yielding an infinite output voltage with $D = 1$. However, the operation of the boost converter contradicts this scenario. With $D = 1$, the active switch remains closed all the time and the diode never conducts. This implies that the output section of the converter is isolated from the source. For this case, the output voltage is zero because no energy is delivered from source to load.

The boost converter becomes inoperative with the unity duty ratio. Thus, the operation with $D = 1$ is the *singular condition* at which the theoretical voltage gain formula breaks down. Apart from the condition with $D = 1$, it is also unrealistic that the voltage gain of the boost converter becomes extremely large as the duty ratio closely approaches unity.

The voltage gain formula, $V_O/V_S = 1/(1 - D)$ is derived under the assumption that all the active and passive components are ideal. The semiconductor switches are considered as the ideal switch and the reactive circuit components are presumed to be totally free from any parasitic resistance. When any non-ideal device property is included, the voltage gain greatly deviates from the ideal formula.

As a simple demonstration, Fig. 4.6 shows a boost converter along with its on-time and off-time subcircuits, where the winding resistance $R_l$ is included to a practical inductor, while the other circuit components are still ideal. The voltage gain expression of the converter with the presence of $R_l$ is now derived to investigate the behavior of the *non-ideal* boost converter.

The volt-sec balance condition of the inductor is formulated as

$$(V_S - \bar{\imath}_L(t)R_l)DT_s = (V_O - \bar{\imath}_L(t)R_l - V_S)(1 - D)T_s \tag{4.28}$$

where $\bar{\imath}_L$ represents the average value of the inductor current: $\bar{\imath}_L = I_L$. The average inductor current $\bar{\imath}_L$ has the same effect as the original inductor current $i_L$, as far as the volt-sec balance condition is concerned. The expression (4.28) is simplified to

$$V_S = I_L R_l + (1 - D)V_O \tag{4.29}$$

On the other hand, the amp-sec balance condition on the output capacitor is written as

$$-\frac{V_O}{R}DT_s = \left(\frac{V_O}{R} - \bar{\imath}_L(t)\right)(1 - D)T_s \tag{4.30}$$

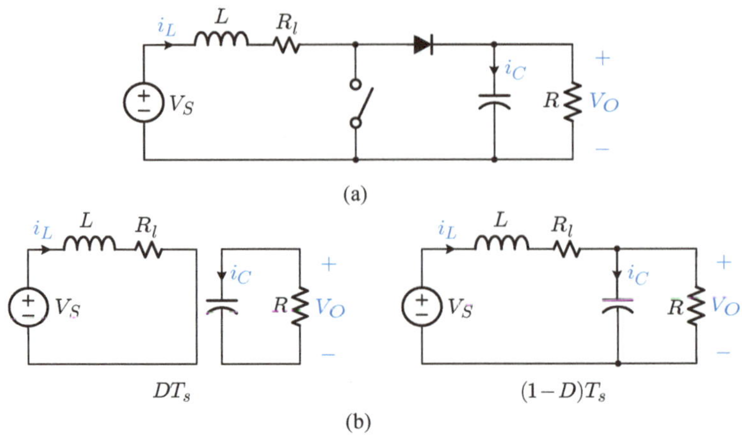


**Figure 4.6** Boost converter with inductor winding resistance. (a) Circuit diagram. (b) On-time and off-time subcircuits.

which is simplified to

$$\frac{V_O}{R} = (1-D)I_L \tag{4.31}$$

The desired expression for the voltage gain is obtained by simultaneously solving (4.29) and (4.31) to eliminate $I_L$. The resulting voltage gain is arranged as the product of the *ideal voltage gain* and the correction factor that accounts for the effect of the winding resistor $R_l$

$$\boxed{\frac{V_O}{V_S} = \frac{1}{1-D}\left(\frac{1}{1+\dfrac{1}{(1-D)^2}\dfrac{R_l}{R}}\right)} \tag{4.32}$$

The key parameter in the correction factor is the ratio of the winding resistance to the load resistance, $R_l/R$. Figure 4.7 shows the voltage gain curves, calculated from (4.32) with different values for $R_l/R$. The voltage gain curves reveal substantial deviations from the ideal case. The deviation intensifies as the $R_l/R$ ratio increases, showing a wide gap between the actual voltage gain and the ideal voltage gain when the duty ratio is large. In particular, all the actual voltage gain curves merge to *zero* as the duty ratio approaches unity, rather than growing boundlessly. This phenomenon is consistent with the operation of the boost converter. With $D = 1$, the output of the converter is never connected to the source, and the output capacitor thus remains uncharged producing the *zero* output voltage.

It is noteworthy that the $R_l/R$ ratio limits the maximum voltage gain that a boost converter could produce. For example, with $R_l/R = 0.02$, Fig. 4.7 indicates that the maximum voltage gain is approximately 3.5 at the duty ratio of $D = 0.875$. If a voltage gain larger than 3.5 is required, $R_l/R$ should be reduced smaller than 0.02.

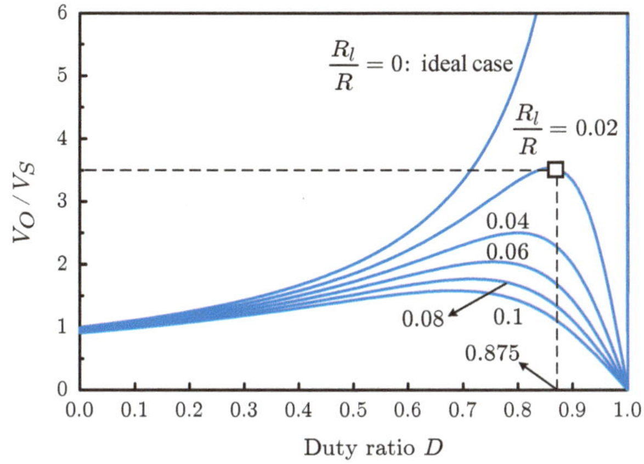


**Figure 4.7** Voltage gain of non-ideal boost converter.

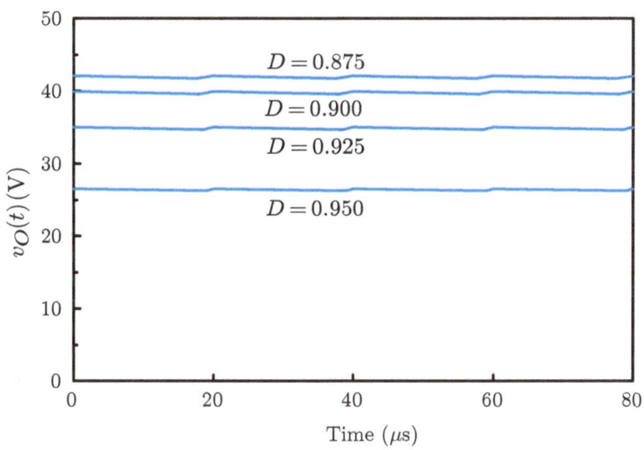


**Figure 4.8** Output voltage waveforms of boost converter with $R_l/R = 0.02$.

## Output voltage with inductor winding resistance EXAMPLE 4.3

This example substantiates the results of the preceding gain analysis. The boost converter introduced in Example 4.2 has the parameters of $V_S = 12\,\text{V}$, $L = 160\,\mu\text{H}$, $C = 400\mu\text{F}$, and $R = 5\Omega$. Now, a $0.1\Omega$ winding resistance is included in the inductor, resulting in $R_l/R = 0.1/5 = 0.02$. Under this condition, a series of simulations is performed with four different values for the duty ratio: $D = 0.875$, 0.900, 0.925, and 0.950. The output voltage waveforms of the boost converter are shown in Fig. 4.8. As expected from Fig. 4.7, the output voltage attains its peak value at $D = 0.875$ and decreases as the duty ratio is further increased. The theoretical predictions of the output voltage from (4.32) are given in Table 4.3.

**Table 4.3** Output Voltage Analysis with Inductor Winding Resistance

| Duty ratio | Output voltage |
|---|---|
| $D = 0.875$ | $V_O = \dfrac{1}{1-D}\left(\dfrac{1}{1+\dfrac{1}{(1-D)^2}\dfrac{R_l}{R}}\right)V_S = \dfrac{1}{1-0.875}\left(\dfrac{1}{1+\dfrac{1}{(1-0.875)^2}\dfrac{0.1}{5}}\right)12 = 42\text{ V}$ |
| $D = 0.900$ | 40 V |
| $D = 0.925$ | 35 V |
| $D = 0.950$ | 27 V |

## EXAMPLE 4.4 PSIM® Voltage gain of boost converter with $R_l$

The voltage gain of the boost converter used in Example 4.3 is computationally evaluated using PSIM simulations. The PSIM model for this analysis is shown in Table 4.4. The operational conditions are the same as the case of Example 4.3. In the PWM block in the PSIM model, the control voltage $V_C$ is swept between $0 < V_C < 1$, in order to produce the duty ratio variation between $0 < D < 1$. This feature is programmed with the four parameter sweep blocks.

While sweeping the duty ratio, a series of time-domain simulations are executed to obtain the output voltage for the range of $0 < D < 1$. The results are converted to the $V_O/V_S$ ratio using the division block in the PSIM model and displayed as a gain curve in Table 4.4. The empirical gain curve exactly reproduces the theoretical gain curve with $R_L/R = 0.02$ in Fig. 4.7.

## 4.2 BUCK/BOOST CONVERTER

The buck/boost converter is a dc-to-dc converter that could either step-up or step-down the input voltage. For this, the buck/boost converter is also called the up-down converter. This section discusses the circuit topology and steady-state operation of the buck/boost converter.

**Table 4.4** Voltage Gain of Boost Converter with $R_l = 0.1\ \Omega$

PSIM model: Boost_Converter_02

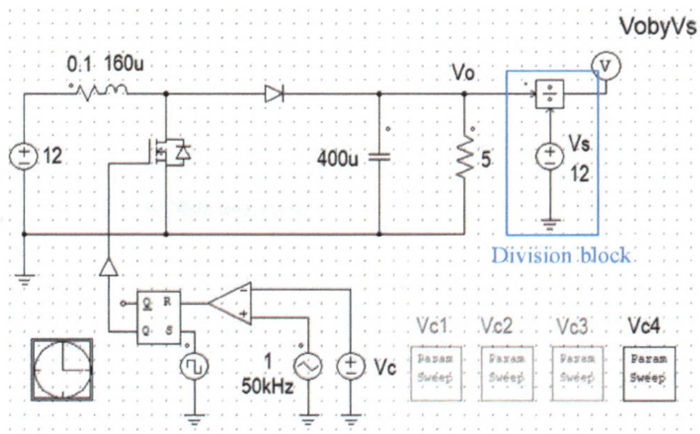


Simulation outcomes

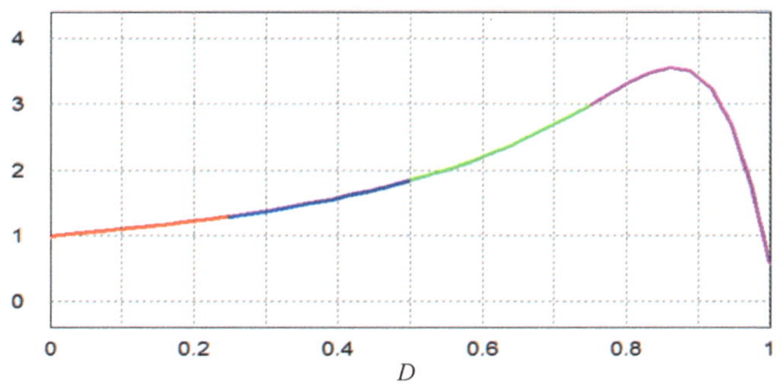


### 4.2.1 Evolution to Buck/Boost Converter

The buck/boost converter is created by cascading the buck converter and boost converter and simplifying the cascaded circuit. Figure 4.9(a) shows the buck converter connected to a boost converter downstream. Figure 4.9(a) is changed to Fig. 4.9(b) by replacing the active-passive switch pairs with SPDT switches. It is now assumed that the two SPDT switches are synchronized so that both the SPDT switches are concurrently held at position **a** for $DT_s$ and position **p** for $(1 - D)T_s$.

It should be noted that the upstream buck converter in Fig. 4.9(b) is normally functioning, even if its output filter capacitor is removed. The output filter capacitor, employed to enhance the filtering performance in a standalone buck converter, becomes redundant when the buck converter is connected to the boost converter having its own output capacitor.

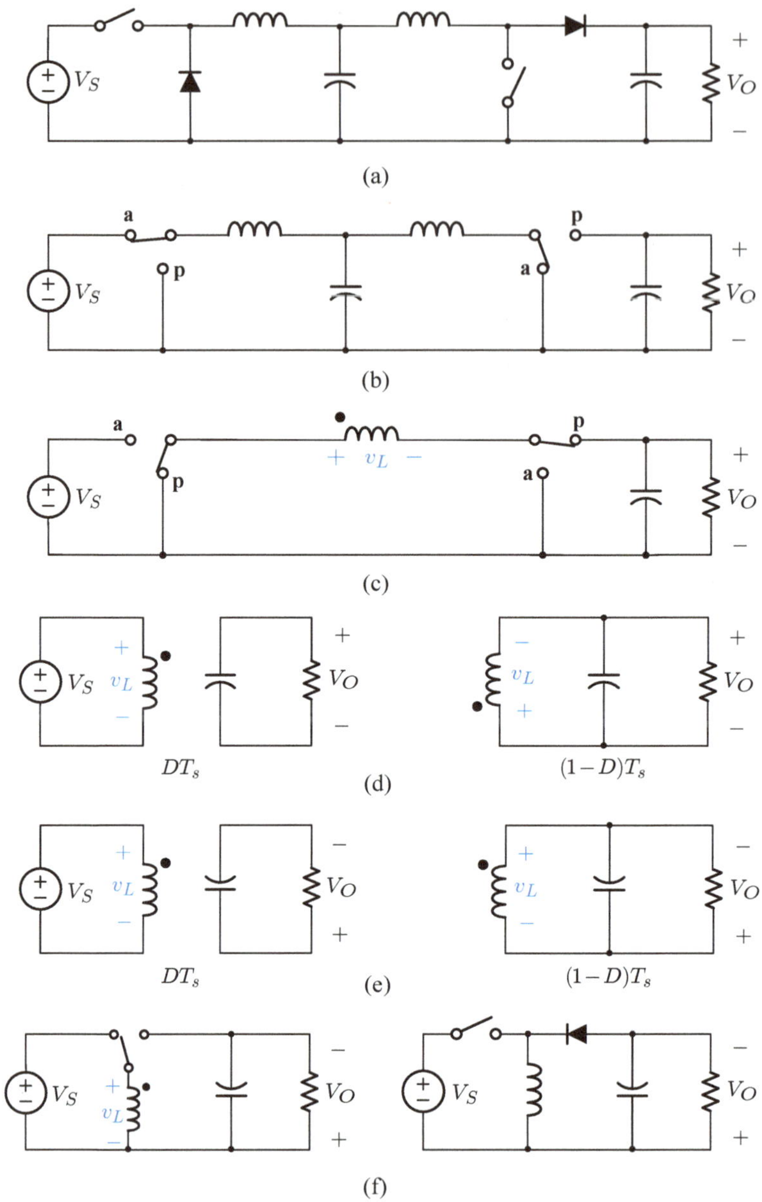


**Figure 4.9** Derivation of buck/boost converter. (a) Cascaded connection of buck converter and boost converter. (b) Alternative representation using two synchronized SPDT switches. (c) Non-inverting buck/boost converter. (d) Subcircuits of non-inverting buck/boost converter. (e) Modification of subcircuits. (f) Buck/boost converter.

Once the output capacitor of the buck converter is removed, the inductor of the buck converter and that of the boost converter merge together, resulting in the circuit shown in Fig. 4.9(c). In Fig. 4.9(c), the left-hand side of the inductor terminal is marked • to highlight the polarity of the inductor voltage. This circuit is referred to as the non-inverting buck/boost converter for the reasons that will become clear shortly.

Application of the volt-sec balance condition to the inductor in Fig. 4.9(c) yields

$$V_S D T_s = V_O (1 - D) T_s \tag{4.33}$$

which is rearranged for the voltage gain of the buck/boost converter

$$\boxed{\frac{V_O}{V_S} = D \frac{1}{1 - D}} \tag{4.34}$$

The first term $D$ originates from the buck converter upstream, while the second term $1/(1 - D)$ comes from the boost converter downstream.

The on-time and off-time subcircuits of the non-inverting buck/boost converter are shown in Fig. 4.9(d). The subcircuits in Fig. 4.9(d) are redrawn into Fig. 4.9(e) keeping the circuit operation unchanged. The vertical flipping of the secondary side of the on-time subcircuit and off-time subcircuit does not affect the circuit operation, but the polarity of the output voltage in Fig. 4.9(e) is inverted from that of Fig. 4.9(d). A new circuit is now synthesized from the two subcircuits in Fig. 4.9(e). The resulting circuit is shown in Fig. 4.9(f) in two different forms: one with the SPDT switch and the other with an active-passive switch pair. It is easy to see that the new circuit in the left-hand side of Fig. 4.9(f) reduces to the two subcircuits in Fig. 4.9(e) with the two different SPDT switch connections.

The converter circuit shown in Fig. 4.9(f) is named the buck/boost converter because the converter is derived from the connection of the buck converter and the boost converter. The polarity of the output voltage of the buck/boost converter is opposite to the input voltage. Meanwhile, in the converter circuit in Fig. 4.9(c), the output voltage has the same polarity as the input voltage, as such, the converter is referred to as the non-inverting buck/boost converter. The non-inverting buck/boost is suited for the applications requiring the positive output voltage. However, the converter requires two pairs of the active-passive switch combination.

### PSIM® Start up response of buck/boost converter EXAMPLE 4.5

This example shows the start-up response of a buck/boost converter operating with a fixed duty ratio $D = V_{con}/V_{ramp} = 1.6/4 = 0.4$ at the switching frequency $f_s = 50$ kHz. The converter operates with $V_S = 12$ V, $L = 160\ \mu$H, $C = 400\ \mu$F, and $R = 3\ \Omega$. The PSIM model and simulation results are given in Table 4.5.

The output voltage $V_O$ is given by

$$V_O = -D \frac{1}{1 - D} V_S = -0.4 \frac{1}{1 - 0.4} 12 = -8 \text{ V}$$

**Table 4.5** Start-up Response of Buck/Boost Converter

PSIM model: Buck_Boost_01

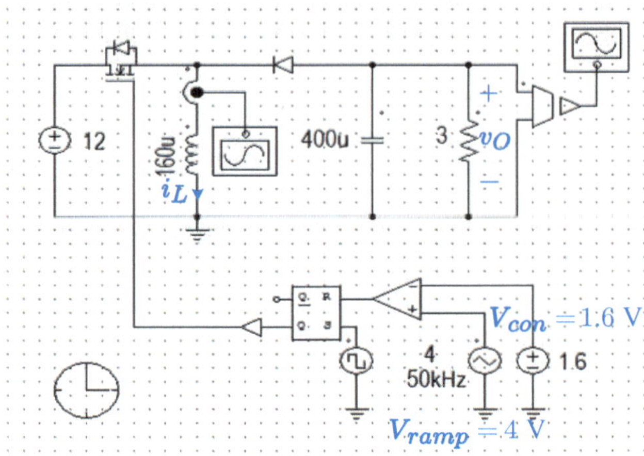


Simulation outcomes

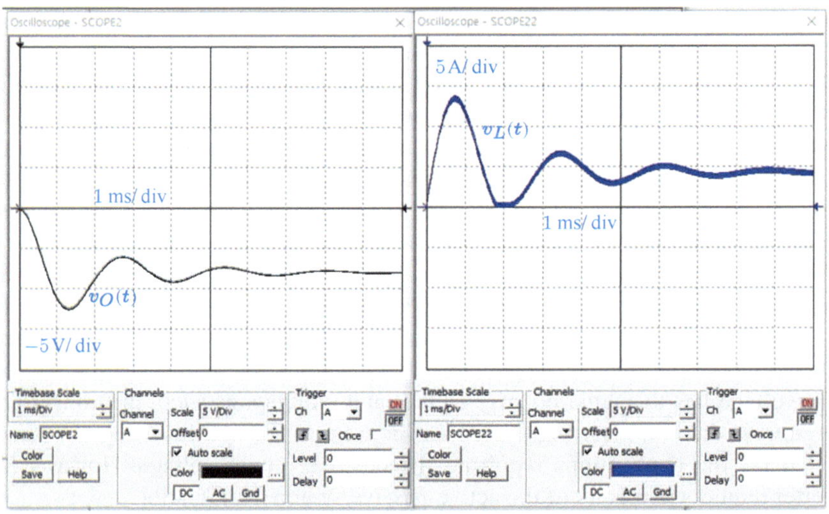


from (4.34). The inversion in the output voltage polarity is apparent in the simulation.

The average value of the inductor current is given by

$$I_L = \frac{1}{1-D}\frac{V_O}{R} = \frac{1}{1-0.4}\frac{8}{3} = 4.4 \text{ A}$$

as will be seen in the next section.

### 4.2.2 Steady-State Analysis in CCM

The steady-state analysis of the buck/boost converter is performed in the same manner as that of the buck converter and boost converter cases. The piecewise linear analysis is combined with the small-ripple approximation, volt-sec balance condition, and amp-sec balance condition.

#### Steady-State Operation in CCM

Figure 4.10 shows the buck/boost converter, on-time and off-time subcircuits, and major circuit waveforms. The steady-state operation of the converter is illustrated with the waveforms in Fig. 4.10(c). Application of the volt-sec balance condition to the inductor

$$V_S D T_s = V_O (1 - D) T_s \tag{4.35}$$

yields the voltage gain of the converter

$$\boxed{\frac{V_O}{V_S} = \frac{D}{1 - D}} \tag{4.36}$$

The gain formula predicts that the output voltage varies from zero to infinity when the duty ratio is varied between $0 < D < 1$. However, as is the case with the boost converter, the voltage gain deviates from the ideal prediction when non-ideal circuit components are considered. It could be shown that the voltage gain is modified as

$$\frac{V_O}{V_S} = \frac{D}{1 - D}\left(\frac{1}{1 + \dfrac{1}{(1 - D)^2}\dfrac{R_l}{R}}\right) \tag{4.37}$$

at the presence of the inductor winding resistance $R_l$. The structure of (4.37) is the same as the case of the practical boost converter in (4.32). Accordingly, the actual buck/boost converter will resemble the behavior of the practical boost converter discussed in Section 4.1.4.

The excursion of the inductor current is given by

$$\boxed{\Delta i_L = \frac{V_S}{L} D T_s = \frac{V_O}{L}(1 - D) T_s} \tag{4.38}$$

The average values of the switch current and diode current are determined as

$$I_Q = D I_L$$

$$I_D = (1 - D) I_L \tag{4.39}$$

Same as the boost converter case, the load current is supported by the diode current. Thus, the average diode current becomes the load current: $I_D = I_O = V_O/R$.

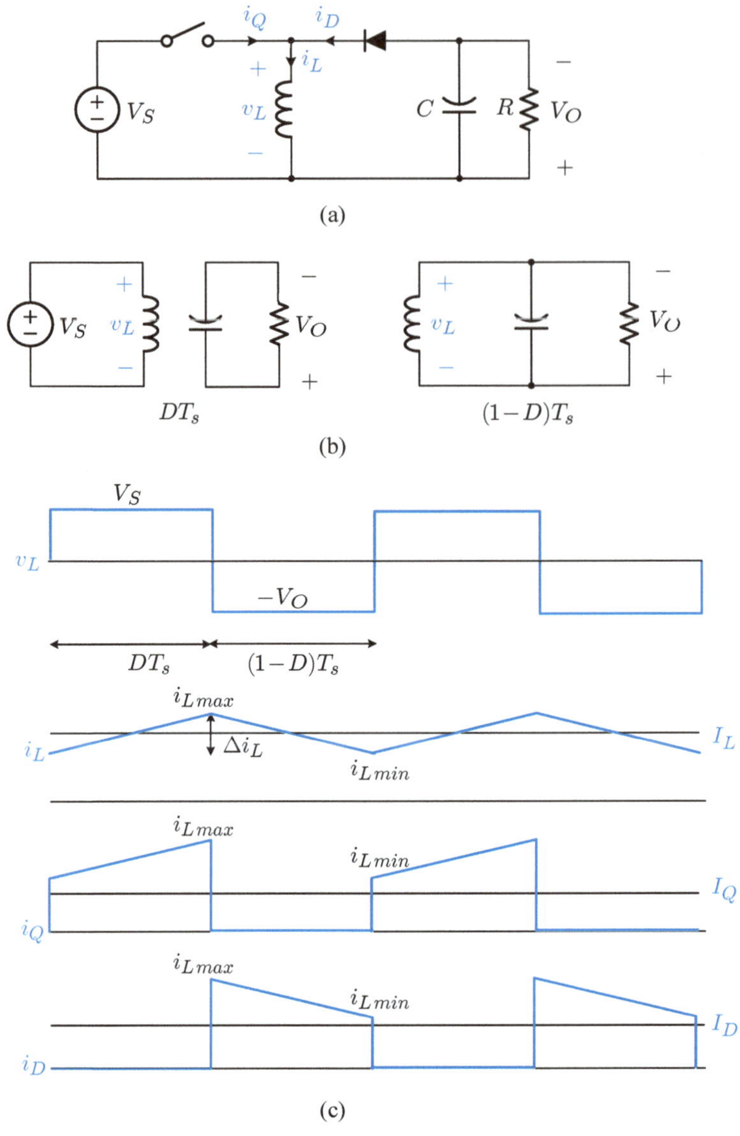


**Figure 4.10** Steady-state analysis of buck/boost converter in CCM. (a) Buck/boost converter. (b) On-time and off-time subcircuits. (c) Major circuit waveforms.

Accordingly, the average value of the inductor current is given as

$$I_L = \frac{1}{1-D} I_D = \frac{1}{1-D} \frac{V_O}{R} \tag{4.40}$$

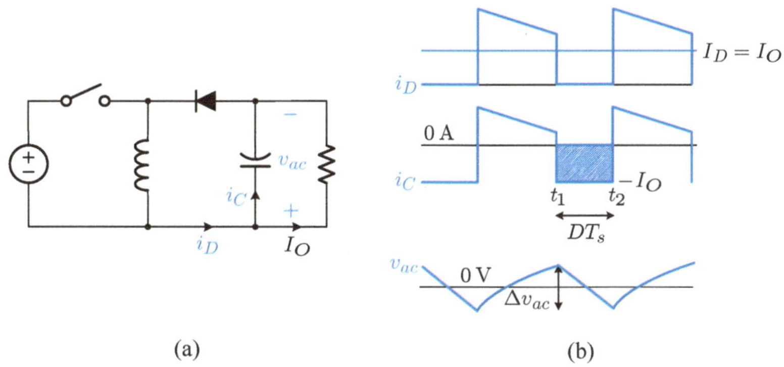


**Figure 4.11** Estimation of output ripple. (a) Circuit diagram. (b) Current waveforms and output ripple.

The maximum value of the inductor current then becomes

$$i_{L\,max} = I_L + \frac{1}{2}\Delta i_L \tag{4.41}$$

The minimum value is given by

$$i_{L\,min} = I_L - \frac{1}{2}\Delta i_L \tag{4.42}$$

## Estimation of Output Voltage Ripple

Figure 4.11 shows the circuit diagram and waveforms used in predicting the magnitude of the output voltage ripple. The waveforms are identical to those of the boost converter in Fig. 4.4. The output stage of the buck/boost converter retains the circuit properties of the boost converter, because the converter is evolved from the connection of the upstream buck converter and downstream boost converter. Consequently, the output voltage ripple of the buck/boost converter is identical to the boost converter

$$\boxed{\Delta v_{ac} = \Delta v_O = \frac{1}{C}\frac{V_O}{R}DT_s} \tag{4.43}$$

This expression only holds true when the converter is in CCM operation and no esr is present in the output capacitor. When the esr of the output capacitor is considered, the output ripple is approximated to

$$\boxed{\Delta v_O \approx \Delta i_D R_c = i_{D\,max} R_c = i_{L\,max} R_c} \tag{4.44}$$

for the buck/boost converter operating in either CCM or DCM.

**Table 4.6** Steady-State Analysis of Buck/Boost Converter

| Circuit variable | Expression |
|---|---|
| $V_O$ | $\frac{D}{1-D}V_S = \frac{0.4}{1-0.4}12 = 8\text{ V}$ |
| $v_{L\,max}$ | $V_S = 12\text{ V}$ |
| $v_{L\,min}$ | $-V_O = -8\text{ V}$ |
| $I_L$ | $\frac{1}{1-D}\frac{V_O}{R} = \frac{1}{1-0.4}\frac{8}{5} = 2.67\text{ A}$ |
| $\Delta i_L$ | $\frac{V_S}{L}DT_s = \frac{12}{160\times10^{-6}}0.4\cdot 20\times10^{-6} = 0.6\text{ A}$ |
| $i_{L\,max}$ | $I_L + \frac{\Delta i_L}{2} = 2.67 + \frac{0.6}{2} = 2.97\text{ A}$ |
| $i_{L\,min}$ | $I_L - \frac{\Delta i_L}{2} = 2.67 - \frac{0.6}{2} = 2.37\text{ A}$ |

**EXAMPLE 4.6 Steady-state operation of buck/boost converter**

This example demonstrates the circuit waveforms of a buck/boost converter. The circuit parameters of the buck/boost converter are $V_S = 12$ V, $L = 160$ $\mu$H, $C =$ 400 $\mu$F, $R = 5\,\Omega$, $T_s = 20$ $\mu$s, and $D = 0.4$. Table 4.6 summarizes the steady-state values of major circuit variables. Figure 4.12 shows the simulated waveforms of important circuit variables.

### 4.2.3 Steady-State Analysis in DCM

For the buck/boost converter, the CCM/DCM borderline condition, $I_L = \Delta i_L/2$, is written as

$$\frac{1}{1-D}\frac{V_O}{R} = \frac{1}{2}\frac{V_S}{L}DT_s \tag{4.45}$$

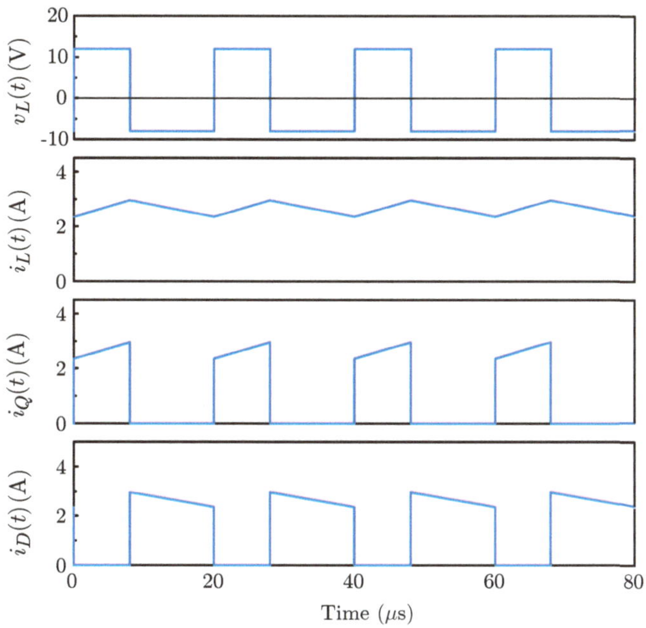


**Figure 4.12** Circuit waveforms of buck/boost converter.

using (4.40) and (4.38). From (4.45) and (4.36), the critical value for the load resistor is given by

$$R_{crit} = \frac{2L}{(1-D)^2 T_s} \tag{4.46}$$

The converter enters DCM operation with the load resistance larger than $R_{crit}$. Otherwise, the converter is in CCM operation.

Figure 4.13 shows the major DCM waveforms of the buck/boost converter. Application of the volt-sec balance condition to the inductor yields

$$V_S D T_s = V_O D_1 T_s \tag{4.47}$$

which is simplified to

$$\frac{V_O}{V_S} = \frac{D}{D_1} \tag{4.48}$$

By equating the average diode current $I_D$ to the load current, it follows that

$$\frac{\frac{1}{2}\left(\frac{V_S}{L} D T_s\right) D_1 T_s}{T_s} = \frac{V_O}{R} \tag{4.49}$$

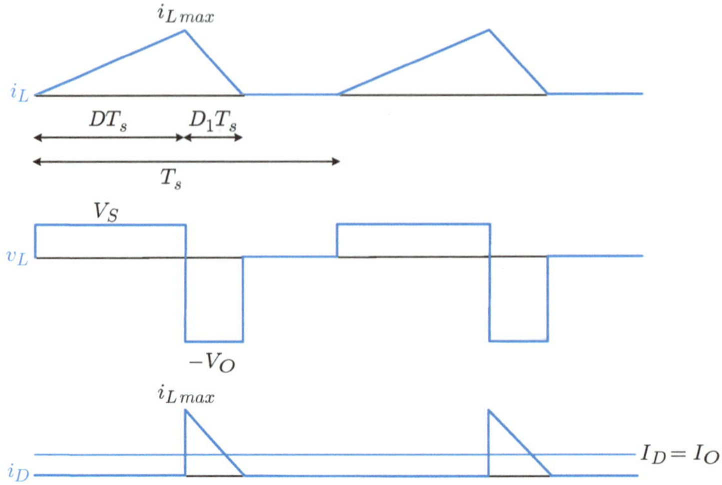


**Figure 4.13** Circuit waveforms of buck/boost converter in DCM operation.

which provides an expression for $D_1$

$$D_1 = \frac{V_O}{V_S}\frac{2L}{RDT_s} \tag{4.50}$$

Finally, by combining (4.48) and (4.50), the DCM voltage gain is given by

$$\boxed{\frac{V_O}{V_S} = D\sqrt{\frac{RT_s}{2L}}} \tag{4.51}$$

As is the case with the buck and boost converters, the voltage gain is proportional to the load resistance. Also, the DCM voltage gain is usually larger than the CCM counterpart for the same duty ratio.

## EXAMPLE 4.7 PSIM® Step load response with voltage mode control

In Chapter 3, we studied the closed-loop PWM control using a buck converter. Referring to Fig. 3.28 in Section 3.6.2, the feedback signal was derived from the output voltage of the converter. The feedback signal is fed to the three-pole two-zero circuit to yield the control signal. The control signal is compared with the ramp signal to produce the pulsewidth modulated switch drive signal.

The control scheme used in Section 3.6.2 is called *voltage mode control* in the sense that the output voltage alone is used as the feedback signal. We demonstrated that voltage mode control provides satisfactory transient responses for the buck converter, provided that the control is correctly designed. Thus, we learned that voltage mode control is a viable scheme for the buck converter and we naturally expect the same for the boost and buck/boost converters. However, as will be explored in the

**Table 4.7** Step Load Response of Buck/Boost Converter

PSIM model: Buck_Boost_02: Voltage Mode Control

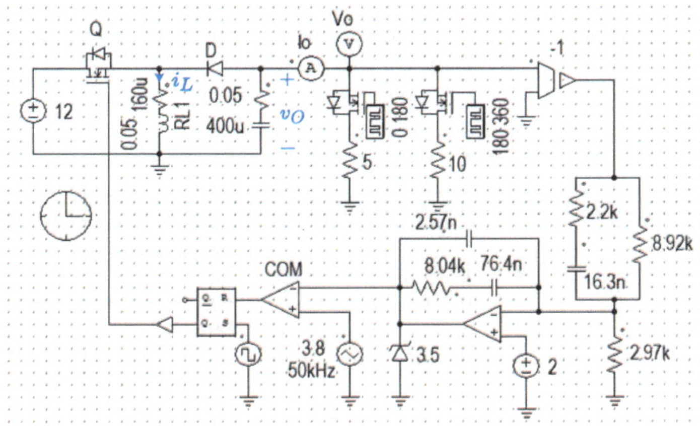


Simulation outcomes

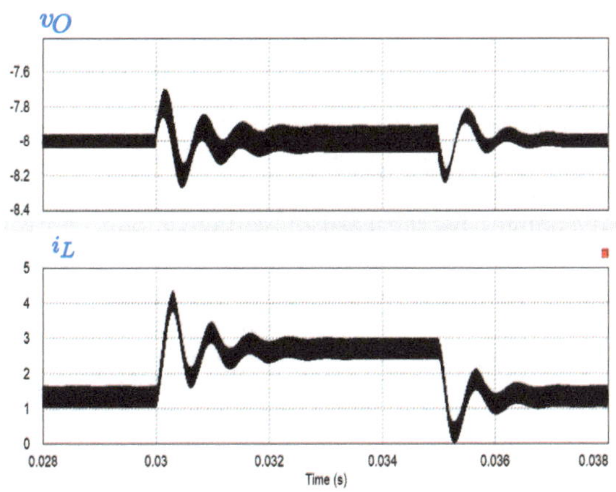


second book of the series, voltage mode control is only suited for the buck converter and does not provide acceptable performance for the boost and buck boost converters. This well-known fact stems from the differences in the dynamic characteristics of the converters.

In this example, we evaluate the performance of the buck/boost converter employing voltage mode control, in an attempt to reveal limitations of voltage mode control. This example supports the necessity of a careful study on the dynamics and control of PWM converters.

Table 4.7 shows the PSIM model of the buck/boost converter employing voltage mode control. The operational details and controller design are not discussed in this example. Yet, it is addressed that the controller is precisely designed to achieve the best possible performance within theoretical constraints of voltage mode control.

Table 4.7 illustrates the transient response of the output voltage $v_O$ and inductor current $i_L$, in response to the step changes between the load resistor $R = 5\ \Omega$ and $10\ \Omega$. Although stable, the responses are very oscillatory and sluggish, both revealing the limits of voltage mode control.

### EXAMPLE 4.8 PSIM® Step load response with current mode control

This example introduces an alternative control scheme, which has been widely adapted to PWM converters to overcome the limits of voltage mode control. In particular, the transient performance of the previous buck/boost converter example will be significantly improved with the alternative control scheme.

The structure of the new control scheme is illustrated in the PSIM model in Table 4.8. As highlighted with a box in the PSIM model, the switch current is sensed with a proportional gain and later mixed with the ramp signal. The resulting composite signal is compared with the output of the voltage feedback circuit, in order to generate the pulsewidth modulated switch drive signal. This control scheme is called *current mode control* in the sense that the control method employs an additional current feedback on top of the existing voltage feedback. Details about current mode control will be addressed in the second volume of the book series.

The superiorities of current mode control are well pronounced in the transient responses of the converter. Table 4.8 shows the output voltage $v_O$ and inductor current $i_L$ of the buck/boost converter with current mode control, in response to the same step load changes as the case of voltage mode control. The merits of current mode control are clearly seen in the transient waveforms. The oscillatory behavior is totally eliminated, thus reducing fluctuations and response time of the circuit waveforms. In fact, current mode control is the prerequisite for the boost and buck/boost converters to obtain acceptable performance. In forthcoming PSIM examples, current mode control will be often employed for PWM converters.

## 4.3 THREE BASIC CONVERTERS

We have studied the three basic dc-to-dc converter topologies: the buck converter, the boost converter, and the buck/boost converter. The structure, operational principles, and voltage gain of these three converters are now reviewed from a more general perspective.

### Structure and Operation of Three Basic Converters

The three basic converters contain a common circuit block, consisting of an SPDT switch and inductor. This common circuit block is shown in Fig. 4.14(a). This circuit

**Table 4.8** Step Load Response of Buck/Boost Converter

PSIM model: Buck_Boost_03: Current Mode Control

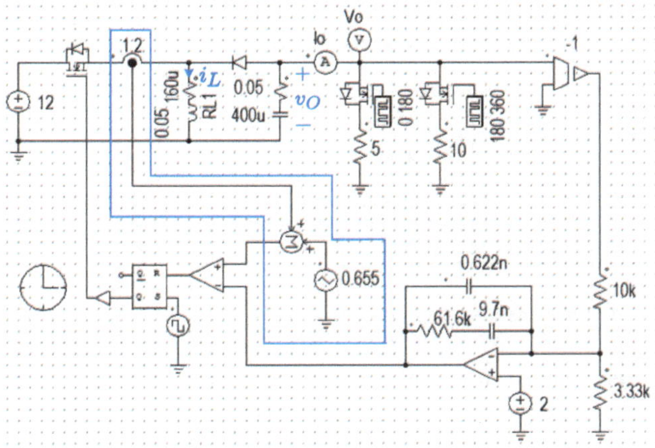


Simulation outcomes

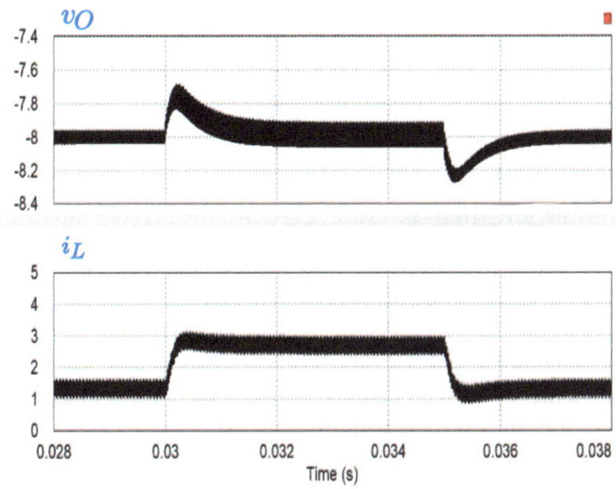


block is referred to as the *three-terminal switch cell.* The node **a** denotes the active switch terminal and **p** indicates the passive switch terminal. The node **c** stands for the common terminal to which the filter inductor is connected. The other end of the filter inductor is denoted as the inductor terminal **i**.

The three-terminal switch cell is inserted between the source and load in three different ways for the three basic converters. As shown in Fig. 4.14, the inductor terminal **i** is directed to the output port in the buck converter, while the inductor

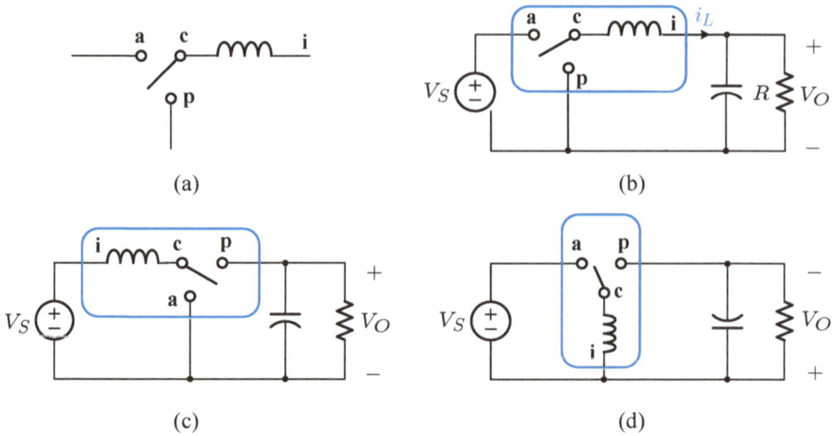


**Figure 4.14** Three-terminal switch cell and basic converters. (a) Three-terminal switch cell. (b) Buck converter. (c) Boost converter. (d) Buck/boost converter.

terminal is linked the input port in the boost converter. In the buck/boost converter, the inductor terminal is connected to the ground.

Despite the different configurations, the three converters have a common feature that enables each circuit to function as a dc-to-dc power converter. The inductor current of all PWM converters exhibits a triangular variation on top of its average value that is determined by the load current requirement, for example $I_L = V_O/R$ for the buck converter in Fig. 4.14(b).

When the common terminal is connected to the active terminal, the inductor current increases with a positive inductor voltage and extra energy is delivered from the source to inductor. On the other hand, when the common terminal is tied to the passive terminal, the inductor current decreases with a negative inductor voltage and the extra energy is released from the inductor to filter capacitor. Thus, the switching action of the three-terminal switch cell provides a balanced energy transfer from the source to load. During the energy transfer, the inductor serves as an intermediate energy storage, while also providing the dc current for the load.

To summarize, the three-terminal switch cell in the three basic dc-to-dc converters is positioned to link the source and load, while allowing a controlled energy transfer between them using the inductor as a transitional energy reservoir. In fact, the buck converter, boost converter, and buck/boost converter are the only three possible topological structures that perform the dc-to-dc power conversion using the three-terminal switch cell.

## Voltage Gain of Three Basic Converters

The three basic dc-to-dc converters offer the different input-to-output voltage gains. Figure 4.15 shows the voltage gain characteristics of the three converters. For each

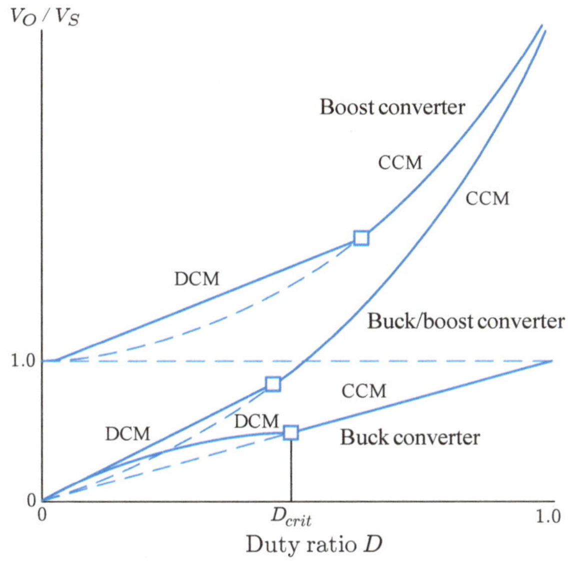


**Figure 4.15** Voltage gain curves of three basic converters.

**Table 4.9** DCM Voltage Gain and $D_{crit}$ of Three Basic Converters

| | DCM voltage gain | $D_{crit}$ |
|---|---|---|
| Buck converter | $\dfrac{2D}{D+\sqrt{D^2+\dfrac{8L}{RT_s}}}$ | $\dfrac{1}{2}(1-D_{crit}) = \dfrac{L}{RT_s}$ |
| Boost converter | $\dfrac{1}{2}\left(1+\sqrt{1+\dfrac{2D^2RT_s}{L}}\right)$ | $\dfrac{1}{2}D_{crit}(1-D_{crit})^2 = \dfrac{L}{RT_s}$ |
| Buck/boost | $D\sqrt{\dfrac{RT_s}{2L}}$ | $\dfrac{1}{2}(1-D_{crit})^2 = \dfrac{L}{RT_s}$ |

converter, the voltage gain is displayed for both CCM operation and DCM operation using $D_{crit}$ as the borderline. The converter operates in DCM when its duty ratio is less than $D_{crit}$ and moves to CCM operation as the duty ratio grows larger than $D_{crit}$. The expressions for the DCM voltage gain and $D_{crit}$ of the three converters are summarized in Table 4.9.

## 4.4 ĆUK CONVERTER

In Section 4.2, we studied the buck/boost converter as a composite converter circuit, formulated by 1) cascading the buck converter and boost converter, and 2) modifying the resultant circuit to eliminate redundant switches. Later in Section 4.3, we viewed the buck/boost converter as one member of the three basic converters, originated from the three different arrangements of the three-terminal switch cell in a generic dc-to-dc power conversion structure. This concept was illustrated in Fig. 4.14.

In late 1970s, Dr. Slobodan Ćuk proposed another innovative amalgamation of the buck converter and boost converter, where the boost converter is positioned upstream and the buck converter is placed downstream. The integration of the two converters in this specific order created a new non-isolated PWM converter circuit [1]. The converter was named as Ćuk converter. In this section, we explore the origin, operation, and waveforms of the Ćuk converter.

### 4.4.1 Evolution of Ćuk Converter

The development of the Ćuk converter is illustrated in Fig. 4.16. First, a boost converter is followed by a buck converter, as shown in Fig. 4.16(a). The composite converter is depicted using two synchronized SPDT switches in Fig. 4.16(b). Figure 4.16(c) portrays the on-time and off-time subcircuits of Fig. 4.16(b). Figure 4.16(c) is rearranged into Fig. 4.16(d) by 1) vertically flipping the secondary side of the on-time and off-time subcircuits and 2) relocating the inductor in the on-time subciruit.

Now, the two subcircuits in Fig. 4.16(d) are merged into one circuit, as shown in Fig. 4.16(e). Finally, the single SPDT switch in Fig. 4.16(e) is replaced with the MOSFET-diode pair to produce the final form of the Ćuk converter in Fig. 4.16(f). The *sign inversion* in the output voltage $V_O$ has naturally arisen during the circuit manipulation, as is the case with the buck/boost converter.

In the Ćuk converter, *the capacitor* $C$ in the middle of the converter is employed as an energy transfer element, instead of *the inductor* in the case of the three basic converters. Referring to Fig. 4.16(e), extra energy is transferred from the source to the capacitor when the pole of the SPDT is connected to **p** terminal. Conversely, when the SPDT switch is linked **a** terminal, the extra energy stored in $C$ is released to the load. The capacitor $C$ operates as an energy reservoir during the power conversion process.

### 4.4.2 Steady-State Analysis in CCM

Figure 4.17 depicts the Ćuk converter, along with its on-time subcircuit and off-time subcircuit. We assume that the two inductors, $L_1$ and $L_2$, are sufficiently large so the converter operates in CCM. It is also presumed that the output capacitor is large enough to ignore the output ripple. Here, we note that the energy transfer capacitor

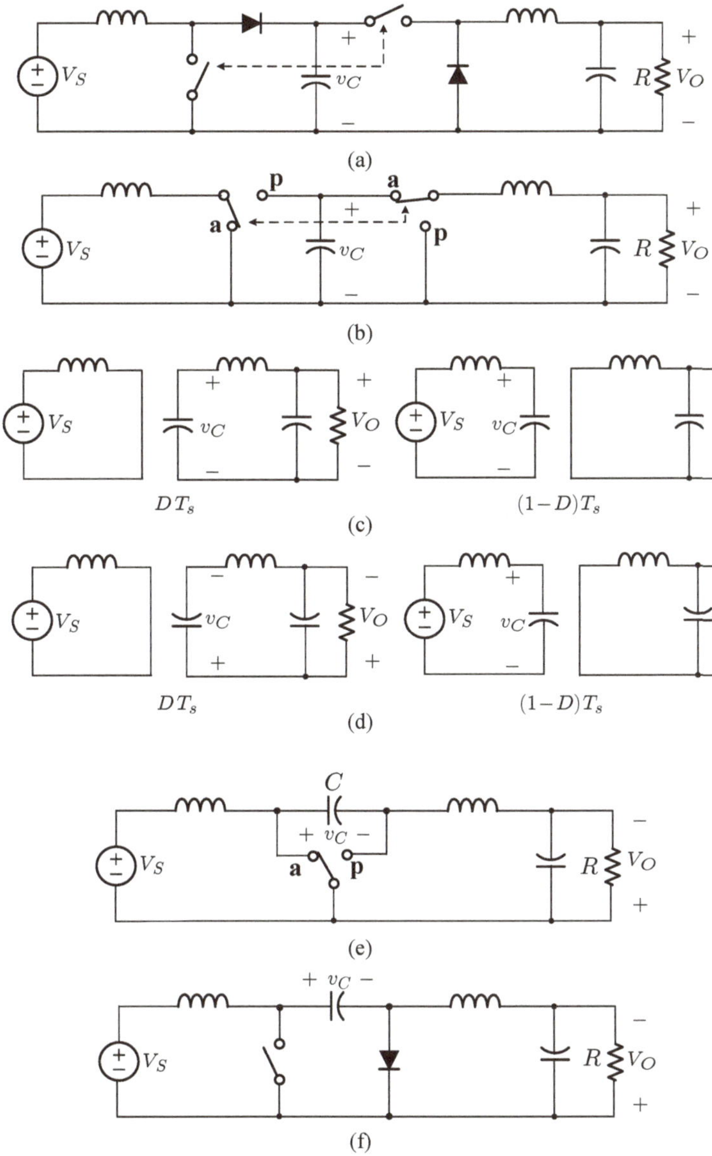


**Figure 4.16** Evolution of Ćuk converter. (a) Cascaded connection of boost converter and buck converter. (b) Alternative representation using two synchronized SPDT switches. (c) On-time and off-time subcircuits. (d) Modified on-time and off-time subcircuits. (e) Ćuk converter with single SPDT switch. (f) Ćuk converter with MOSFET-diode pair.

$C$ is charged with a positive voltage $V_C$ in steady state, as will be seen later in this section.†

† The structure and operation of the Ćuk converter are analogous to those of the capacitive switching circuit studied in Example 2.7.

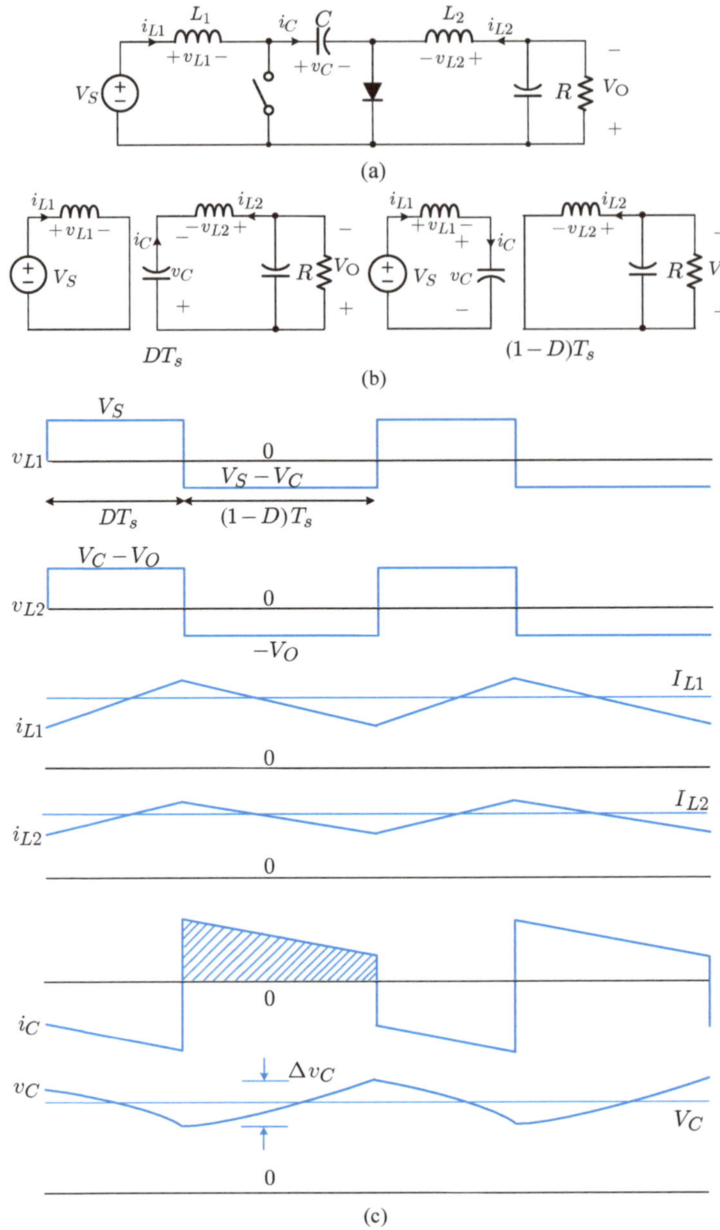


**Figure 4.17** Steady-state analysis in CCM. (a) Ćuk converter. (b) On-time subcircuit and off-time subcircuit. (c) Circuit waveforms.

When the active switch is closed during an on-time period, the capacitor voltage $V_C$ reverse-biases the diode $D$, thus resulting in the on-time subcircuit in Fig. 4.17(b).

When the active switch is opened, the inductor current $i_{L2}$ turns on the diode, yielding the off-time subcircuit in Fig. 4.17(b).

Figure 4.17(c) portraits the steady-state waveforms of the circuit variables specified in Fig. 4.17(b). The facts $V_C > V_S$ and $V_C > V_O$ are used in Fig. 4.17(c). The operation of the Ćuk converter will prove that the two conditions $V_C > V_S$ and $V_C > V_O$ always hold true.

Application of the volt-sec valance condition on the inductor voltage $L_1$ leads to

$$V_S DT_s = (V_C - V_S)(1 - D)T_s \quad \Rightarrow \quad V_C = \frac{1}{1 - D} V_S \tag{4.52}$$

Analogously, the volt-sec balance condition on $L_2$ yields

$$(V_C - V_O)DT_s = V_O(1 - D)T_S \quad \Rightarrow \quad V_O = DV_C \tag{4.53}$$

From (4.52) and (4.53),[†] the voltage gain of the converter is expressed as

$$\frac{V_O}{V_S} = \frac{D}{1 - D} \tag{4.54}$$

which is the product of the voltages gains of the boost converter and buck converter. The conditions $V_C > V_S$ and $V_C > V_O$ are apparent from (4.52) and (4.53).

The average values of the second inductor current $i_{L2}$ is determined as

$$I_{L2} = \frac{V_O}{R} \tag{4.55}$$

From the power balance condition, the average value of the first inductor current $i_{L1}$ is found as

$$V_S I_{L1} = V_O I_{L2} \quad \Rightarrow \quad I_{L1} = \frac{I_{L2} V_O}{V_S} = \frac{V_O^2}{RV_S} \tag{4.56}$$

The excursions of the inductor currents are given as

$$\Delta i_{L1} = \frac{V_S}{L_1} DT_s \tag{4.57}$$

and

$$\Delta i_{L2} = \frac{V_O}{L_2}(1 - D)T_s \tag{4.58}$$

The capacitor current is given by $i_C = -i_{L2}$ for the on-time period $DT_s$ and $i_C = i_{L1}$ for the off-time period $(1 - D)T_s$. The charge balance condition on $i_C$

$$I_{L2} DT_s = I_{L1}(1 - D)T_S \tag{4.59}$$

[†]Relations (4.52) and (4.53) can be immediately seen from Fig. 4.16(a). The $V_C$ is the output voltage of the upstream boost converter, $V_C = V_S/(1 - D)$, and $V_C$ is also the input voltage of downstream buck converter, $V_O = DV_C$.

Table 4.10 Dc Analysis of Ćuk onverter

| Circuit variable | Expression |
|---|---|
| $V_O$ | $\frac{D}{1-D}V_S = \frac{0.25}{1-0.25} \cdot 24 = 8\text{ V}$ |
| $I_{L1}$ | $\frac{V_O^2}{RV_S} = \frac{8^2}{1 \cdot 24} = 2.67\text{ A}$ |
| $I_{L2}$ | $\frac{V_O}{R} = \frac{8}{1} = 8\text{ A}$ |
| $V_C$ | $\frac{1}{1-D}V_S = \frac{1}{1-0.25}24 = 32\text{ V}$ |

results in

$$\frac{I_{L1}}{I_{L2}} = \frac{D}{1-D} \tag{4.60}$$

From (4.52), the average value of the capacitor voltage $v_C$ is determined as

$$V_C = \frac{1}{1-D}V_S \tag{4.61}$$

The capacitor voltage decreases during the on-time period in which the capacitor current is negative. Conversely, the capacitor voltage increases during the off-time period with a positive capacitor current. The excursion of the capacitor voltage $\Delta v_C$ is found by evaluating the area of the the shaded rectangle in Fig. 4.17(c) and dividing the outcome by the capacitance of $C$.

### EXAMPLE 4.9 PSIM® Start-up response of Ćuk converter

This example simulates the start-up response of the Ćuk converter operating with $V_S = 24$ V, $T_s = 20$ $\mu$s, and $D = 0.25$. The circuit components of the converter are $L_1 = L_2 = 200$ $\mu$H, $C = 500$ $\mu$F, and $R = 1$ $\Omega$. The dc values of the important circuit variables are calculated in Table 4.10. The PSIM model and start-up response of the converter are given in Table 4.11.

### EXAMPLE 4.10 PSIM® Steady-state response of Ćuk converter

The state-state analysis of the Ćuk converter used in Example 4.9 are shown in Table 4.12. The PSIM simulations of the converter are displayed in Table 4.13. The

**Table 4.11** Start-up Response of Ćuk Converter

PSIM model: Ćuk_Converter_01

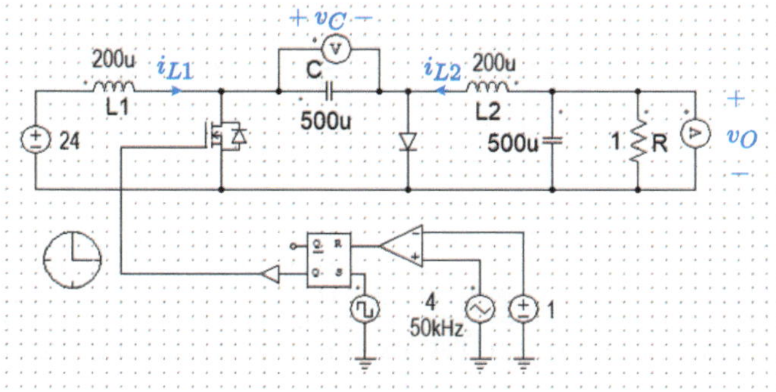


Simulation outcomes

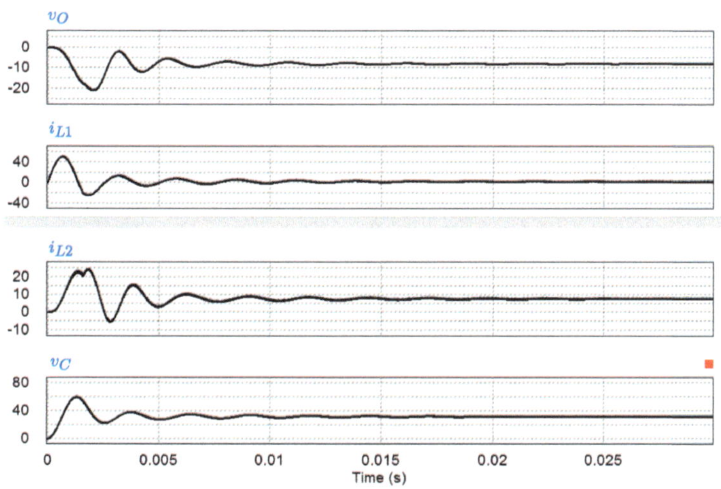


simulation outcomes validate the accuracy of the theoretical predictions in Table 4.12.

## 4.5 BIDIRECTIONAL BOOST CONVERTER

In the previous chapters, the PWM power conversion was discussed only considering the unidirectional energy transfer from the source section to load section. However, there are emerging demands for the bidirectional energy transfer, where the roles of the source section and load section are frequency interchanged over operational cycles.

**Table 4.12** Steady-State Analysis of Ćuk Converter

| Circuit variable | Expression |
|---|---|
| $\Delta i_{L1}$ | $\frac{V_S}{L_1} DT_s = \frac{24}{200 \times 10^{-6}} 0.25 \cdot 20 \times 10^{-6} = 0.6$ A |
| $i_{L1\,max}$ | $I_{L1} + \frac{1}{2}\Delta i_{L1} = 2.67 + 0.3 = 2.97$ A |
| $i_{L1\,min}$ | $I_{L1} - \frac{1}{2}\Delta i_{L1} = 2.67 - 0.3 = 2.37$ A |
| $\Delta i_{L2}$ | $\frac{V_O}{L_2}(1-D)T_s = \frac{8}{200 \times 10^{-6}}(1-0.25) \cdot 20 \times 10^{-6}$ |
| | $= 0.6$ A |
| $i_{L2\,max}$ | $I_{L2} + \frac{1}{2}\Delta i_{L2} = 8 + 0.3 = 8.3$ A |
| $i_{L2\,min}$ | $I_{L2} - \frac{1}{2}\Delta i_{L2} = 8 - 0.3 = 7.7$ A |
| $i_{C\,max}$ | $i_{L1\,max} = 2.97$ A |
| $i_{C\,min}$ | $-i_{L2\,max} = -8.3$ A |
| $\Delta v_C$ | $\frac{1}{C}\left((1-D)T_s\, i_{L1\,min} + \frac{1}{2}(1-D)T_s\, \Delta i_{L1}\right)$ |
| | $= \frac{0.75 \cdot 20 \times 10^{-6} \cdot 2.37 + 0.5 \cdot 0.75 \cdot 20 \times 10^{-6} \cdot 0.6}{500 \times 10^{-6}}$ |
| | $= 0.08$ V |
| $v_{C\,max}$ | $V_C + \frac{1}{2}\Delta v_C = 32.04$ V |
| $v_{C\,min}$ | $V_C - \frac{1}{2}\Delta v_C = 31.96$ V |

Such examples include photovoltaic power systems, automotive power systems, and space power systems, which employ batteries and other subsidiary energy sources together. These power systems employ an assembly of battery cells, referred to as the battery unit, as the primary energy source and also utilize another resource as the secondary energy source.

**Table 4.13** Steady-State Response of Ćuk Converter

PSIM model: Cuk_Converter_01

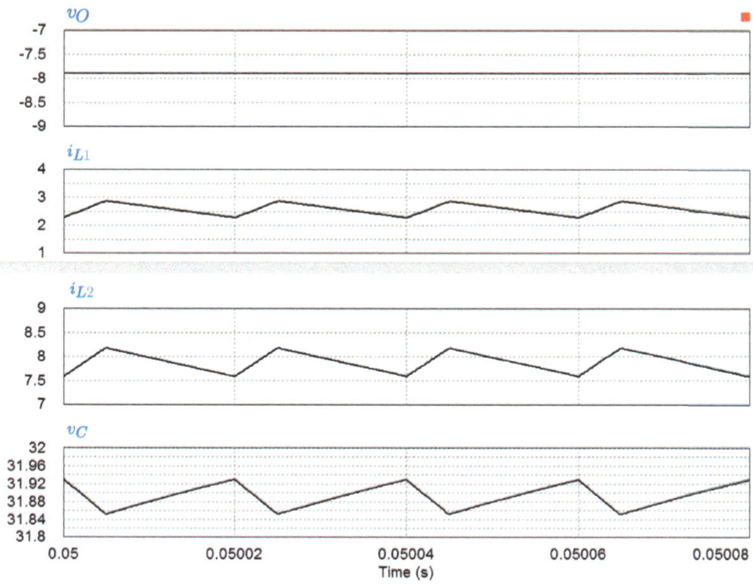


* The capacitor voltage $v_C$ is not fully settled into the steady state yet. The $v_C$ will eventually attain the values given in Table 4.12, as the simulation time advances.

Figure 4.18 shows the structure of a simple photovoltaic power system. In Fig. 4.18, the battery unit is the primary voltage source that delivers most electrical energy to the load via a bidirectional power converter. Between the battery unit and load, the bidirectional power converter is located which could could deliver carry electrical energy in both the forward and reverse directions. As the secondary energy source, a solar panel is connected to the output port of the bidirectional converter, where the energy-consuming load is located.

The load mainly receives electrical power from the battery unit, yet it also acquires additional energy from the solar panel. Over operational cycles, the battery unit will be overly discharged, whereas the energy influx from the solar panel becomes greater than the energy consumption in the load. Under this condition, the bidirectional power converter reverses the direction of energy flow, in order to recharge the battery unit using the surplus energy from the solar panel.

In Section 4.1, we viewed the boost converter as a *reverse* buck converter, which is developed from the buck converter by interchanging the the positions of the source section and the load section, and rearranging the active and passive switches to allow the converter circuit to deliver energy in the reverse direction. A logical extension of

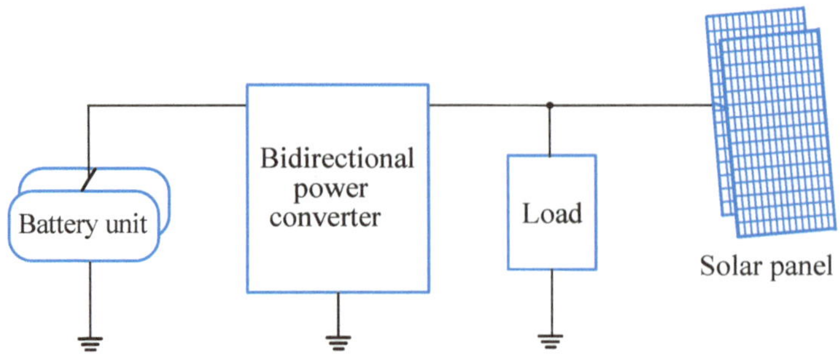


Figure 4.18 Simple photovoltaic power system.

this perspective directs that the boost converter could be employed as a bidirectional converter which delivers electrical energy in both the forward and reverse directions.

The boost converter, which processes electrical energy in the forward direction with a step-up function, could transfer energy in the reverse direction with a step-down operation, provided that the switches could carry the current in the forward and reverse directions. The boost converter engineered for such feature is referred to as the *bidirectional boost converter.*

In this section, we study the structure and operation of the bidirectional boost converter, employed in a battery-powered power system with an auxiliary energy source. An example of such cases was introduced in Section 1.4.2, where the eclipse mode operation of a spacecraft power system was addressed using in Fig. 1.7. Figure 4.18 is the simplest illustration of such operation.

### 4.5.1 Bidirectional Boost Converter in Photovoltaic Power System

This section presents the configuration and operation of a bidirectional boost converter employed in a photovoltaic power system in Fig. 4.18. We first review the operation of a conventional unidirectional boost converter and later deal with the case of the bidirectional boost converter in the photovoltaic power system.

#### Conventional Boost Converter

Figure 4.19 shows the conventional boost converter. The MOSFET symbol in Fig. 4.19 includes the parasitic anti-parallel body diode. The origin of the parasitic body diode was discussed in Section 2.1.1. In the conventional boost converter, as well as all other unidirectional PWM converters, the body diode is always turned-off and does not play any role. Thus, the MOSFET only functions as a unidirectional switch, delivering a positive current from the drain terminal to source terminal via the conduction channel.

The conventional boost converter employs the diode as a passive unidirectional switch. When the MOSFET switch is turned off, the diode turns on to create the freewheeling path for the inductor current. The boost converter employs a closed-

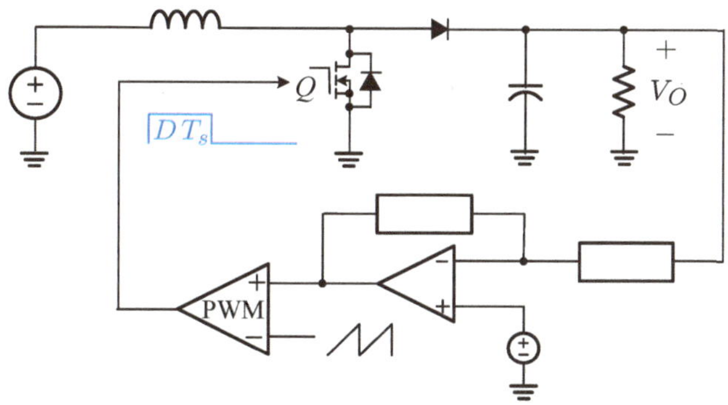


**Figure 4.19** Unidirectional boost converter.

loop PWM control to regulate the output voltage $V_O$. Details about the closed-loop regulation were presented in Section 3.6.1.

## Bidirectional Boost Converter

Figure 4.20 shows the structure and operational environment of a photovoltaic power system. Figure 4.20(a) exhibits the system configuration. The system consists of a rechargeable battery unit, bidirectional boost converter, load, and solar panel. The system employs the battery unit as the primary energy source and the solar panel as the secondary resource. The bidirectional boost converter is located between the battery unit and the load. The output voltage of the bidirectional boost converter is labeled as $V_{DC}$, in order to highlight its role as a constant dc bus. The bidirectional boost converter ensures a tight regulation of $V_{DC}$ through a closed-loop PWM control.

As an auxiliary energy source, the solar panel transforms the solar energy into the dc current. The output of the solar panel is tied to the output port of the bidirectional boost converter, where the load is also connected. The load draws a rated dc current $I_{\text{Load}}$ from the boost converter and solar array. The load current is given by $I_{\text{Load}} = V_{DC}/R_{\text{Load}}$ with $R_{\text{Load}}$ being the load resistance for the case of a pure resistive load.

The bidirectional boost converter should carry out two functions. The first is the regulation of the output voltage $V_{DC}$. This task is the same as the previous conventional boost converter. The second mission is the energy flow control that offers a reliable and seamless system operation, while meeting the energy balance condition, under certain changes in the operational environment. For this second task, the bidirectional boost converter is employed in Fig. 4.20.

The bidirectional boost converter has a couple of features, not found in the conventional boost converter. First, the passive diode switch, previously used in the conventional boost converter, is replaced with the active MOSFET switch. With this change, the two MOSFET switches both operate as a bidirectional switch, as will be seen soon. The bidirectional MOSFET switches are labeled as $S_1$ and $S_2$ in Fig. 4.20, in order to differentiate them from the conventional MOSFET switch $Q$ in Fig. 4.19.

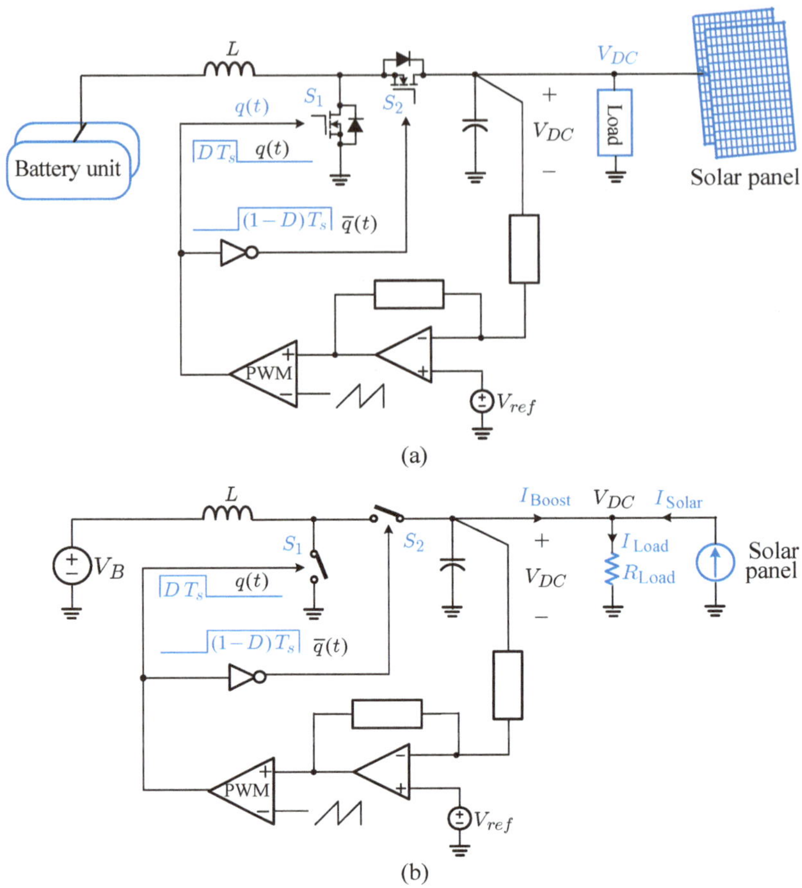


**Figure 4.20** Photovoltaic power system and bidirectional boost converter. (a) System configuration. (b) Functional circuit diagram.

Second, the bidirectional switches, $S_1$ and $S_2$, are individually excited by two complementary switch drive signals, $q(t)$ and $\bar{q}(t)$ in Fig. 4.20. The conduction channel in the MOSFET $S_1$ is created for the time period $DT_s$ and is removed for the period $(1-D)T_s$. Conversely, $S_2$ is turned-on for $(1-D)T_s$ and turned-off for $DT_s$. With this complementary PWM scheme, the bidirectional switches $S_1$ and $S_2$ carry the current in the forward direction or reverse direction when the switches are turned-on, as demanded for the bidirectional power transfer mission.

A simplified system diagram is given in Fig. 4.20(b). The battery unit is denoted as a voltage source $V_B$ and the load is simplified as a pure resistor, $R_{\text{Load}}$. The solar panel, which produces a dc current from the solar energy, is represented as a current source.

The current flows at the output port of the boost converter are illustrated in Fig. 4.20(b). The load resistor $R_{\text{load}}$ pulls a fixed current, $I_{\text{Load}}$, from two individual current flows. The first flow is the current coming from the bidirectional boost converter, $I_{\text{Boost}}$. The other is the dc current originating from the solar panel, $I_{\text{Solar}}$. The relation among these current flows is expressed as

$$I_{\text{Load}} = I_{\text{Boost}} + I_{\text{Solar}} \quad \Rightarrow \quad I_{\text{Boost}} = I_{\text{Load}} - I_{\text{Solar}} \tag{4.62}$$

We assume that the load current $I_{\text{Load}}$ is fixed. On the other hand, the solar panel current $I_{\text{Solar}}$ will vary depending on the amount of available solar energy. Accordingly, the boost converter current $I_{\text{Boost}}$ can be positive, negative, or zero, depending on the status of the solar panel. This indicates that the bidirectional boost converter will operate in one of the following three modes.

1) Battery Discharging Mode with $I_{\text{Solar}} < I_{\text{Load}}$: This mode occurs when the dc current from the solar panel is less than the load current. The bidirectional boost converter delivers the current difference, $I_{\text{Boost}} = I_{\text{Load}} - I_{\text{Solar}}$, from the battery unit. The bidirectional boost converter operates in usual step-up mode with the input voltage of $V_B$.

2) Regulation Mode with $I_{\text{Solar}} = I_{\text{Load}}$: When $I_{\text{Solar}}$ is identical to $I_{\text{Load}}$, the average value of the boost converter current, $I_{\text{Boost}}$, reduces to zero. Under this condition, the bidirectional boost converter is still operative and regulates its output voltage at $V_{DC} = V_B/(1-D)$, but the converter does not deliver any net current to the load or source.

3) Battery Charging Mode with $I_{\text{Solar}} > I_{\text{Load}}$: When the solar panel produces more energy than necessary, the extra energy is used to recharge the battery unit. Under this condition, the bidirectional boost converter delivers the current $I_{\text{Boost}} = I_{\text{Solar}} - I_{\text{Load}}$ in the reverse direction. As far as the energy flow is concerned, the output port now constitutes the source section and the battery unit becomes the load. The bidirectional boost converter now operates in step-down mode, which is functionally equivalent to the buck converter. Even so, the bidirectional converter still regulates $V_{DC}$ at $V_B/(1-D)$, the same as the other two operational modes.

Figure 4.21 illustrates the current distributions in the three operational modes, defined in the preceding discussions. The converter is in the discharging mode when $I_{\text{Solar}}$ falls below $I_{\text{Load}}$. The converter enters the regulation mode when $I_{\text{Solar}}$ rises to and remains being equal to $I_{\text{Load}}$. Finally, the converter goes to the charing mode with the condition $I_{\text{Solar}} > I_{\text{Load}}$. The boost converter always operates in continuous conduction mode (CCM), due to the presence of the bidirectional switches, regardless of the operational mode of the system. Details about steady-state and transitional responses of the bidirectional boost converter are studied in the next section.

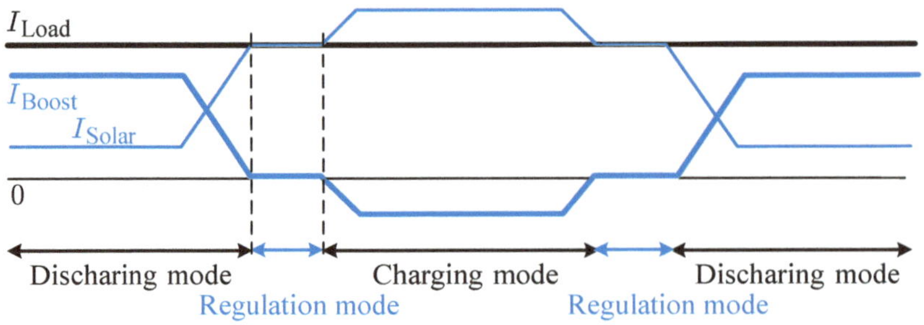


**Figure 4.21** Operational modes of bidirectional boost converter.

### 4.5.2 Operation of Bidirectional Boost Converter

Figure 4.22 shows the operational details about the bidirectional boost converter. Referring to Fig. 4.22, we explore the steady-state operation.

#### Operational Properties

From the previous discussions, we conclude that the following facts always hold true for the bidirectional converter in Fig. 4.22(a) in all three operational modes.

***$V_{DC}$ Regulation*** : The volt-sec balance condition on the inductance $L$

$$V_B D T_s = (V_{DC} - V_B)(1 - D)T_s \tag{4.63}$$

is alway valid. From this equation, it follows that

$$V_{DC} = \frac{1}{1 - D} V_B \tag{4.64}$$

regardless of the operational mode or the direction of inductor current. We further know that $V_{DC}$ is tightly regulated by the closed-loop PWM control. The feedback scheme in Fig. 4.20 forces $V_{DC}$ to track $V_{ref}$ for all the operational modes, irrespective of the apparent role of $V_{DC}$ as the output voltage in the discharging mode or the input voltage in the charging mode.

***Dc Current Distribution*** : The following relations among the dc currents

$$I_{\text{Boost}} = I_{\text{Load}} - I_{\text{Solar}} \tag{4.65}$$

and

$$I_L = \frac{1}{1 - D} I_{\text{Boost}} \tag{4.66}$$

are always valid.

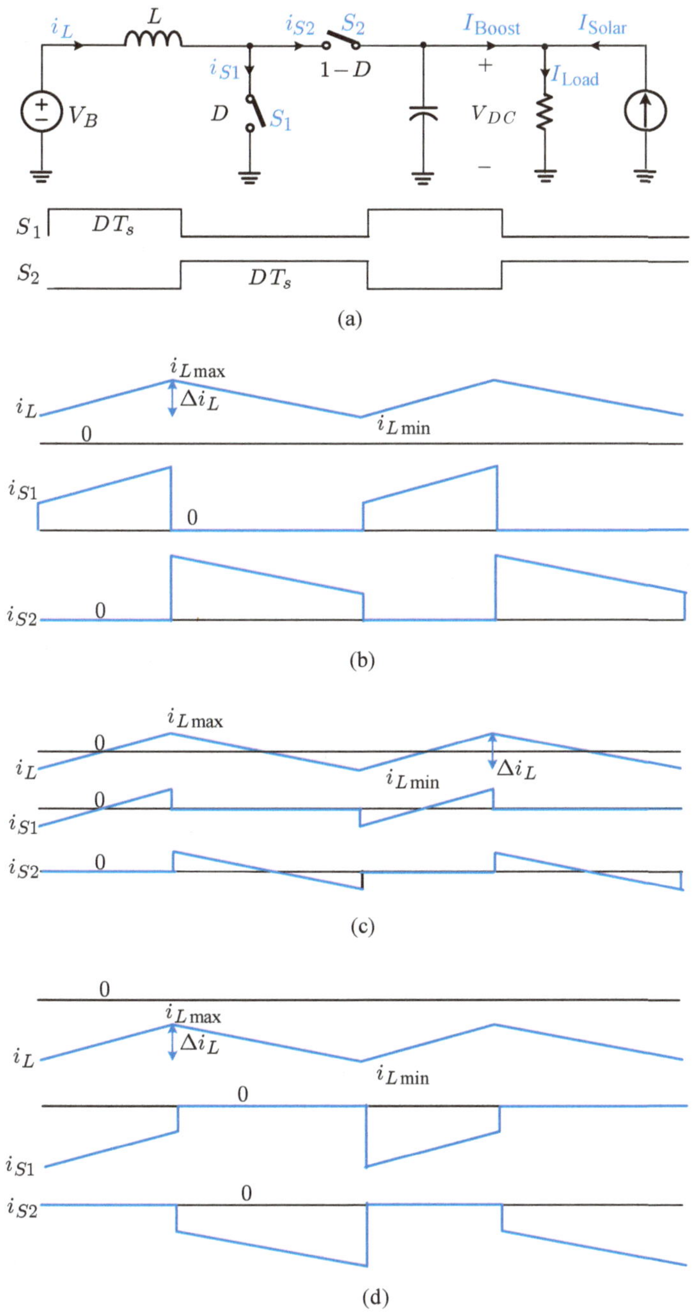


**Figure 4.22** Bidirectional boost converter. (a) Circuit configuration and switch drive signals. (b) Deep battery discharging mode operation. (c) Regulation mode operation. (d) Deep battery charging mode operation.

***Inductor Current Excursion*** : The excursion of the inductor current is given by

$$\Delta i_L = \frac{V_B}{L} DT_s = \frac{V_{DC} - V_B}{L}(1 - D)T_s \tag{4.67}$$

for the entire operational cycle.

***CCM Operation*** : The bidirectional switches ensure the continuous conduction mode (CCM) operation of the converter. During the period $DT_s$, the bidirectional switch $S_1$ delivers the positive current via the conduction channel and the negative current through the conduction channel and the body diode .

Likewise, $S_2$ carries the positive current via the conduction channel and body diode, while delivering the negative current through the conduction channel, during the period $(1 - D)T_s$.

## Converter Operation in Three Operational Modes

The converter operation in three different operational modes are illustrated in Figs. 4.22(b)-(d), which shows the inductor and switch currents in the ***deep*** battery discharging/charging modes and regulation mode.[†]

1) Deep Battery Discharging Mode: Figure 4.22(b) shows the current waveforms of the converter in the deep battery discharging mode. The converter operation is the same as the conventional boost converter. The inductor current flows through $S_1$ for $DT_s$ and $S_2$ for $(1 - D)T_s$. The inductor current remains positive all the time. The $S_1$ function as the active switch in the conventional boost converter; for this reason, $S_1$ is often referred to as the boost switch.

2) Regulation Mode: The current waveforms in the regulation mode are shown in Fig. 4.22(c). While the inductor and switch currents change their directions within the switching period, their average values are just balanced at zero. Thus, the bidirectional converter does not deliver any net current to the load or source. It just regulates $V_{DC}$ at $V_B/(1 - D)$.

3) Deep Battery Charging Mode: The converter delivers the net current from the $V_{DC}$ bus to the battery unit. The inductor and switch currents remain negative all the time and the converter functions as a buck converter in the reverse direction. Here, $S_2$ behaves as the active switch in the conventional buck converter. Thus, $S_2$ is sometimes called the buck switch.

## Converter Operation in Mode Transition Periods

Operations of the bidirectional boost converter during the mode transition periods are illustrated in Fig. 4.23. Figure 4.23(a) shows the inductor and switch currents of

[†]In the *deep* battery discharging/charging modes, the inductor and switch currents remain positive or negative for the entire switching period. Conversely, the currents in the *narrow* battery discharging/charging modes can be partially both positive and negative in one operational period.

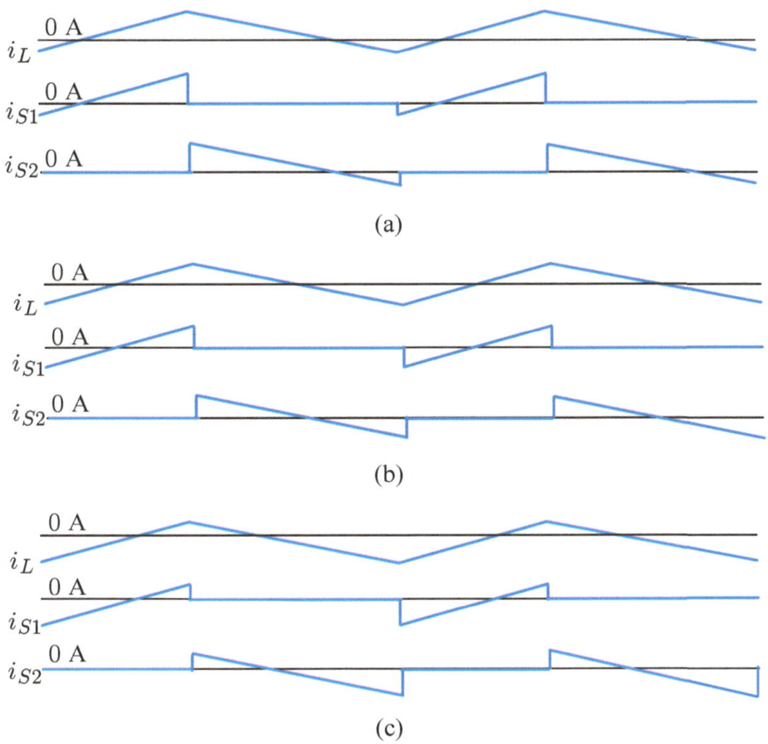


Figure 4.23 Transitions between operational modes. (a) Transition from narrow battery discharging mode to regulation mode. (b) Regulation mode. (c) Transition from regulation mode to narrow battery charging mode.

the converter undergoing a transition from the ***narrow*** battery discharging mode to regulation mode. The two bidirectional switches deliver both the positive and negative currents when the respective switch is turned-on. However, the net current of each bidirectional switch is positive and the converter in effect carries the current in the forward direction.

The situation in the regulation mode is shown in Fig. 4.23(b). The two bidirectional switches deliver the positive and negative current in an exact equal amount. Thus, the converter does not deliver any net current and only regulates $V_{DC}$ at a fixed value.

The transition from the regulation mode to the narrow battery charging mode is shown in Fig. 4.23(c). Here, the net values of the inductor and switch currents are negative and the bidirectional converter charges the partially-depleted battery unit.

As illustrated in Fig. 4.23, the bidirectional converter secures a seamless operation over mode transition periods via harmonious operations of the two bidirectional switches.

**Table 4.14** Steady-State Response in Battery Discharging Mode

PSIM model: Bidirectional_Boost _Converter_01

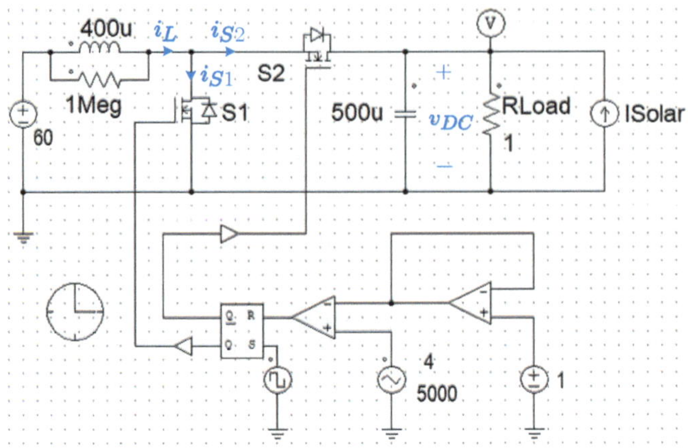


Simulation outcomes

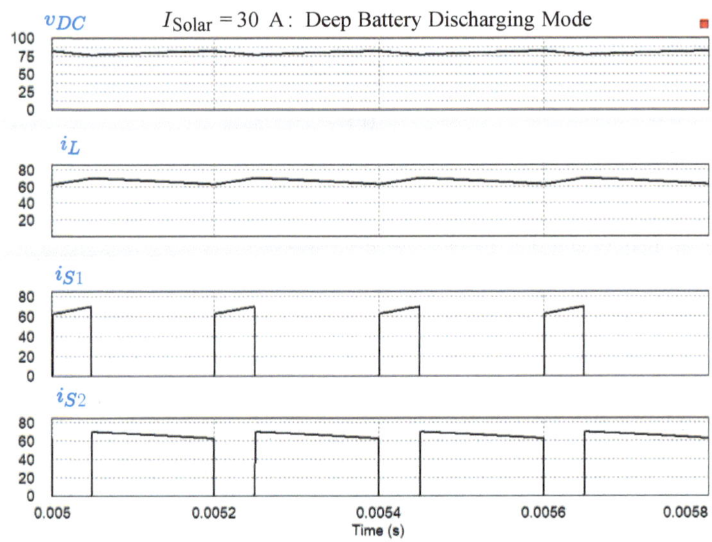


## EXAMPLE 4.11 PSIM® Bidirectional boost converter in steady-state

This example illustrates the steady-state response of the bidirectional boost converter. The converter operates with $V_S = 60$ V, $f_s = 5$ kHz, and $D = 0.25$. The circuit components of the converter are $L = 400\ \mu$H, $C = 500\ \mu$H, and $R_{\text{Load}} = 1\ \Omega$.

The PSIM®model is shown in Table 4.14. In the PSIM model, a 1 MΩ resistor is connected across the inductor $L$, in order to avoid an *open-circuit error* in the

**Table 4.15** Steady-State Analyses of Bidirectional Boost Converter

| Circuit variable | Expression |
|---|---|
| $V_{DC}$ | $\frac{1}{1-D}V_S = \frac{1}{1-0.25}60 = 80$ V |
| $\Delta i_L$ | $\frac{V_S}{L}DT_s = \frac{60}{400\times10^{-6}}0.25\cdot200\times10^{-6} = 7.5$ A |
| Deep Battery Discharging Mode with $I_{Solar} = 30$ A | |
| $I_{\text{Boost}}$ | $I_{\text{Load}} - I_{\text{Solar}} = 80 - 30 = 50$ A |
| $I_L$ | $\frac{1}{1-D}I_{\text{Boost}} = \frac{1}{1-0.25}50 = 66.7$ A |
| Regulation Mode with $I_{\text{Solar}} = 80$ A | |
| $I_{\text{Boost}}$ | 0 |
| $I_L$ | 0 |
| Deep Battery Charging Mode with $I_{Solar} = 130$ A | |
| $I_{\text{Boost}}$ | $I_{\text{Load}} - I_{\text{Solar}} = 80 - 130 = -50$ A |
| $I_L$ | $\frac{1}{1-D}I_{\text{Boost}} = -\frac{1}{1-0.25}50 = -66.7$ A |

model. We assume that the bidirectional boost converter provides a tight regulation at $V_{DC} = 80$ V. To emulate this operational environment, the PSIM simulations are done under the open-loop condition with $D = 0.25$. This simulation set-up is functionally equivalent to a tight regulation of the output voltage at

$$V_{DC} = \frac{1}{1-D}V_S = \frac{1}{1-0.25}60 = 80 \text{ V}$$

The load current $I_{\text{Load}}$ is determined as

$$I_{\text{Load}} = \frac{80}{1} = 80 \text{ A}$$

**Table 4.16** Steady-State Response in Regulation and Battery Charging Modes

Simulation outcomes

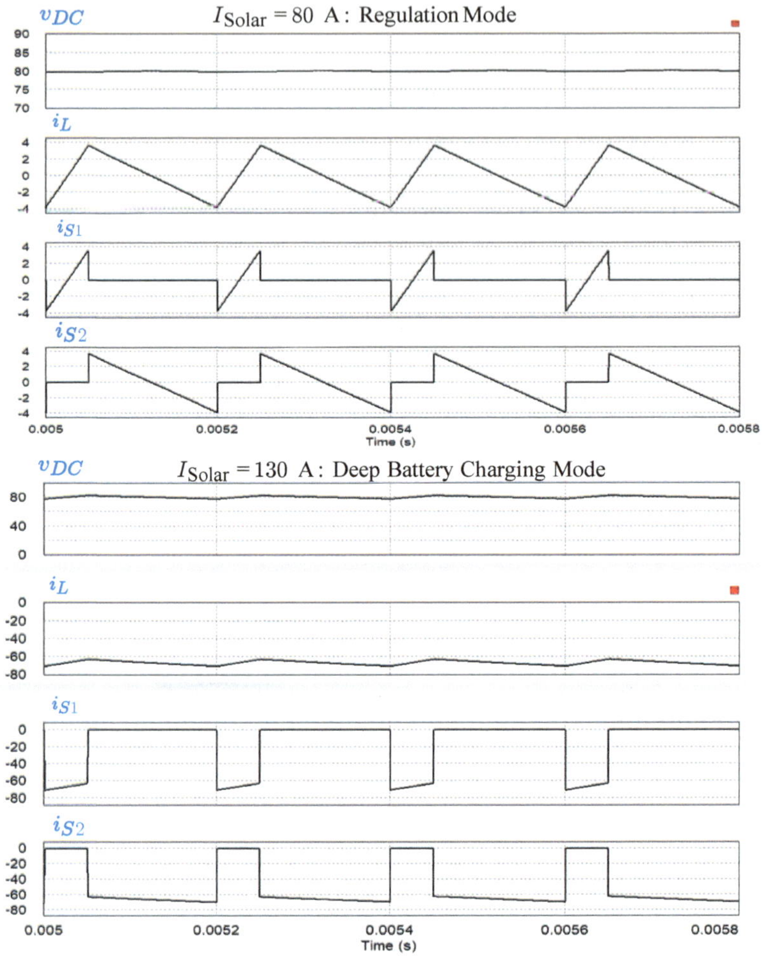


The bidirectional boost converter is in a battery discharging mode with $I_{Solar} <$ 80 A, in the regulation mode with $I_{Solar} = 80$ A, and in the battery charging mode with $I_{Solar} > 80$ A.

Steady-state analyses of the converter in the three operational modes with different $I_{Solar}$ values, the deep battery discharging mode with $I_{Solar} = 30$ A, the regulation mode $I_{Solar} = 80$ A, and the deep battery charging mode with $I_{Solar} = 130$ A, are given in Table 4.15.

The steady-state current waveforms in the deep battery discharging mode are shown in Table 4.14. The $V_{DC}$ shows a relatively large voltage ripple, due to the wide current fluctuations in $S_1$ and $S_2$. The waveforms show a good match with the predictions in Table 4.15.

Table 4.16 displays the waveforms in the regulation mode and deep battery charging mode. In the regulation mode, the inductor current is symmetrical with respect to *zero* value. The converter is does not transfer any net energy between the load and source. A finite energy circulates within the converter while regulating $V_{DC}$. The output voltage ripple is reduced to the minimum value with the least current fluctuations in $S_1$ and $S_2$.

The waveforms in the deep battery charging mode illustrates the operation of the bidirectional converter as a *reverse buck converter*. The battery unit is charged with the extra energy delivered from the solar panel. The circuit waveforms, operational principles, and power transfer are all *reversely* symmetrical to those of the conventional boost converter.

**PSIM® Bidirectional boost converter in transient EXAMPLE 4.12**

This example illustrates the operation of the bidirectional boost converter during the transitional periods among the battery discharging mode, regulation mode, and battery charging modes. Table 4.17 shows the PSIM model and simulation outcomes. The time-varying solar panel current is generated in the PSIM model using the time-varying current source, $I_{\text{Solar}}$. The current distribution at the output of the bidirectional boost converter is shown in the format of Fig. 4.21. Because we employed a 1 Ω resistor as the load, the current $I_{\text{Load}}$ also represents the dc bus voltage $V_{DC}$.

The converter is in the battery discharging mode for the period $0 < t < 0.025$, in the regulation mode for $0.025 < t < 0.03$, and in the battery charging mode after $t = 0.03$. The converter goes through a transition from the narrow battery discharging mode to regulation mode in the period near $t = 0.025$ and a transition from the regulation mode to narrow battery charging mode in the time interval around $t = 0.03$, as highlighted with circles in the current waveforms in Table 4.17. The system shows the smooth transitions among the three operational modes, as predicted from the preceding analyses.

## 4.6 FLYBACK CONVERTER: TRANSFORMER-ISOLATED BUCK/BOOST CONVERTER

For many dc-to-dc converter applications, electrical isolation between the input and output ports is required by law for practical reasons such as safety. One simple way to provide such an input-output isolation is to insert an isolation transformer in the mid-section of dc-to-dc converters. Indeed, many isolated converter topologies are proliferated by adding a transformer to the three basic dc-to-dc converters. The flyback converter is one prime example of them.

The flyback converter is evolved from the buck/boost converter by adding an isolation transformer and simplifying the resulting circuit. The flyback converter has

**Table 4.17** Transitional Response of Bidirectional Boost Converter

PSIM model: Bidirectional_Boost _Converter_02

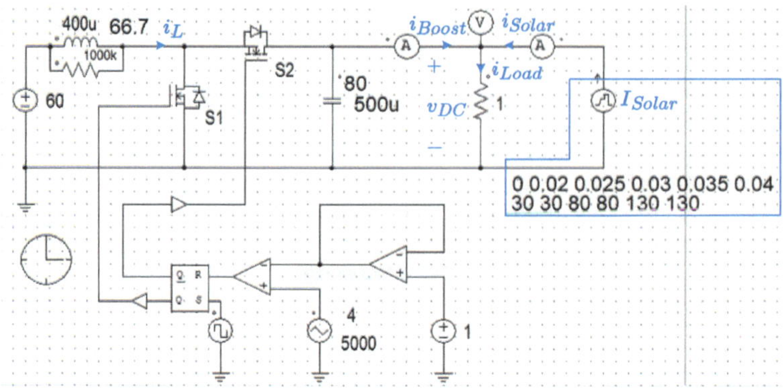


Simulation outcomes

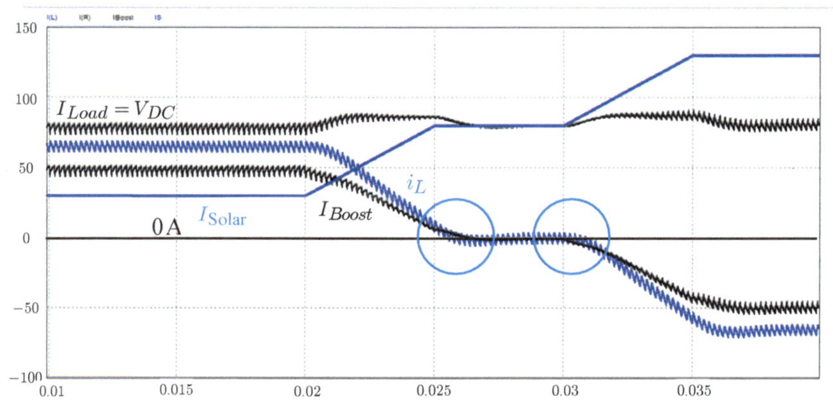


* The simulation of this example requires the professional version of PSIM® software

a very simple structure with minimal component count, while providing the required input-output isolation. Accordingly, the flyback converter is widely used as a cost-effective converter for consumer electronics. This section presents the topological origin and steady-state operation of the flyback converter.

### 4.6.1 Evolution to Flyback Converter

Figure 4.24 shows the evolution of the flyback converter from the buck/boost converter. The buck/boost converter is shown in Fig. 4.24(a). A transformer is inserted in the middle of the buck/boost converter, resulting in the circuit shown in Fig. 4.24(b).

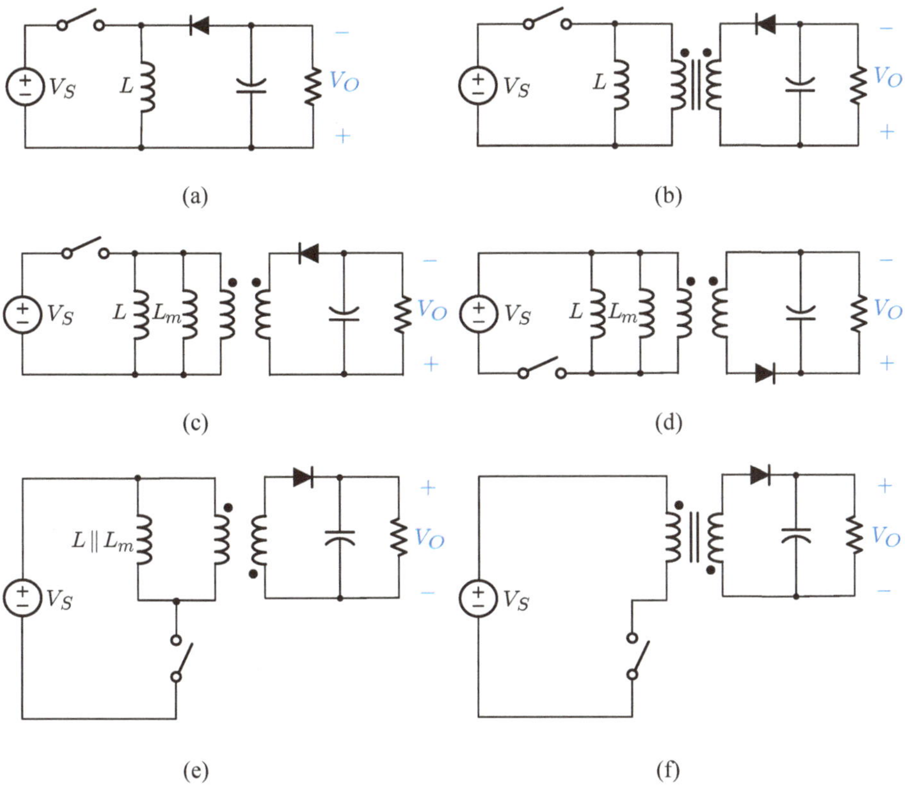


**Figure 4.24** Evolution to flyback converter. (a) Buck/boost converter. (b) Insertion of isolation transformer. (c) Circuit model. (d) Equivalent circuit model. (e) Modified circuit model. (f) Flyback converter.

After replacing the transformer with a combination of the magnetizing inductance $L_m$ and ideal transformer, as shown in Fig. 4.24(c), the active switch and diode are relocated without affecting the converter operation, yielding Fig. 4.24(d). Figure 4.24(d) is then changed to Fig. 4.24(e) by combining the inductor of the buck/boost converter and the magnetizing inductance of the transformer, and vertically flipping the circuitry in the secondary side of the ideal transformer. The final form of the flyback converter is shown in Fig. 4.24(f), where the composite inductance, $L \| L_m$, is considered as the magnetizing inductance of the isolation transformer.

The flyback converter utilizes the magnetizing inductance of the isolation transformer as its *functional inductance*. Therefore, the isolation transformer should be fabricated in such a way that could offer a controllable magnetizing inductance. One easy solution is to create a gap in the magnetic path of the transformer. This is commonly implemented by introducing an air gap between the magnetic cores, which effectively determines the magnetizing inductance of the isolation transformer.

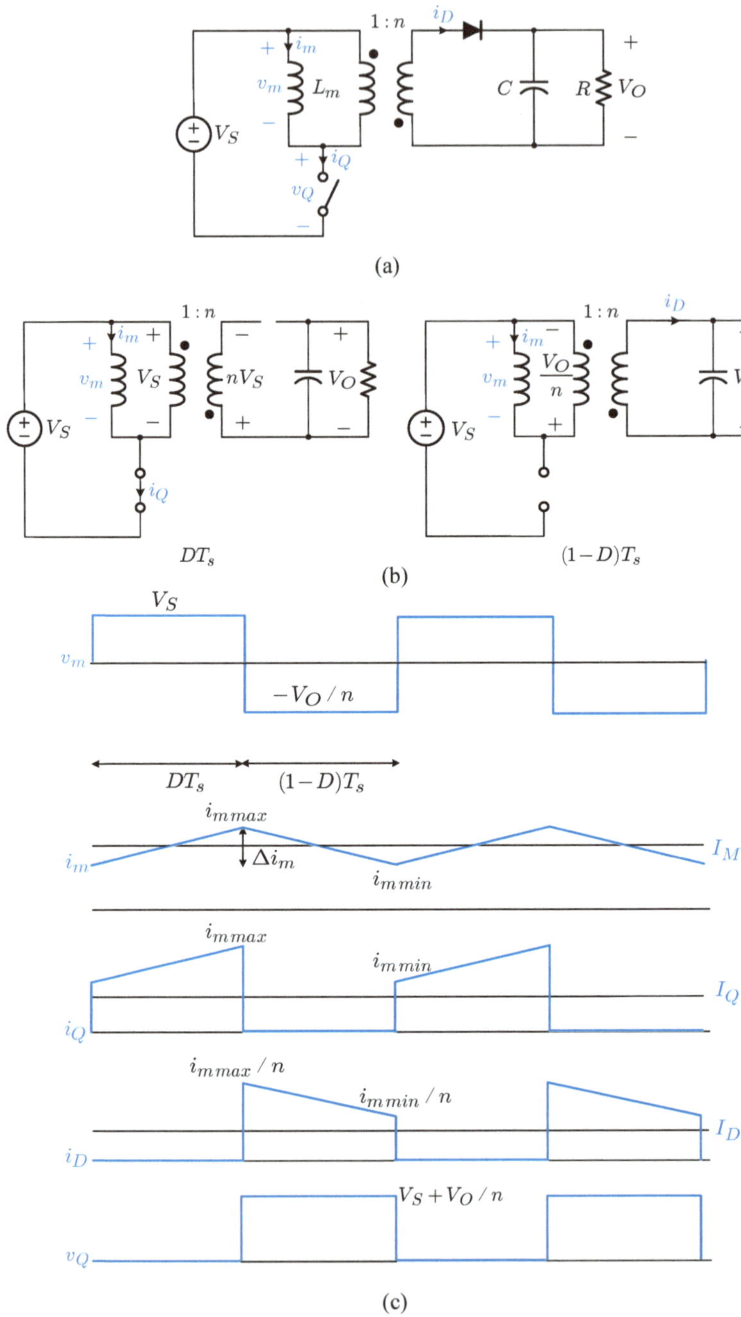


**Figure 4.25** Steady-state analysis of flyback converter with $1:n$ transformer. (a) Flyback converter. (b) On-time and off-time subcircuits. (c) Major circuit waveforms.

### 4.6.2 Steady-State Analysis in CCM

Functionally speaking, the flyback converter is a buck/boost converter with transformer isolation. Thus, the flyback converter closely mimics the buck/boost converter in operation. Especially, the magnetizing inductance of the transformer executes the role of the regular inductor in the buck/boost converter. Namely, the magnetizing inductance carries a continuous dc-biased triangular current, as is the case of the regular inductor in all other PWM converters, including the buck/boost converter. For this reason, we call the current through the magnetizing inductance as the *inductor current or magnetizing inductor current*, not the magnetizing current.

Figure 4.25 illustrates the operation of the flyback converter in CCM, where a $1 : n$ turns ratio is assumed for the transformer. The transformer is represented by the parallel connection of the magnetizing inductance and ideal transformer.

During an on-time period in which the active switch is closed, the diode is reverse-biased by the sum of the output voltage and the voltage across the secondary transformer winding: $V_O + nV_S$. Accordingly, the on-time subcircuit for $DT_s$ is identified as shown in Fig. 4.25(b).

The operation of the converter during the on-time period $DT_s$ is explained as follows. The primary winding of the ideal transformer does not carry any current because the secondary winding is open-circuited. The voltage across the magnetizing inductance, $v_m$ in Fig. 4.25, is identical to the input voltage $V_S$. The magnetizing inductor current $i_m$ increases from its minimum value with the slope $V_S/L_m$. The inductor current $i_m$ flows through the switch, $i_m = i_Q$, while both the primary and secondary currents of the ideal transformer are zero.[†]

When the active switch is turned-off, the inductor current $i_m$ directs towards the *undotted* terminal of the primary winding of the transformer, and this pushes a current flow from the *undotted* terminal of the secondary winding towards the diode at the secondary side of the transformer. Accordingly, the diode is turned-on and the converter circuit now presents the off-time subcircuit in Fig. 4.25(b) for $(1 - D)T_s$. In this period, the inductor current $i_m$ runs into the *undotted* terminal of the primary winding, while the diode current $i_D$ leaves from the *undotted* terminal of the secondary winding. The diode current in this period is given by $i_D = i_m/n$, according to the current equation of the $1 : n$ ideal transformer.

During the off-time period, the output voltage $V_O$ is reflected through the $1 : n$ transformer and the reflected voltage is applied to the magnetizing inductance $L_m$ in the negative polarity. Thus, the inductor current $i_m$ ramps down from its maximum value with a slope $(-V_O/n)/L_m$. Figure 4.25(c) shows the inductor current $i_m$, constructed based on the operational principles explained above. The magnetizing inductor current $i_m$ shows triangular excursions from its average value $I_M$, in the exact same manner as the regular inductor current in other PWM converters.

[†] Operation of the flyback converter in the on-time period is similar to that of the capacitor charging circuit studied in Section 2.3.2.

Application of the volt-sec balance condition to the magnetizing inductance $L_m$ yields

$$V_S DT_s = \frac{V_O}{n}(1 - D)T_s \tag{4.68}$$

which leads to the voltage gain for the flyback converter

$$\boxed{\frac{V_O}{V_S} = \frac{D}{1 - D}n} \tag{4.69}$$

The voltage gain is given by the product of the voltage gain of the buck/boost converter and the turns ratio of the transformer.

The magnetizing current swing is given by

$$\boxed{\Delta i_m = \frac{V_S}{L_m}DT_s = \frac{V_O/n}{L_m}(1 - D)T_s} \tag{4.70}$$

The average value of the diode current is equal to the load current because the diode is connected to the load

$$I_D = \frac{V_O}{R} \tag{4.71}$$

The average diode current $I_D$ is related to the average inductor current $I_M$ by the following equation

$$I_D = (1 - D)I_M\frac{1}{n} \tag{4.72}$$

From (4.71) and (4.72), the average inductor current is given by

$$\boxed{I_M = \frac{1}{1 - D}nI_D = \frac{1}{1 - D}n\frac{V_O}{R}} \tag{4.73}$$

The maximum value of the inductor current is determined as

$$i_{m\,max} = I_M + \frac{1}{2}\Delta i_m \tag{4.74}$$

and the minimum value is given by

$$i_{m\,min} = I_M - \frac{1}{2}\Delta i_m \tag{4.75}$$

The voltage across the active switch, $v_Q$ in Fig. 4.25(a), is zero during the on-time period. During the off-time period, $v_Q$ is given by the sum of the input voltage and the reflected output voltage

$$\boxed{v_Q = V_S + \frac{V_O}{n}} \tag{4.76}$$

The operation of the flyback converter is virtually identical to that of the buck/boost converter, yet providing an input-to-output isolation, positive output voltage, and more flexible voltage gain.

**Table 4.18** Steady-State Analysis of Flyback Converter

| Circuit variable | Expression |
|---|---|
| $V_O$ | $\frac{D}{1-D}nV_S = \frac{0.4}{1-0.4}0.5 \cdot 24 = 8\text{ V}$ |
| $v_{m\,max}$ | $V_S = 24\text{ V}$ |
| $v_{m\,min}$ | $-\frac{V_O}{n} = -\frac{8}{0.5} = -16\text{ V}$ |
| $I_M$ | $\frac{1}{1-D}n\frac{V_O}{R} = \frac{1}{1-0.4}0.5\frac{8}{2.5} = 2.67\text{ A}$ |
| $\Delta i_m$ | $\frac{V_S}{L_m}DT_s = \frac{24}{160\times10^{-6}}0.4\cdot 20\times10^{-6} = 1.2\text{ A}$ |
| $i_{m\,max}$ | $I_M + \frac{1}{2}\Delta i_m = 2.67 + \frac{1.2}{2} = 3.27\text{ A}$ |
| $i_{m\,min}$ | $I_M - \frac{1}{2}\Delta i_m = 2.67 - \frac{1.2}{2} = 2.07\text{ A}$ |
| $i_{Q\,max}$ | $i_{mmax} = 3.27\text{ A}$ |
| $i_{D\,max}$ | $\frac{i_{mmax}}{n} = \frac{3.27}{0.5} = 6.54\text{ A}$ |
| $v_{Q\,max}$ | $V_S + \frac{V_O}{n} = 24 + \frac{8}{0.5} = 40\text{ V}$ |

## Steady-state operation of flyback converter — EXAMPLE 4.13

This example shows the steady-state analysis and circuit waveforms of a flyback converter. The operational conditions and power stage parameters of the flyback converter are $V_S = 24\,\text{V}$, $L_m = 160\,\mu\text{H}$, $n = 0.5$, $C = 400\,\mu\text{F}$, $R = 2.5\,\Omega$, $T_s = 20\,\mu\text{s}$, and $D = 0.4$. The results of the steady-state analysis are summarized in Table 4.18. The simulated circuit waveforms are shown in Fig. 4.26.

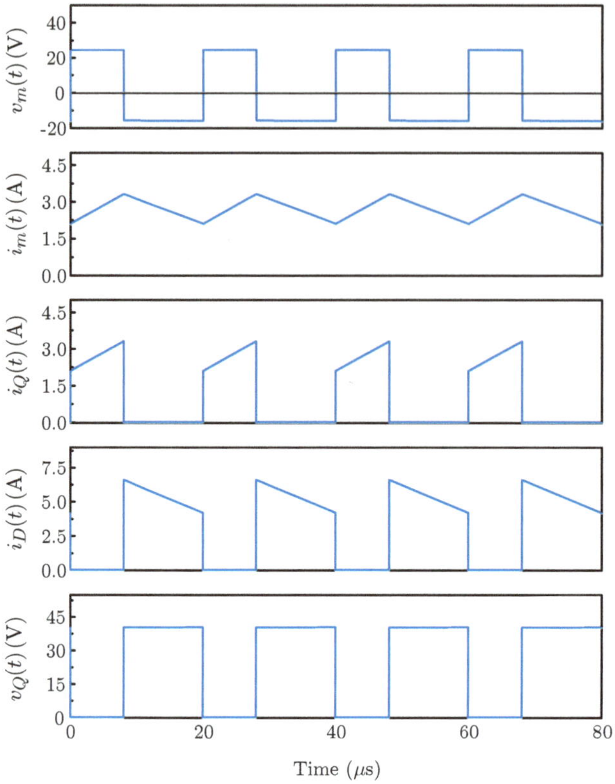


**Figure 4.26** Circuit waveforms of flyback converter.

### 4.6.3 Steady-State Analysis in DCM

For the flyback converter, the magnetizing inductor current decides the operational mode and the CCM/DCM borderline is $I_M = \Delta i_m/2$. The CCM/DCM borderline condition $I_M = \Delta i_m/2$ is expressed as

$$\frac{1}{1-D} n \frac{V_O}{R} = \frac{1}{2} \frac{V_S}{L_m} DT_s \tag{4.77}$$

referring to (4.73) and (4.70). By incorporating (4.69) into (4.77), the critical load resistance is found as

$$\boxed{R_{crit} = 2n^2 \frac{L_m}{(1-D)^2 T_s}} \tag{4.78}$$

When the load resistance is increased beyond $R_{crit}$, the converter encounters DCM operation.

Figure 4.27 shows the major waveforms of the flyback converter in DCM operation. When $i_m$ becomes zero, the diode at the secondary side of the transformer

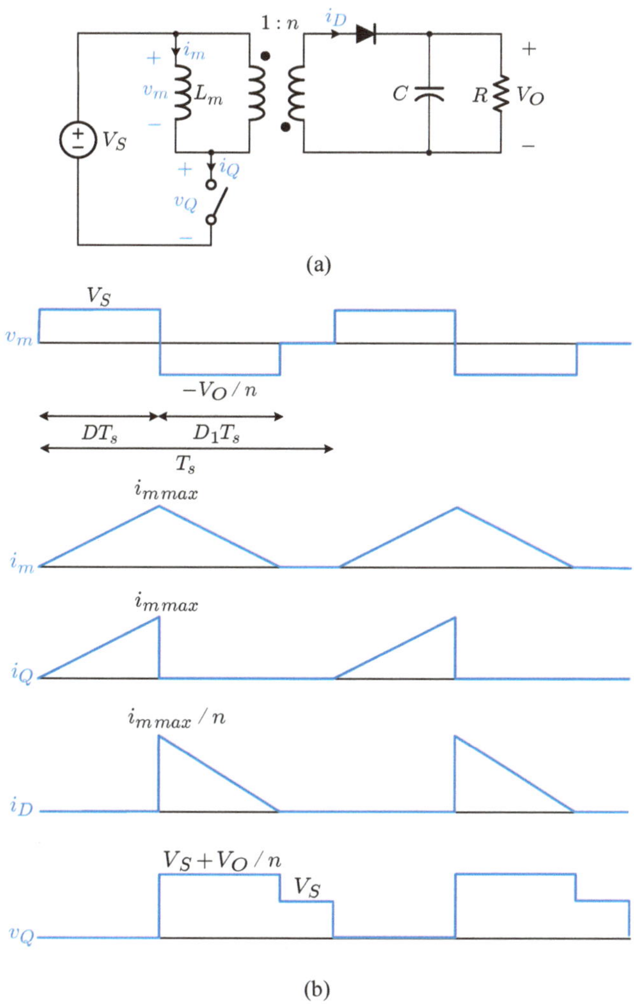


**Figure 4.27** Steady-state analysis of flyback converter in DCM. (a) Circuit diagram. (b) Major circuit waveforms.

is turned-off. Under this condition, $i_m$ remains at zero and thus $v_m$ also becomes zero. Accordingly, the voltage across the active switch equals to the input voltage, $v_Q = V_S$, as shown in the bottom curve in Fig. 4.27(b).

Application of the volt-sec balance condition to the magnetizing inductance $L_m$ yields

$$V_S DT_s = \frac{V_O}{n} D_1 T_s \tag{4.79}$$

which leads to

$$\frac{V_O}{V_S} = \frac{D}{D_1}n \tag{4.80}$$

By equating the average value of the diode current to the load current

$$\frac{\frac{1}{2} i_{mmax} \frac{1}{n} D_1 T_s}{T_s} = \frac{\frac{1}{2}\left(\frac{V_S}{L_m} DT_s\right)\frac{1}{n} D_1 T_s}{T_s} = \frac{V_O}{R} \tag{4.81}$$

an expression for $D_1$ is obtained

$$D_1 = \frac{V_O}{V_S}\frac{2L_m}{RDT_s}n \tag{4.82}$$

Finally, by combining (4.80) and (4.82), the DCM voltage gain is given by

$$\boxed{\frac{V_O}{V_S} = D\sqrt{\frac{RT_s}{2L_m}}} \tag{4.83}$$

The DCM voltage gain is independent of the transformer turns ratio $n$, and is the same as that of the buck/boost converter.

## EXAMPLE 4.14 PSIM® Flyback converter with current mode control

This example illustrates the transient response of a closed-loop controlled flyback converter during the start-up period. The PSIM model and simulation results are shown in Table 4.19. The operational conditions of the converter are shown in the PSIM model. The converter employs the current mode control scheme introduced in Example 4.8. The circuitry associated with current mode control is highlighted with a box. As shown in the simulation results, the output voltage of the converter settles into the desired value $V_O = V_{ref} = 8$ V. The steady-state duty ratio of the converter is found as $D = 0.4$ by solving the equation

$$\frac{V_O}{V_S} = \frac{D}{1-D}n$$

with $V_O = 8$ V, $V_S = 24$ V, and $n = 0.5$. The average value of the magnetizing current converges to the theoretical value

$$I_M = \frac{1}{1-D}n\frac{V_O}{R} = \frac{1}{1-0.4}0.5\frac{8}{2.5} = 2.67 \text{ A}$$

## EXAMPLE 4.15 PSIM® Flyback converter with current mode control

The transient response of the flyback converter in Example 4.14 is illustrated in this example. Using the PSIM model in Table 4.20, the transient behavior of the

**Table 4.19** Start-up Response of Flyback Converter

PSIM model: Flyback_Converter_01: Current Mode Control

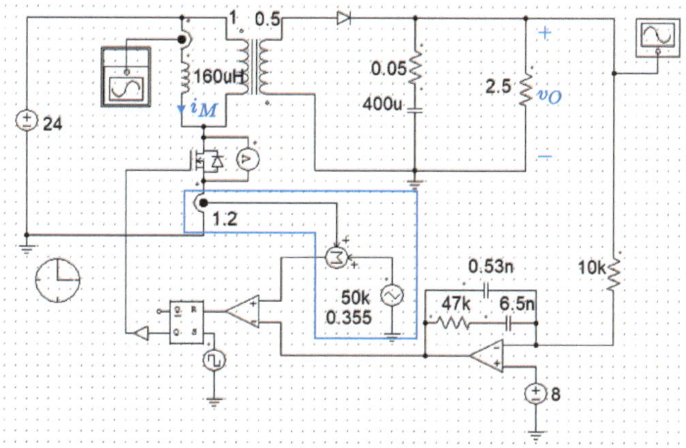


Simulation outcomes

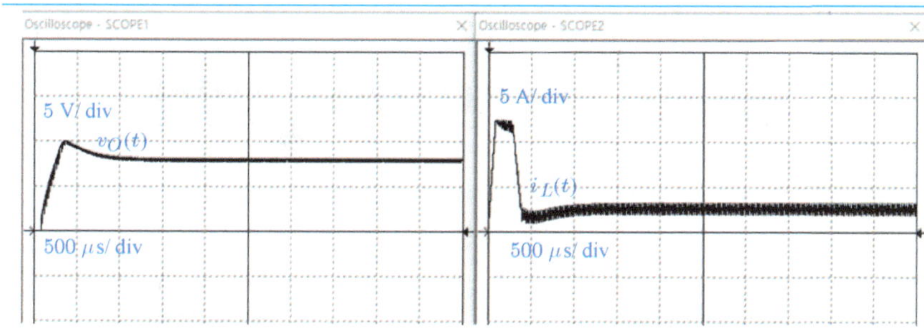


converter is simulated due to the step load changes between $R = 5\ \Omega$ and $2.5\ \Omega$. The converter responses are shown at the bottom of the table. The output voltage reveals an excellent transient behavior. The magnetizing inductor current with $R = 2.5\ \Omega$ was found as $I_M = 2.67$ A. With $R = 5\ \Omega$, the inductor current becomes half the previous case, $I_M = 1.33$ A.

## PSIM® Off-line flyback converter — EXAMPLE 4.16

Power supplies for the consumer and industrial electronics often receive an ac voltage from the utility line and provide a dc voltage to the load. The converters for these applications are referred to as the off-line converter. The flyback converter is most widely adapted for off-line applications. The flyback converter provides the power stage isolation with minimal component count.

**Table 4.20** Step Load Response of Flyback Converter

PSIM model: Flyback_Converter_02: Current Mode Control

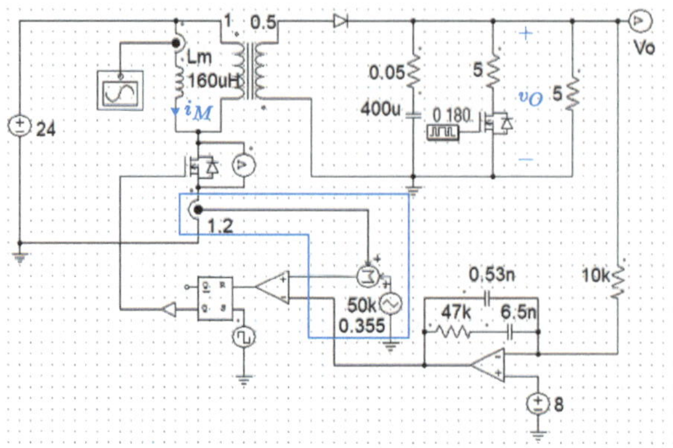


Simulation outcomes

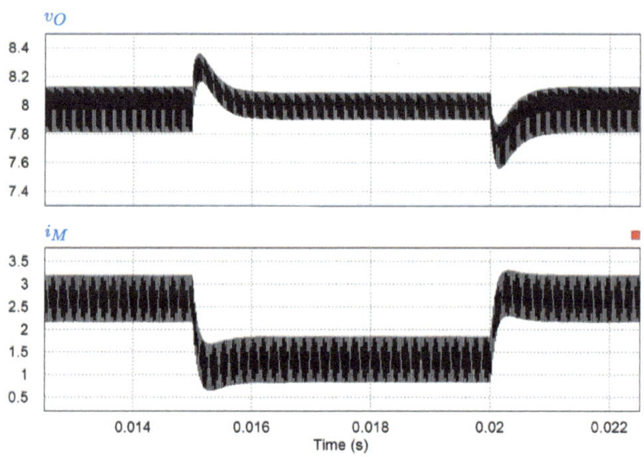


The off-line converter should offer the galvanic isolation in the feedback controller, as well as in the power stage. For the feedback controller isolation, the controller typically uses an optocouler, which couples the infrared diode and photo-transistor in order to transmit the control signal with electrical isolation.

This example illustrates the operation and performance of the off-line flyback converter employing the optocupler-isolated current mode control. The PSIM model in Table 4.21 shows the structure, operational environment, and circuit components of the off-line flyback converter. Current mode control is implemented with the two resistors shown inside the box in the power stage model. The optocoupler in the feedback controller is also highlighted with a box. The optocoupler provides a functional connection between the voltage feedback circuit and the PWM block.

**Table 4.21** Start-up Response of Off-Line Flyback Converter

PSIM model: Flyback_Converter_03: Current Mode Control

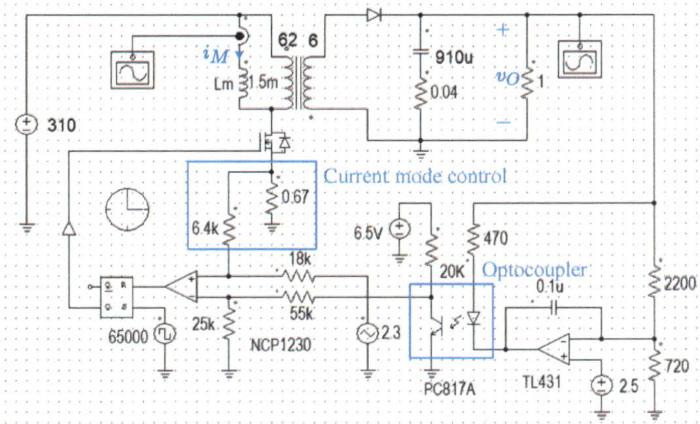


Simulation Outcomes

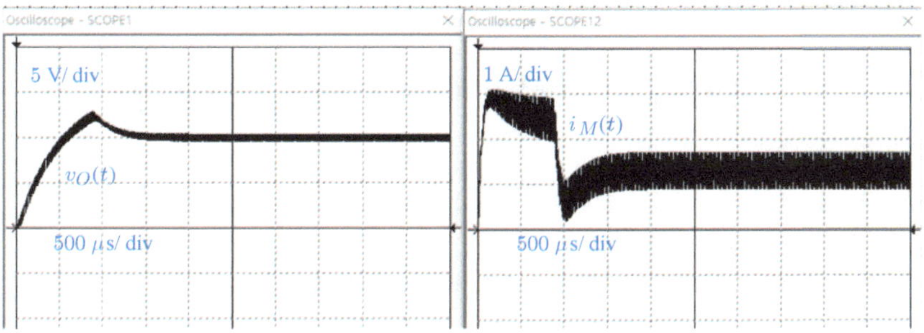


The input voltage is a 310 V dc source – the dc source obtained by rectifying a 220 V ac utility line. Details about the controller design and dynamic analysis of the converter will be addressed in the second book of the series.

Table 4.21 displays the start-up response of the converter. Based on the steady-state analysis, it can be shown that the output of the converter is regulated at $V_O =$ 10.1 V with the duty ratio $D = 0.25$. The average inductor current is determined as $I_M = 1.3$ A.

Table 4.22 displays the transient response of the off-line flyback converter. The upper part of the table shows the PSIM model and the lower part are the simulation results with 100% changes in the load current: $R = 0.5\ \Omega \Rightarrow 1\ \Omega\ \Rightarrow 0.5\ \Omega$.

**Table 4.22** Step Load Response of Off-Line Flyback Converter

PSIM model: Flyback_Converter_04: Current Mode Control

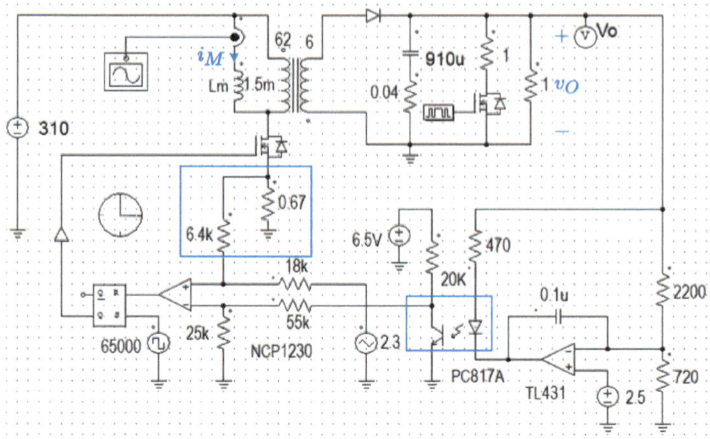


Simulation outcomes

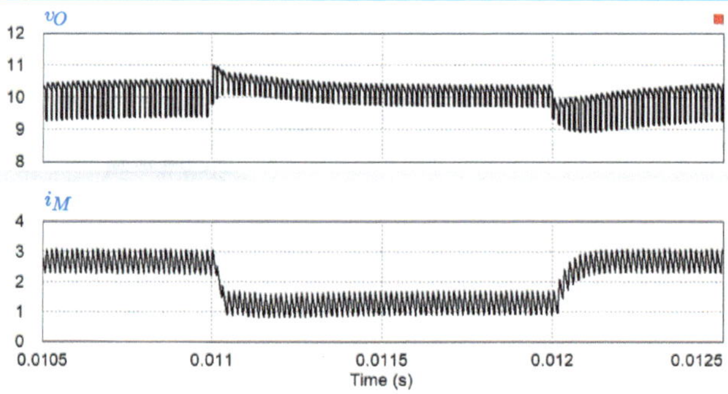


## 4.7 BUCK-DERIVED ISOLATED DC-TO-DC CONVERTERS

Numerous isolated converters have been created to provide the input-to-output isolation and, more importantly, to enhance the utility and performance of non-isolated converters. In the previous section, we studied the flyback converter that was derived from the buck/boost converter by adding an isolation transformer and utilizing the magnetizing inductance of the isolation transformer as the functional inductor.

In principle, a single non-isolated converter produces many different isolated converter topologies with distinct methods of transformer isolation. Such isolated topologies have their own merits and respective application area. Exhaustive discussions about various isolated converter topologies were presented in [1]-[3].

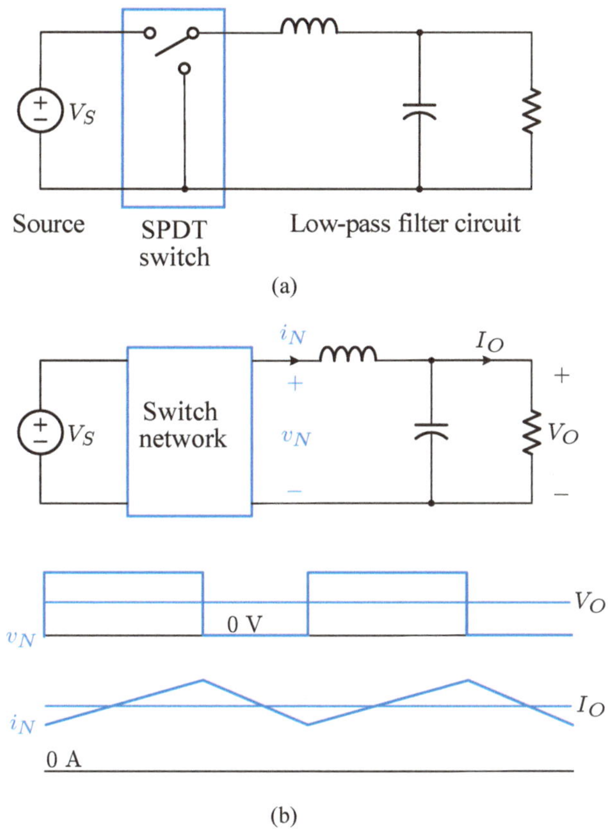


**Figure 4.28** Derivation of isolated dc-to-dc converter from buck converter. (a) Buck converter. (b) Buck-derived isolated dc-to-dc converter and switch network waveforms.

Among various isolated PWM converter topologies, the dc-to-dc converters derived from the buck converter have been widely used in medium/high power applications. Here, we study the buck-derived isolated PWM converter topologies, aiming to provide general insight about the proliferation and operation of different families of isolated converters.

This section provides basic concept and fundamental knowledge, required to understand the structures and operation of buck-derived isolated converters. We first discuss the topological structure of the converters. Then, we review the circuit models for multi-winding transformers, which are frequently employed as an isolation transformer in buck-derived isolated PWM converters. The current section is followed by other two sections, Section 4.8 and Section 4.9, which cover the details about two most prevalent buck-derived converter families – the forward converter and bridge-type converter.

### 4.7.1 Buck-Derived Isolated Converters and Switch Network

Figure 4.28 is a conceptual illustration of deriving isolated converters from the buck converter. In Fig. 4.28(a), the buck converter is viewed as an integration of three functional blocks: the source, SPDT switch, and low-pass filter circuit. The roles of the SPDT switch and low-pass filter were explained in Section 3.2.

Isolated dc-to-dc converters are derived from the buck converter in Fig. 4.28(a), by replacing the SPDT switch with a *switch network* which performs the same function while providing the input-output isolation. This concept is illustrated in Fig. 4.28(b). The requirements of the switch network are summarized below.

1) The switch network should provide a rectangular voltage waveform, $v_N$ in Fig. 4.28(b), as the input voltage to the low-pass filter circuit. The top value of $v_N$ is proportionate to the source voltage $V_S$ and the bottom value is zero. The low-pass filter circuit smooths $v_N$, thereby producing the average value of $v_N$ as the output voltage of the converter: $\bar{v}_N(t) = V_O$.

2) The switch network should deliver a dc-biased triangular current waveform, $i_N$ in Fig. 4.28(b), to the input terminal of the low-pass filter circuit. In particular, the switch network should provide a freewheeling path for $i_N$ when $v_N$ drops to zero. The average value of $i_N$ is the output current of the converter: $\bar{i}_N(t) = I_O$.

3) The switch network should offer the input-output isolation. Most switch networks employ a transformer for the input-output isolation.

### 4.7.2 Circuit Models for Multi-Winding Transformers

The switch network discussed in Fig. 4.28 usually contains a multi-winding transformer. We discuss the circuit models for multi-winding transformers before investigating the structure and operation of transformer-isolated buck-derived converters. Figure 4.29(a) shows a two-winding transformer and its circuit model. The circuit model contains the magnetizing inductance and ideal transformer. Details about the transformer model were covered in Section 2.2.3. From the transformer circuit model, the following relationships are written

$$\boxed{v_P(t) : v_S(t) = 1 : n} \tag{4.84}$$

$$\boxed{i_P(t) = i_m(t) + n\, i_S(t)} \tag{4.85}$$

with

$$\boxed{n = \frac{N_S}{N_P}} \tag{4.86}$$

where $N_P$ is the turns of the primary winding and $N_S$ is that of the secondary winding. The magnetizing current $i_m$ is given by

$$i_m(t) = \frac{1}{L_m} \int v_P(t)\, dt \tag{4.87}$$

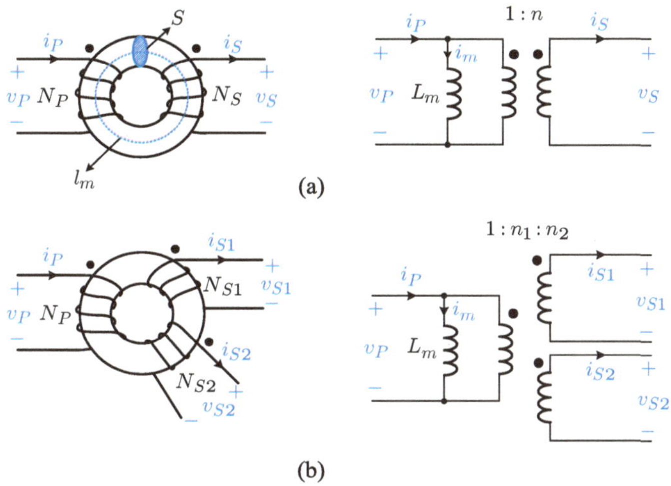


**Figure 4.29** Practical transformers and circuit models. (a) Two-winding transformer. (b) Three-winding transformer.

The magnetizing inductance $L_m$ is expressed as

$$L_m = \mu_r \mu_o \frac{S}{l_m} N_P^2 \tag{4.88}$$

in terms of the transformer parameters shown in Fig. 4.29(a). The $\mu_r$ is the relative permeability of the core material and $\mu_o = 4\pi \times 10^{-7}$ is the permeability of the free space.

Figure 4.29(b) shows a three-winding transformer and its circuit model. The circuit model consists of the magnetizing inductance and ideal three-winding transformer. While the circuit model could be derived from the electro-magnetic phenomena inside the three-winding transformer, it also could be considered as an extension of the two-winding transformer model. Equations (4.84) and (4.85) are modified for the three-winding transformer

$$\boxed{v_P(t) : v_{S1}(t) : v_{S2}(t) = 1 : n_1 : n_2} \tag{4.89}$$

$$\boxed{i_P(t) = i_m(t) + n_1 i_{S1}(t) + n_2 i_{S2}(t)} \tag{4.90}$$

with

$$\boxed{n_1 = \frac{N_{S1}}{N_P} \quad \text{and} \quad n_2 = \frac{N_{S2}}{N_P}} \tag{4.91}$$

where $N_P$, $N_{S1}$, and $N_{S2}$ represent the respective turns of each transformer winding. Expressions for the magnetizing current $i_m$ and magnetizing inductance $L_m$ are the same as those of the previous two-winding transformer. Polarity/direction of the winding voltage/current is consistent with Lenz's law and *dot convention*, all discussed in Section 2.2.1 and 2.2.3. The circuit model and equations for the three-winding transformer could be extended for the transformers with four or more windings.

When the magnetizing inductance $L_m$ is assumed to be infinite, the circuit model reduces to an ideal three-winding transformer. The voltage equation of the ideal three-winding transformer is the same as (4.89). On the other hand, the current equation is modified to

$$\boxed{i_P(t) = n_1 i_{S1}(t) + n_2 i_{S2}(t)} \tag{4.92}$$

because $i_m$ in (4.87) becomes zero with $L_m = \infty$.

## 4.8 FORWARD CONVERTERS

The forward converter is an important family of buck-derived isolated dc-to-dc converters. The forward converter has unique circuit characteristics, not found in other isolated converters. Features of the forward converter will be apparent as we study the operation of the converter.

### 4.8.1 Basic Operational Principles

Functional basics of the forward converter are first explained in Fig. 4.30, using a conceptual converter configured with an ideal transformer. Figure 4.30(a) is the circuit diagram of a forward converter that employs a 1 : 1 ideal transformer. The converter has the configuration of the buck-derived isolated converter, shown in Fig. 4.28(b). The switch network contains one active switch, one ideal two-winding transformer, and two diodes. This switch network is the simplest among all the switch networks used in buck-derived isolated converters.

The operation of the conceptual forward converter is illustrated in Fig. 4.30(b). When the switch is closed, the positive voltage appears at the *dotted* terminal of the primary and secondary transformer windings. During this period, the diode $D_1$ is turned-on and $D_2$ is reverse-biased by the secondary winding voltage $v_S$. Therefore, the switch network current $i_N$ flows through $D_1$ and the switch network voltage $v_N$ equals the input voltage $V_S$. During this period, energy is delivered in the *forward* direction through the ideal transformer.

When the switch is opened, both the primary current and secondary current are identically zero according to the current equation of the ideal two-winding transformer. In this period, the ideal transformer is functionally isolated from the converter circuit. The switch network current $i_N$ now freewheels through $D_2$ and $v_N$ thus becomes zero. The switch network produces the required waveforms for $v_N$ and $i_N$, and the output voltage is given by

$$\boxed{V_O = \bar{v}_N(t) = DV_S} \tag{4.93}$$

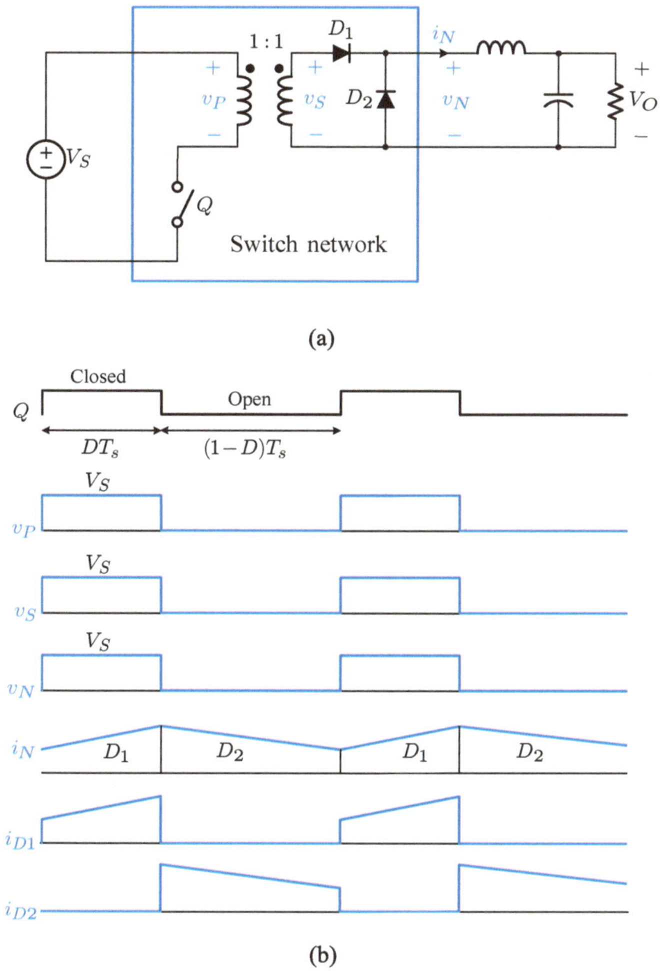


**Figure 4.30** Conceptual forward converter with ideal transformer. (a) Circuit diagram. (b) Major circuit waveforms.

and the output current is given by

$$\boxed{I_O = \bar{\imath}_N(t) = \frac{V_O}{R}} \tag{4.94}$$

The conceptual forward converter is functionally identical to the buck converter.

### 4.8.2 Reset Problem and Reset Circuits

When the ideal transformer is replaced with a practical transformer, the conceptual forward converter in Fig. 4.30 immediately becomes inoperative. Figure 4.31(a)

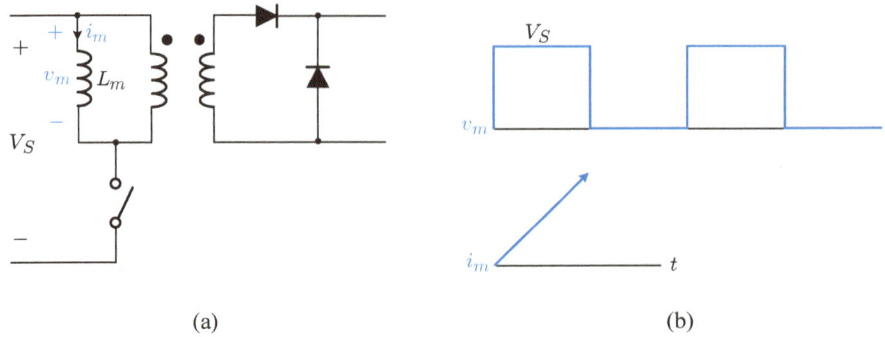


**Figure 4.31** Switch network with practical transformer. (a) Circuit diagram. (b) Reset problem.

shows the switch network including the magnetizing inductance of a practical transformer. Now, the problem of the converter is clear with the presence of the magnetizing inductance $L_m$. As shown in Fig. 4.31(b), the voltage across the magnetizing inductance $v_m$ does not meet the volt-sec balance requirement. More seriously, the magnetizing current, elevated by the equation $i_m(t) = (V_S/L_m)t$ during an on-time period, instantly loses its path when the active switch is opened. The only way for $i_m$ to maintain the continuity is to divert into the *undotted* terminal of the primary winding of the ideal transformer. If this is the case, the secondary current must run into the *dotted* terminal of the secondary winding. However, this current flow is not possible because the diode cannot deliver any current from cathode to anode.

The problem associated with the volt-sec balance condition on the magnetizing inductance $L_m$ or the continuity of the magnetizing current $i_m$ is termed as the reset problem, indicating the problem of resetting the magnetic flux inside the core to its original state.

A variety of auxiliary circuits are employed as a means of meeting the volt-sec balance condition on $L_m$, while providing a continuous path for $i_m$. The auxiliary circuits, applied to practical forward converters to solve the reset problem, are called the reset circuits. In this section, we study several reset circuits in preparation for analyses of various forward converters.

## Switch Network with Zener Diode Reset

The Zener diode is often employed to resolve the reset problem. Figure 4.32 shows the Zener diode reset circuit, which utilizes a pair of Zener diode and regular diode across the primary winding of the transformer. During an on-time period, the regular diode is reverse-biased and the Zener diode reset circuit does not interfere with the circuit operation. The magnetizing current linearly increases with a slope of $V_S/L_m$ during this period.

During an off-time period, the magnetizing current $i_m$ activates the Zener diode and circulates through the loop formed by the magnetizing inductance, Zener diode,

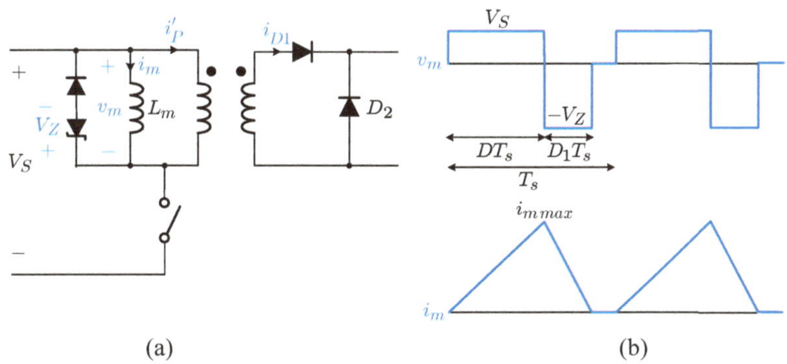


**Figure 4.32** Switch network with Zener diode reset. (a) Circuit diagram. (b) Major circuit waveforms.

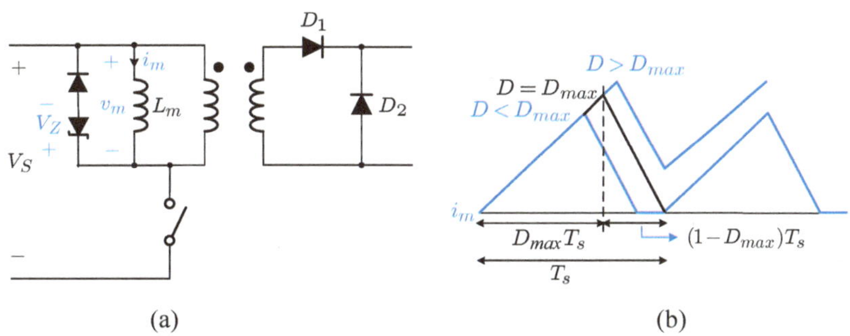


**Figure 4.33** Maximum allowable duty ratio. (a) Circuit diagram. (b) Magnetizing current.

and regular diode. In this period, the primary and secondary currents of the ideal transformer are both zero: $i'_P = i_{D1} = 0$. The breakdown voltage of the Zener diode, $V_Z$, is applied to $L_m$ in the negative polarity and $i_m$ thus linearly decreases with a slope $-V_Z/L_m$. When $i_m$ is reduced to zero, the regular diode turns off and remains off until the onset of the next off-time period. The waveforms for $v_m$ and $i_m$ are shown in Fig. 4.32(b). The volt-sec balance condition on $L_m$ yields

$$V_S DT_s = V_Z D_1 T_s \tag{4.95}$$

where $D_1T_s$ is the fraction of the off-time period in which the magnetizing current is present. The magnetizing current $i_m$ attains its peak value at the end of the on-time period and returns to zero during the off-time period, thus resetting the magnetic flux to its initial state.

The Zener diode reset circuit in Fig. 4.32 has a constraint on the maximum allowable duty ratio, $D_{\max}$. This feature is illustrated in Fig. 4.33. Figure 4.33(b) shows three cases of the magnetizing current $i_m$. The case with $D = D_{\max}$ is shown in black line. The other two cases are the magnetizing current $i_m$ with $D < D_{\max}$ and $D > D_{\max}$.

The expression of $D = D_{\max}$ for the given $V_S$ and $V_Z$ is found from (4.95) with the condition $D_{\max}$ and $D_1 = 1 - D_{\max}$

$$V_S D_{\max} T_s = V_Z(1 - D_{\max})T_s \tag{4.96}$$

which leads to

$$D_{\max} = \frac{1}{1 + \dfrac{V_S}{V_Z}} \tag{4.97}$$

The expression (4.97) specifies the upper limit to the allowable duty ratio. For example, when $V_S = V_Z$, the duty ratio should be lower than or equal to 0.5 $(0 < D \leq 0.5)$ for a proper converter operation. When the duty ratio exceeds 0.5, the magnetizing current increases cycle-by-cycle and the magnetizing inductance will be saturated, leading to a fatal circuit failure.

While the Zener diode reset circuit is simple in structure and operation, it only provides a lossy reset. The energy transferred to $L_m$ during the on-time period is dissipated at the Zener diode during the off-time period. Accordingly, the Zener diode reset circuit is rarely used for the applications where efficiency is an important consideration.

Many alternative reset circuits are available for forward converters. Some reset circuits result in a lossy reset while others operate losslessly. All these reset circuits do not alter the operation of the forward converter, yet only function during the off-time period to meet the volt-sec balance condition on $L_m$ while maintaining the continuity of $i_m$. In other words, the forward converter follows the simple operation of the conceptual converter shown in Fig. 4.30, while bypassing the reset problem with an appropriate reset circuit.

### Switch Network with RCD Clamp Reset

Figure 4.34 shows another switch network, which employs a different dissipative reset scheme. As shown in Fig. 4.34(a), the reset circuit employs a capacitor $C_c$, resistor $R_c$, and diode $D_c$. This reset scheme is referred to as the RCD clamp reset circuit due to its circuit structure and operation. The capacitor $C_c$ is known as the *clamp capacitor*. The voltage across the clamp capacitor $C_c$ is referred to as the clamp voltage $v_C$. The function of the RCD clamp reset circuit is illustrated in Fig. 4.34(b).

As will be seen soon, the capacitor $C_c$ is charged at a fixed clamp voltage in steady state. Here, we assume that $C_c$ is charged at $V_C$ and analyze the reset circuit using Fig. 4.34.

During the on-time period, the diode $D_c$ is reverse-biased by $V_S + V_C$ and the magnetizing current $i_m$ linearly increases by the equation

$$i_m = \frac{V_S}{L_m} t \tag{4.98}$$

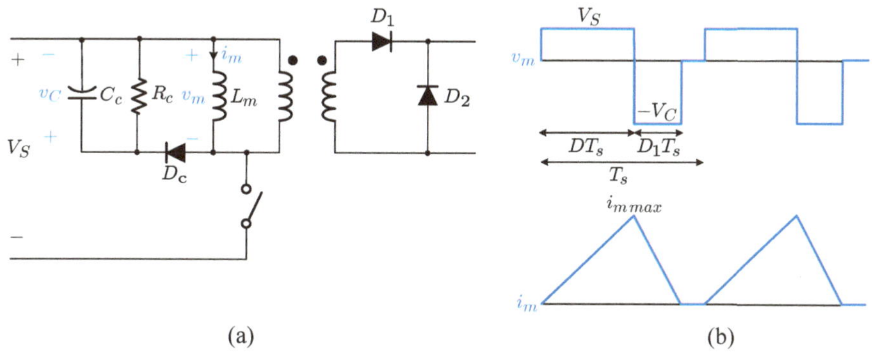


**Figure 4.34** Switch network with RCD Clamp reset. (a) Circuit diagram. (b) Major circuit waveforms.

and attains the maximum value

$$i_{m\,max} = \frac{V_S}{L_m} DT_s \tag{4.99}$$

at the end of the on-time period. At $t = DT_s$, the diode $D_c$ conducts and $i_m$ decreases with the slope $-V_C/L_m$ for the period $D_1T_s$ and reduces to zero at $t = (D+D_1)T_s$.

The magnetic energy stored in $L_m$ during the on-time period

$$E_m = \frac{1}{2} L_m {i^2}_{m\,max} \tag{4.100}$$

is mostly transfered to the capacitor $C_c$ during the off-time period and partially consumed in the resistor $R_c$. We assume that the clamp capacitor $C_c$ is sufficiently large so that $v_C$ remains practically constant over the operational period.

For the periodic circuit operation, the amount of the energy given by (4.100) must be consumed in the resistor $R_c$ in the form of

$$E_R = \frac{V_C^2}{R_c} T_s \tag{4.101}$$

in a cycle-by-cycle fashion. By equating (4.100) and (4.101), it follows that

$$\frac{1}{2} L_m {i^2}_{m\,max} = \frac{V_C^2}{R_c} T_s \tag{4.102}$$

By combining (4.102) and (4.99), the expression of the average value of the clamp capacitor voltage is found as

$$V_C = \sqrt{\frac{R_c T_s}{2L_m}} DV_S \tag{4.103}$$

Same as the previous Zener diode reset circuit, the RCD clamp reset circuit has a limit to an upper limit for the acceptable duty ratio, $D_{\max}$. The expression for

$D_{\max}$ is derived as follows. First, the volt-sec balance condition on $L_m$ is set up with D=$D_{\max}$

$$V_S D_{\max} T_s = V_C (1 - D_{\max}) T_s \tag{4.104}$$

which yields

$$D_{\max} = \frac{1}{1 + \dfrac{V_S}{V_C}} \tag{4.105}$$

This expression is incomplete because the clamp voltage $V_C$ varies as a function of duty ratio $D$.

As the second step, the expression (4.103) is evaluated with $D = D_{\max}$

$$V_C = \sqrt{\frac{R_c T_s}{2L_m}} D_{\max} V_S \tag{4.106}$$

The expressions (4.106) and (4.105) are simultaneously solved to find another equation for the maximum duty ratio

$$D_{\max} = 1 - \sqrt{\frac{2L_m}{R_c T_s}} \tag{4.107}$$

This expression will be more convenient for circuit design purposes.

Now, we discuss the development of a fixed clamp voltage $V_C$ in steady state. Over the circuit operation, a fixed magnetic energy of (4.100) is periodically transfered to the capacitor. In the early stage of circuit operation, only a fraction of the transfered energy is consumed at the resistor $R_c$ by the equation (4.101), because $v_C$ is very small in magnitude. Thus, the following condition prevails

$$\frac{1}{2} L_m i^2{}_{m\,max} > \frac{v_C^2}{R_c} T_s \tag{4.108}$$

before the circuit establishes a steady-state operation. This indicates that the capacitor voltage $v_C$ increases cycle-by-cycle until it settles into the steady-state value $V_C$ that meets the equation (4.103).

Expressions (4.103),(4.105), and (4.107) provide design equations for the RCD clamp reset circuit. It is apparent that the RCD clamp reset circuit only provides a lossy reset, because the energy accumulated in the magnetizing inductor during the period $DT_s$ is dissipated at $R_c$ during the period $D_1 T_s$. Further details about the RCD clamp reset circuit will be given in Section 4.8.5, which covers the design and analysis of an RCD clamp reset forward converter.

### Switch Network with Tertiary Winding Reset

It is also possible to reset the transformer using an additional winding. This alternative reset scheme is shown in Fig. 4.35. The reset is performed by the tertiary winding wound on the same magnetic core on which the original transformer is built. Figure 4.35(a) shows a switch network employing the tertiary winding reset. The structure

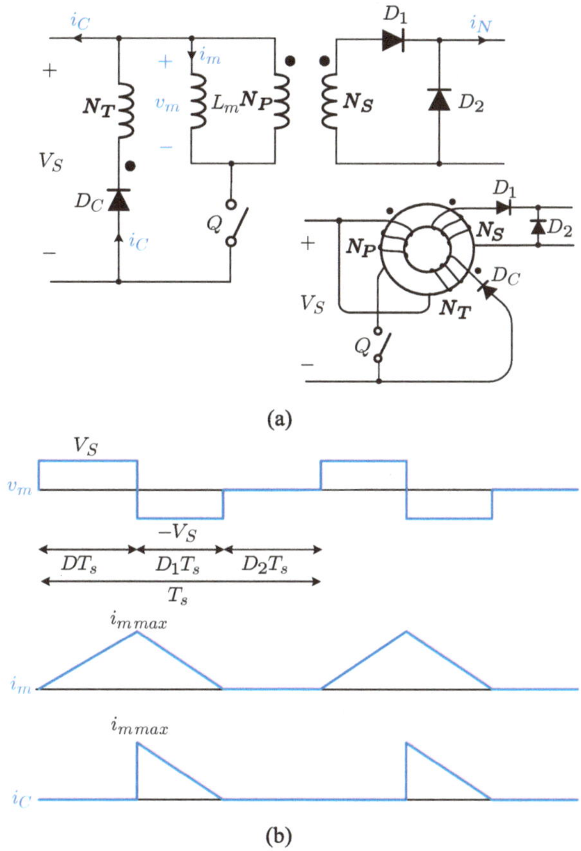


**Figure 4.35** Switch network with tertiary winding reset. (a) Circuit diagram. (b) Major circuit waveforms.

of the three-winding transformer is also shown. For simplicity of discussions, a turns ratio of $N_P : N_S : N_T = 1:1:1$ is assumed for the three-winding transformer. Figure 4.35(b) shows the major circuit waveforms of the switch network. Referring to Fig. 4.35, the operational cycle of the switch network is divided into three periods.

On-Time Period $DT_s$: This period corresponds to the interval in which the active switch $Q$ is closed. The input voltage $V_S$ appears at all the three transformer windings in the positive polarity – with the positive voltage at the *dotted* winding terminal. Accordingly, the diode $D_1$ is forward-biased while the other two diodes are reverse-biased. In particular, the diode at the tertiary winding, the *reset diode* $D_C$, is reversed-biased by $2V_S$ – the input voltage $V_S$ plus the voltage across the tertiary winding, which is $V_S$ for the $1:1:1$ transformer turns ratio. The output current of the switch network $i_N$ flows out of $D_1$ while the same current runs into the *dotted* terminal of the primary winding. The magnetizing current $i_m$

increases linearly with a slope $V_S/L_m$. The sum of the magnetizing current and primary winding current runs through the switch $Q$. Energy in the magnetizing inductance gradually increases and reaches the peak value at the end of this period.

Reset Period $D_1T_s$: This period starts when the active switch is opened, and ends when the magnetizing current $i_m$ reduces to zero. Upon the disappearance of the current path through the active switch $Q$, the reset diode $D_C$ linked to the tertiary winding turns on automatically, because the conduction of $D_C$ allows the continuous flow of the elevated magnetizing current $i_m$, while meeting the current equation of the $1:1:1$ ideal transformer. Details about this function are explained below.

Opening the active switch $Q$ triggers the reset diode $D_C$ to conduct and to deliver $i_C$ through the tertiary winding. The tertiary winding current $i_C$ comes out of the negative terminal of the voltage source $V_S$, located at the left-hand side of the switch network, and flows into the positive terminal of the voltage source $V_S$. The turn-on of $D_C$ also forces all the three windings to take $V_S$ in the negative polarity – with the negative voltage at the *dotted* terminal. Accordingly, $D_1$ is reverse-biased and $i_N$ now freewheels through $D_2$. The magnetizing current $i_m$ circulates through the loop formed by the magnetizing inductance and primary transformer winding. This current is coming out of the *dotted* terminal of the primary winding. The continuous flow of $i_m$ is possible because the same current flows into the *dotted* terminal of the tertiary winding, $i_C = i_m$, while the secondary winding carries *zero* current. This distribution meets the current equation of the 1:1:1 ideal transformer.

The input voltage $V_S$ is applied to the magnetizing inductance $L_m$ in the negative polarity and $i_m$ thus linearly decreases with a slope of $-V_S/L_m$. As $i_m$ decays from its peak, energy stored in $L_m$ is transferred back to the voltage source, located at the left-hand side of the switch network. The $i_m$ continues to decay linearly until it becomes zero. Thus, the circuit provides a lossless reset; all the energy delivered to the magnetizing inductance $L_m$ during the on-time period is fully recovered by the voltage source during the reset period.

Recess Period $D_2T_s$: When the condition $i_C = i_m = 0$ is reached, the reset diode $D_C$ turns off and remains off until the next reset period. All the currents and voltages associated with the switch network are zero in this period.

With the $1:1:1$ transformer turns ratio, the rising slope of $i_m$ during the on-time period is the same as the decaying slope during the reset period. This implies that the maximum duty ratio of the active switch should be limited to 0.5 ($0 < D < 0.5$) for the proper operation of the converter. Otherwise, the magnetizing inductance of the transformer will saturate, thus making the circuit inoperative.

The upper limit of the duty ratio for general cases is evaluated using Fig. 4.36, which illustrates the operation of reset circuit with $N_P : N_S : N_T = 1 : n_1 : n_2$. Figure 4.36(b) shows the voltage $v_m$ and current $i_m$ associated with the magnetizing

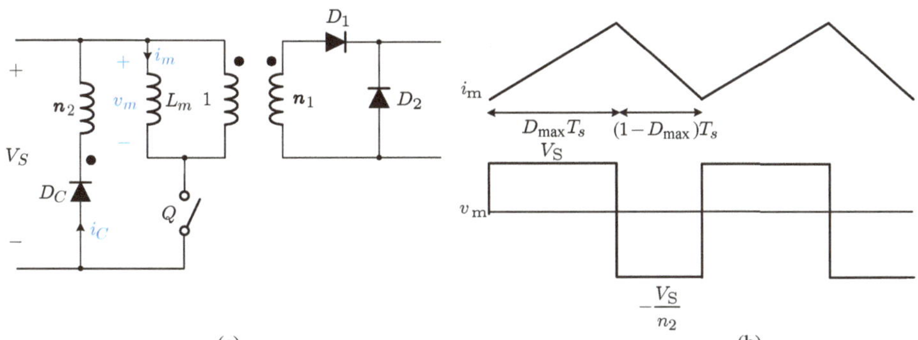


Figure 4.36 Tertiary-winding reset network (a) Circuit diagram. (b) Magnetizing inductance waveforms.

inductance $L_m$ of the isolation transformer. The volt-sec balance condition on $L_m$ with the maximum possible duty ratio, $D_{max}$, is given by

$$V_S D_{max} T_s = \frac{V_S}{n_2}(1 - D_{max})T_s \tag{4.109}$$

which is solved for $D_{max}$ for the given $n_2 = N_T/N_P$

$$D_{max} = \frac{1}{1+n_2} = \frac{1}{1+\dfrac{N_T}{N_P}} \tag{4.110}$$

For the given $n_2$, the duty ratio should lie between $0 < D < 1/(1+n_2)$. For an example, with $n_2 = N_T/N_P = 36/12 = 3$, the maximum allowable duty ratio becomes $D_{max} = 1/(1+3) = 0.25$.

### 4.8.3 Tertiary-Winding Reset Forward Converter

In this section, we study the forward converter that is configured by combining the tertiary-winding reset circuit and low-pass filter circuit. The resulting converter circuit is referred as the tertiary-winding reset forward converter, which offers a lossless reset. Two other lossy reset forward converters – the Zener diode reset forward converter and the RCD clamp reset forward converter – will be covered in the ensuing two sections.

The tertiary-winding reset forward converter is shown in Fig. 4.37. In the tertiary-winding reset circuit in Fig. 4.37(a), the primary and tertiary windings retain the $N_P : N_T = 1:1$ turns ratio, yet the primary and secondary windings have an $N_P : N_S = 1:n$ turns ratio. The tertiary winding is called the *reset winding*. Figure 4.37(a) also shows the typical magnetizing current $i_m$. Figure 4.37(b) depicts three subcircuits of the converter, each representing the converter circuit during the on-time period, reset period, or recess period. Expressions of major circuit waveforms, derived from these three subcircuits, are listed in Table 4.23.

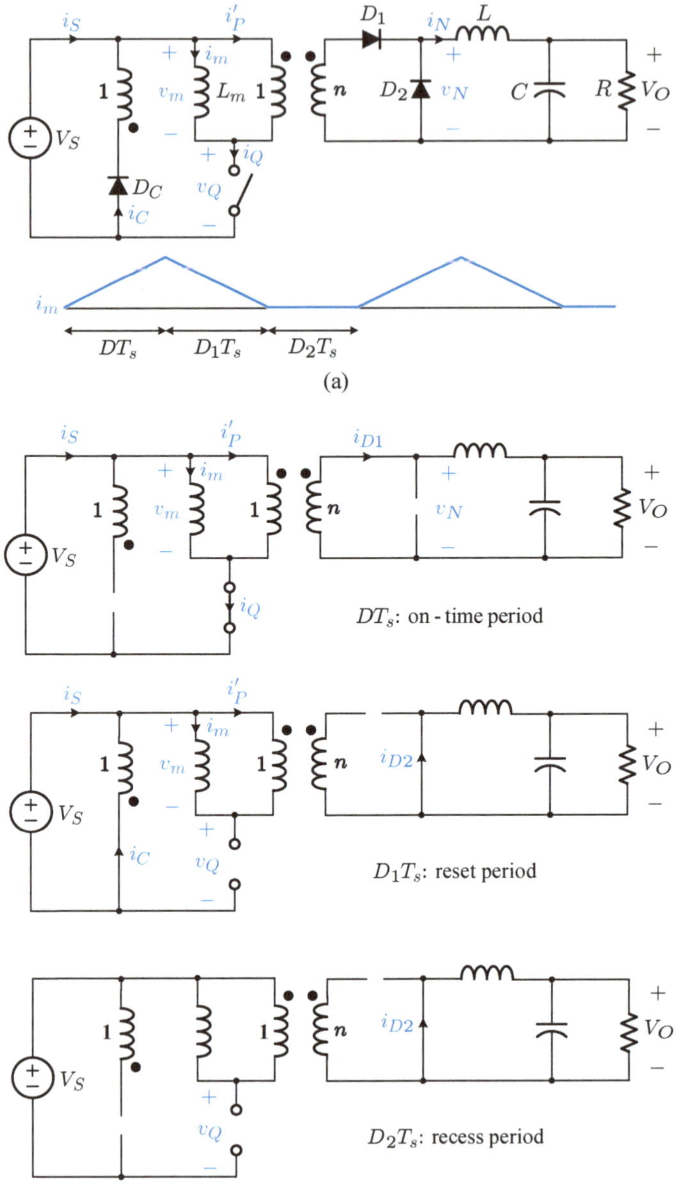


**Figure 4.37** Tertiary-winding reset forward converter. (a) Circuit diagram and magnetizing current. (b) Three subcircuits.

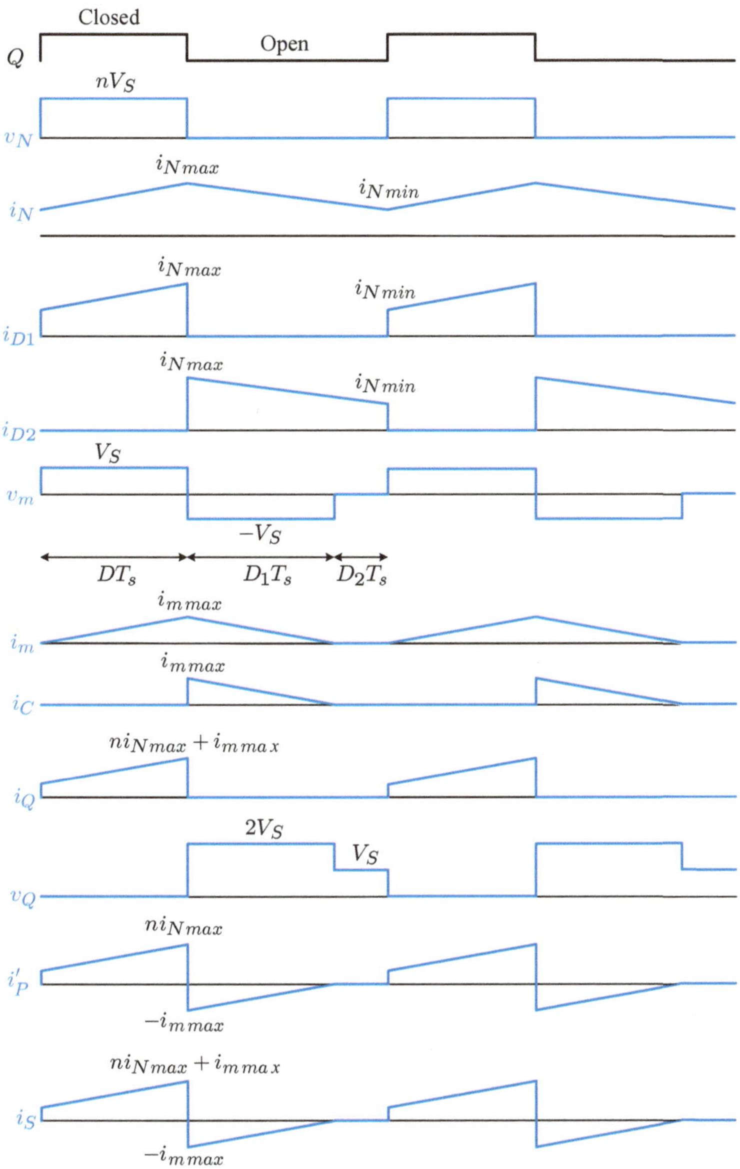


**Figure 4.38** Waveforms of tertiary-winding reset forward converter.

**Table 4.23** Expressions for Circuit Variables of Tertiary-Winding Reset Forward Converter

| Circuit variable | On-time period | Reset period | Recess period |
|---|---|---|---|
| $v_N$ | $nV_S$ | 0 | 0 |
| $v_m$ | $V_S$ | $-V_S$ | 0 |
| $i_m$ | $\frac{V_S}{L_m}t$ | $-\frac{V_S}{L_m}t$ | 0 |
| $i_C$ | 0 | $-\frac{V_S}{L_m}t$ | 0 |
| $i_N$ | $\frac{nV_S - V_O}{L}t$ | $-\frac{V_O}{L}t$ | $-\frac{V_O}{L}t$ |
| $i_{D1}$ | $i_N$ | 0 | 0 |
| $i_{D2}$ | 0 | $i_N$ | $i_N$ |
| $i'_P$ | $ni_N$ | $-i_m$ | 0 |
| $i_Q$ | $i_m + i'_P$ | 0 | 0 |
| $i_S$ | $i_m + i'_P$ | $-i_C$ | 0 |
| $v_Q$ | 0 | $2V_S$ | $V_S$ |

Figure 4.38 shows the circuit waveforms, constructed using the equations in Table 4.23. The waveforms indicate that the operation of the forward converter is identical to that of the conceptual converter in Fig. 4.30, except for the waveforms associated with the magnetizing inductance and reset winding. The output voltage of the converter is given by

$$\boxed{V_O = \bar{v}_N(t) = nDV_S} \tag{4.111}$$

with the $1 : n$ turns ratio between the primary and secondary windings.

**Table 4.24** Steady-State Analysis of Tertiary-Winding Reset Forward Converter

| Circuit variable | Expression |
|---|---|
| $V_O$ | $nDV_S = 0.5 \cdot 0.25 \cdot 64 = 8\ \mathrm{V}$ |
| $v_{Nmax}$ | $nV_S = 0.5 \cdot 64 = 32\ \mathrm{V}$ |
| $I_N$ | $\frac{V_O}{R} = \frac{8}{1} = 8\ \mathrm{A}$ |
| $\Delta i_N$ | $\frac{nV_S - V_O}{L} DT_s = \frac{0.5 \cdot 64 - 8}{40 \times 10^{-6}} 0.25 \cdot 20 \times 10^{-6} = 3.0\ \mathrm{A}$ |
| $i_{Nmax}$ | $I_N + \frac{1}{2}\Delta i_N = 8.0 + \frac{1}{2}3.0 = 9.5\ \mathrm{A}$ |
| $i_{Nmin}$ | $I_N - \frac{1}{2}\Delta i_N = 8.0 - \frac{1}{2}3.0 = 6.5\ \mathrm{A}$ |
| $v_{mmax}$ | $V_S = 64\ \mathrm{V}$ |
| $v_{mmin}$ | $-V_S = -64\ \mathrm{V}$ |
| $i_{mmax}$ | $\frac{V_S}{L_m} DT_s = \frac{64}{200 \times 10^{-6}} 0.25 \cdot 20 \times 10^{-6} = 1.6\ \mathrm{A}$ |
| $v_{Qmax}$ | $2V_S = 2 \cdot 64 = 128\ \mathrm{V}$ |

The peak value of $i_m$ is given by

$$i_{m\,max} = \frac{V_S}{L_m} DT_S \tag{4.112}$$

The maximum value for the the switch network current $i_N$ is determined as

$$i_{N\,max} = \frac{V_O}{R} + \frac{1}{2}\frac{nV_S - V_O}{L} DT_s \tag{4.113}$$

The minimum value of $i_N$ is given by

$$i_{N\,min} = \frac{V_O}{R} - \frac{1}{2}\frac{nV_S - V_O}{L} DT_s \tag{4.114}$$

From these values, the peak and valley values of the other current waveforms are evaluated.

The tertiary-winding reset forward converter is one of most popular converter topologies. Advantages of this converter include the lossless reset with minimal component count. Due to the lossless reset, efficiency of the converter is high, compared with other forward converters with a lossy reset scheme. This converter uses only one active switch, while handling the power level comparable to that of the half-bridge converter.

There are many other variations of the forward converter with lossless reset. One broadly-used converter topology with lossless reset is the forward converter that employs two active switches and one two-winding transformer, known as the two-switch forward converter. The operation of this converter is described in a later section.

### EXAMPLE 4.17 Steady-state operation of forward converter

This example illustrates the steady-state analysis and major circuit waveforms of a tertiary-winding reset forward converter. The operational conditions and circuit parameters of the converter are $V_S = 64\text{V}$, $L_m = 200\mu\text{H}$, $N_P : N_S : N_T = 1 : 0.5 : 1$, $L = 40\mu\text{H}$, $C = 400\mu\text{F}$, $R = 1\Omega$, $T_s = 20\mu\text{s}$, and $D = 0.25$. For the given power stage parameters and operating conditions, steady-state values of major circuit variables are evaluated using the previous analysis results. The steady-state values are listed in Table 4.24 and the simulated circuit waveforms are given in Fig. 4.39.

### EXAMPLE 4.18 PSIM® Tertiary-winding reset forward converter

This example describes the steady-state response of another tertiary-winding reset forward converter. The converter has the same operational conditions and circuit parameters as those of Example 4.17, except for the the turns of the three winding three-winding transformer. The turns of the three transformer windings are $N_P : N_S : N_T = 10 : 15 : 5$.

Referring to (4.110), the upper limit of the duty ratio of the converter is given by

$$D_{\text{max}} = \frac{1}{1 + \dfrac{N_T}{N_P}} = \frac{1}{1 + \dfrac{5}{10}} = \frac{2}{3}$$

The operation with $D_{\text{max}} = 2/3$ produces the output voltage

$$V_O = D_{\text{max}} \frac{N_S}{N_P} V_S = \frac{2}{3} \frac{15}{10} 64 = 64 \text{ V}$$

The PSIM model for this analysis is shown in Table 4.25. The PWM block is programmed to produce the duty ratio $D = 2/3$. The steady-state waveforms of the major circuit variables are shown in the table. The converter produces the output voltage $V_O = 64$ V. The magnetizing current indicates that the converter operates at

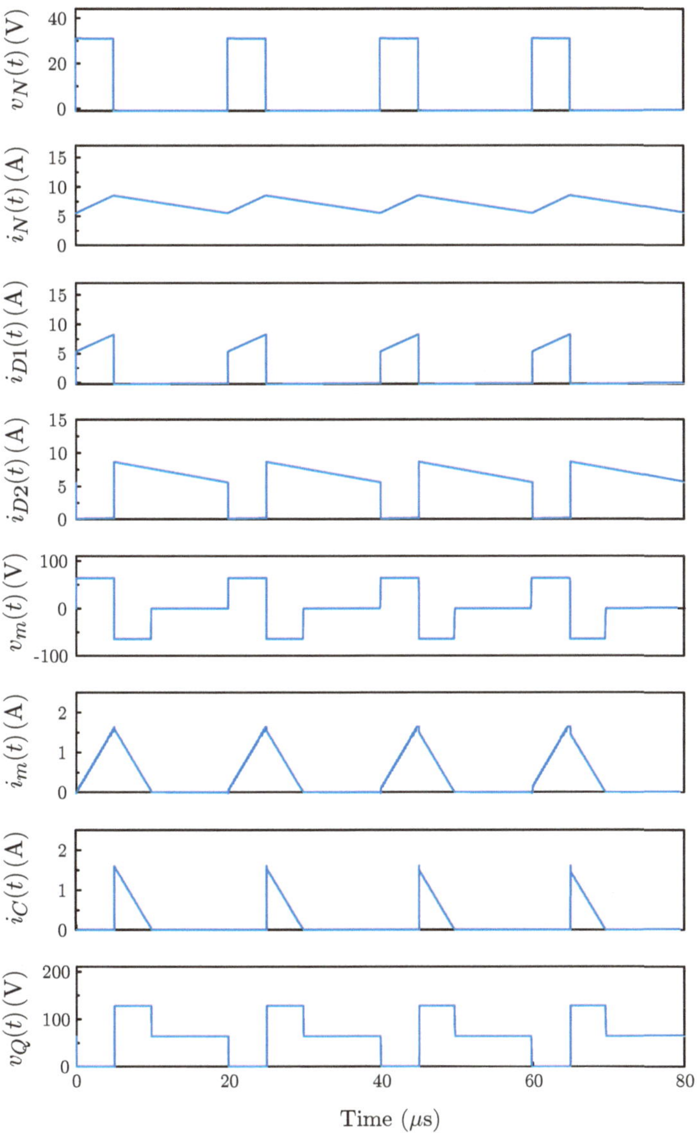


**Figure 4.39** Circuit waveforms of tertiary-winding reset forward converter.

Table 4.25 Steady-State Response of Tertiary-Winding Reset Forward Converter

PSIM model: Tertiary_Winding_Reset_ Forward_Converter_01

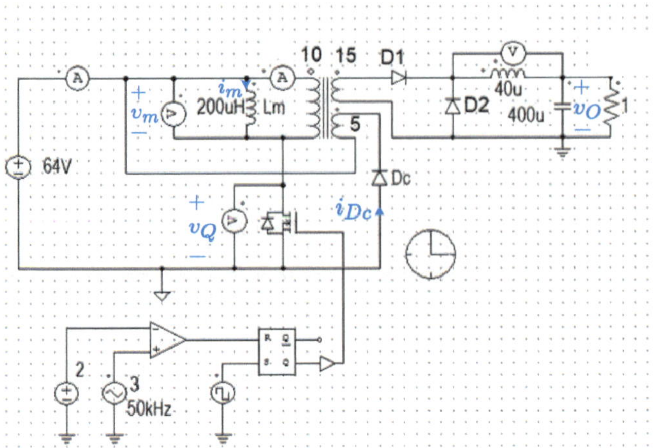


Simulation outcomes

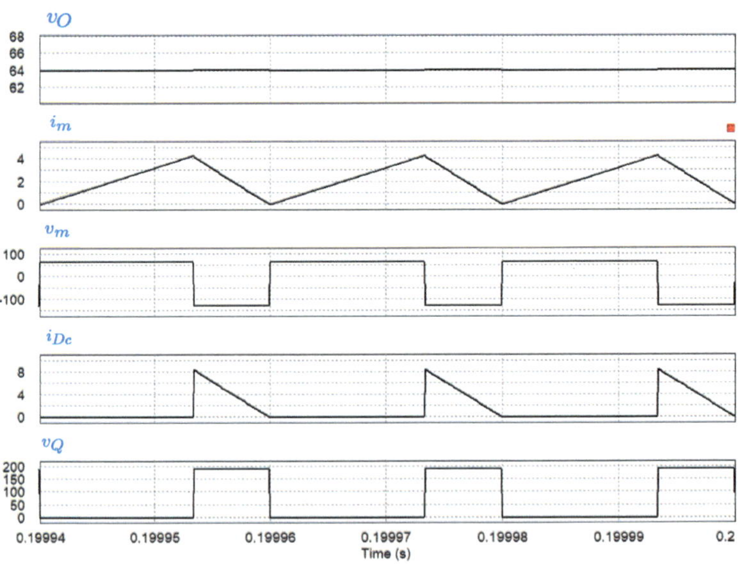


the maximum possible duty ratio. If the duty ratio is increased beyond the current value, the transformer will saturate, thus leading to a fatal failure to the converter.

### 4.8.4 Zener Diode Reset Forward Converter

This section deals with the Zener diode reset forward converter, which combines the Zener diode reset switch network, discussed in Section 4.8.2, and low-pass filter. Figure 4.40 shows the circuit diagram of the converter along with its three subcircuits in different operational periods: the on-time period, off-time period, and recess period. The structures of the converter circuit and its subcircuits are very similar to those of the tertiary-winding reset forward converter. Accordingly, the analysis of the Zener diode reset forward converter is practically the same as the case of the tertiary-winding reset forward converter.

The only differences between the two converters lie in the waveforms associated with the reset circuit. The reset circuit waveforms are shown in Fig. 4.41. The reset winding voltage in the previous converter is replaced with the Zener breakdown voltage in the current converter.

Operations and waveforms of Zener diode rest forward converter are illustrated using PSIM®simulations in Example 4.19. Readers are recommended to adaptively adopt the information given in Fig. 4.38, Table 4.23, and Table 4.24 for comprehensions of simulation results.

**PSIM® Zener diode reset forward converter** **EXAMPLE 4.19**

This example illustrates the steady-state response of Zener diode reset forward converter. The converter operates with $V_S = 64$ V and $f_s = 50$ kHz. The circuit parameters are selected as $L = 40\ \mu$H, $C = 400\ \mu$F, and $R = 1\ \Omega$. The turns of the transformer windings are $N_P : N_S = 10 : 15$ and the magnetizing inductance of the transformer is $L_m = 200\ \mu$H. The Zener breakdown voltage is $V_Z = 32$ V.

In this example, the duty ratio is selected to produce the largest possible value for the output voltage. Using the expression (4.97), the duty ratio for the maximum output voltage is determined as

$$D_{\max} = \frac{1}{1 + \dfrac{V_S}{V_Z}} = \frac{1}{1 + \dfrac{64}{32}} = \frac{1}{3}$$

First, we investigate the start-up response of the converter. The duty ratio of the converter is selected as $D = 1/3$ to produce the maximum possible output voltage of

$$V_O = D_{\max} n V_S = \frac{1}{3} \cdot 1.5 \cdot 64 = 32 \text{ V}$$

The PSIM model of the converter and the start-up response of the converter are shown in Table 4.26. The output voltage settles into the target voltage of $V_O = 32$ V. The inductor current reach at

$$I_L = \frac{V_O}{R} = \frac{32}{1} = 32 \text{ A}$$

The steady-state waveforms of the Zener diode reset forward converter are shown in Table 4.27. The upper figure exhibits the response of the primary circuit wave-

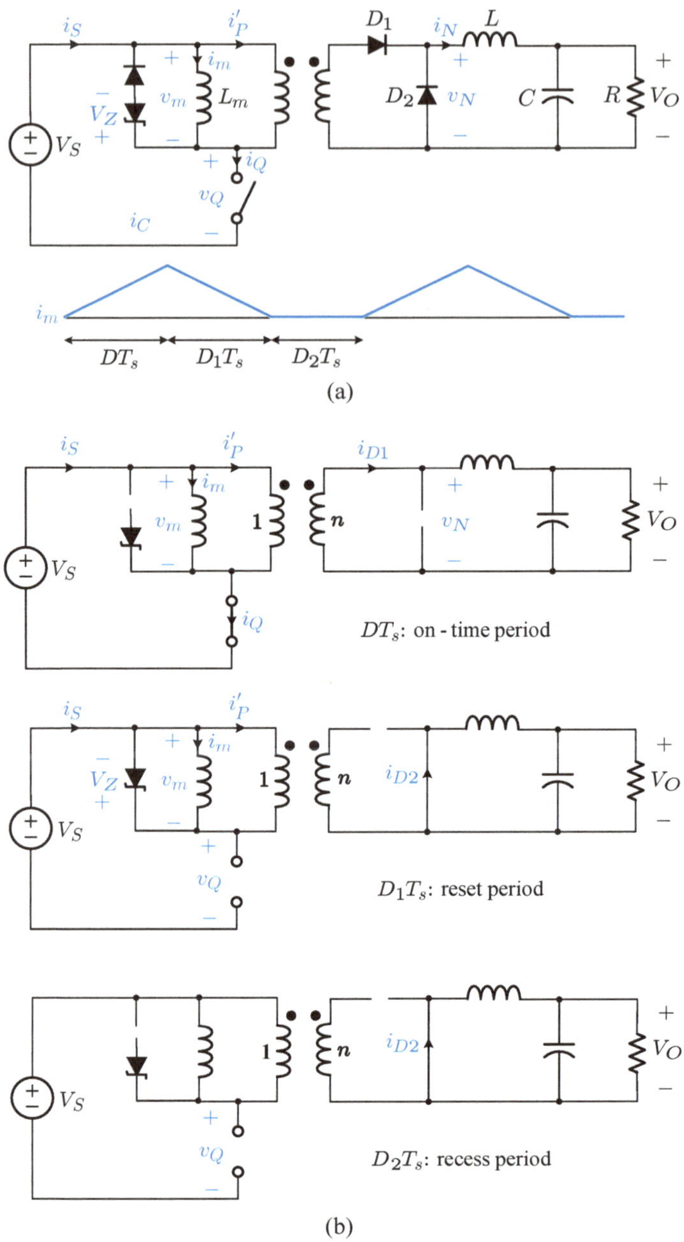


**Figure 4.40** Zener diode reset forward converter. (a) Circuit diagram and magnetizing current. (b) Three subcircuits.

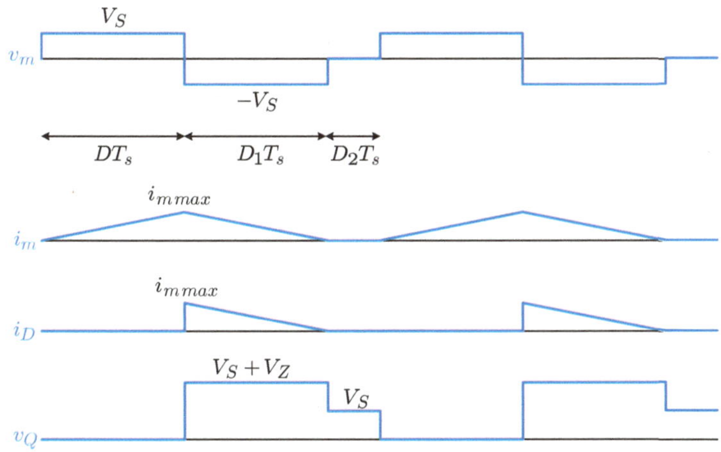


**Figure 4.41** Reset waveforms of Zener diode reset forward converter.

forms of the converter. The magnetizing current with $D = 1/3$ confirms that converter produces a continuous $i_m$ waveform with a *zero* recess period.

The secondary circuit waveforms are displayed in the lower figure. The converter circuit produces the largest possible output voltage of $V_O = D_{\max} n V_S = (1/3) \cdot 1.5 \cdot 64 = 32$ V. It is left as exercises for reader to validate the simulation results in Table 4.27, referring to Fig. 4.38, Table 4.23, and Table 4.24 with appropriate modification.

### 4.8.5 RCD Clamp Reset Forward Converter

When the RCD clamp reset circuit covered in Section 4.8.2 is connected to a low-pass filter, another lossy forward converter is created. The resulting converter is shown in Fig. 4.42. The converter is called the RCD clamp reset forward converter.

The structure, three subcircuits, and operation of the converter are practically identical to those of the previous two forward converters. As one exception, the voltage across the clamp capacitor $C_c$, referred to the clamp voltage $v_C$ in earlier discussions, is determined by the circuit components and operational conditions of the converter. As shown in Section 4.8.2, the steady-state value of the clamp capacitor is given by

$$V_C = \sqrt{\frac{R_c T_s}{2L_m}} D V_S \tag{4.115}$$

The reset circuit waveforms are shown in Fig. 4.43. The clamp voltage $V_C$ affects the switch voltage during the reset and recess periods.

**Table 4.26** Start-up Response of Zener Diode Reset Forward Converter

PSIM model: Zener_Reset Forward_Converter_01

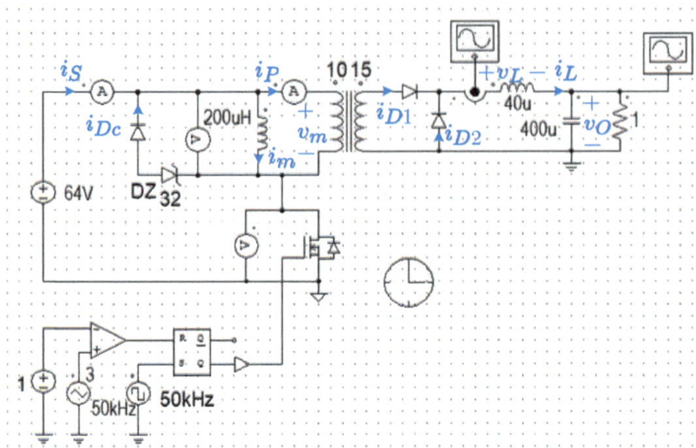


Simulation outcomes

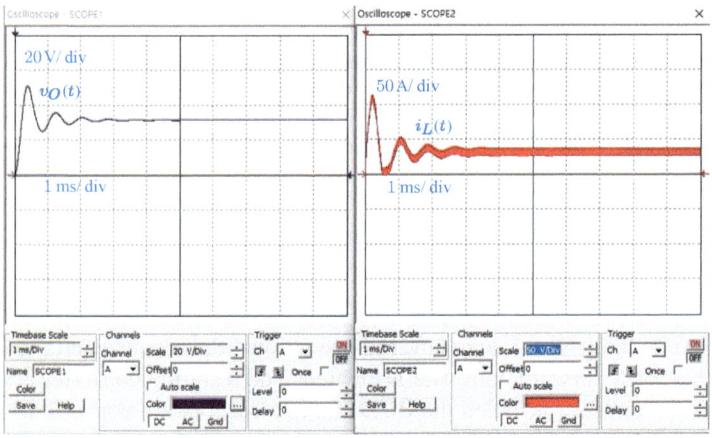


The earlier analysis in Section 4.8.2 showed that the maximum allowable duty ratio is determined as

$$D_{\max} = \frac{1}{1 + \dfrac{V_S}{V_C}} \tag{4.116}$$

By simultaneously solving (4.115) and (4.116) with the condition $D = D_{\max}$, another expression for $D_{\max}$ is given as

$$D_{\max} = 1 - \sqrt{\frac{2L_m}{R_c T_s}} \tag{4.117}$$

**Table 4.27** Steady-State Response of Zener Diode Reset Forward Converter

| PSIM model: Zener_Reset Forward_Converter_02 |
|---|
| Simulation outcomes: Primary circuit waveforms |
| 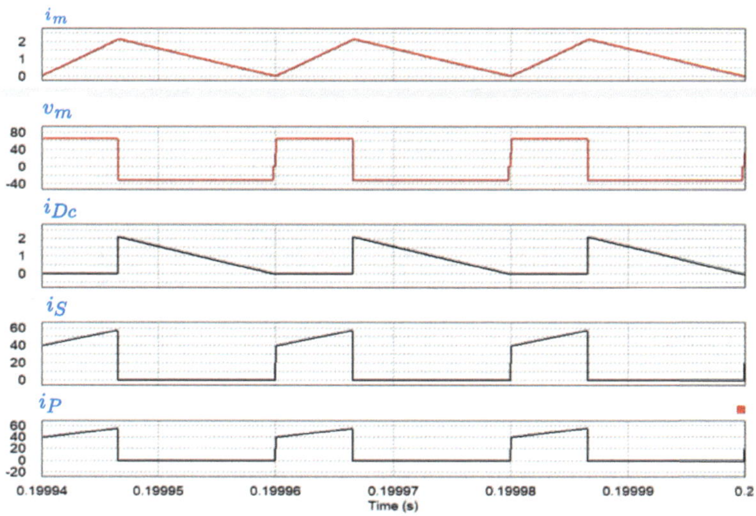 |
| Simulation outcomes: Secondary circuit waveforms |
| 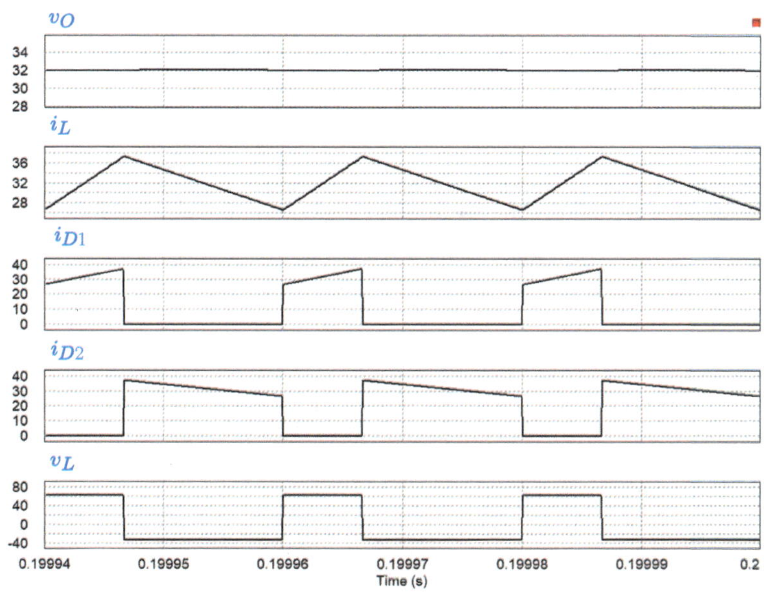 |

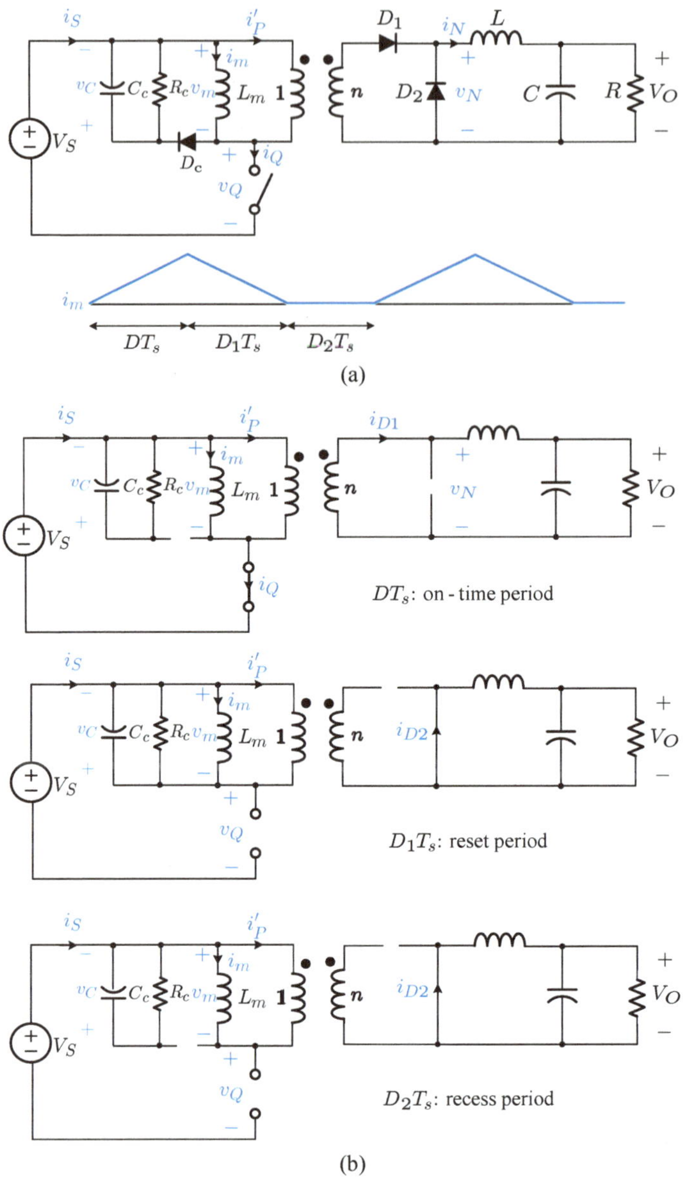


**Figure 4.42** RCD clamp reset forward converter. (a) Circuit diagram and magnetizing current. (b) Three subcircuits.

The design and analysis of the RCD reset forward converter are done be using the expressions (4.115) through (4.117). An example is given.

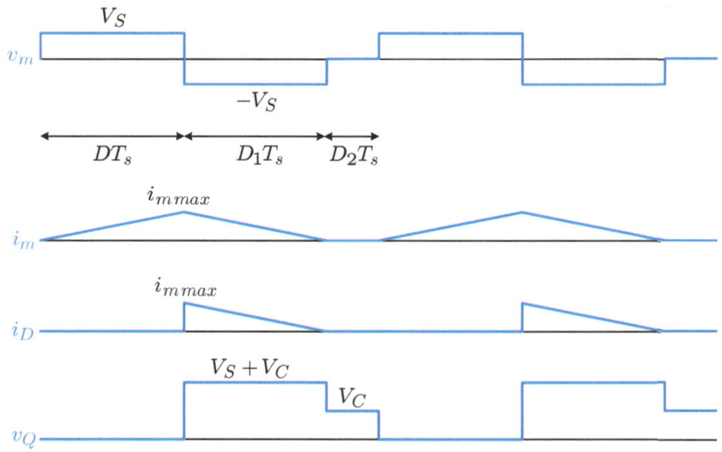


**Figure 4.43** Reset waveforms of RCD clamp reset forward converter.

## PSIM® RCD clamp reset forward converter — EXAMPLE 4.20

This example illustrates the design and analysis of an RCD clamp reset forward converter. The operational conditions and circuit components of the converter are as follows.

- Converter specification:
  - Input voltage: $V_S = 24 - 36$ V
  - Output voltage: $V_O = 5$ V
  - Output power: $P_O = 100$ W
  - Switching frequency: $f_s = 100$ kHz.
- Output filter components: $L = 40\mu$H, $C = 400\mu$F, $R = V_O^2/P_O = 5^2/100 = 0.25\Omega$
- Transformer parameters:
  - Turns ratio: $N_p : N_s = 20 : 10 \Rightarrow n = 10/20 = 0.5$
  - Magnetizing inductance: $L_m = 200\ \mu$H.

For the given operational conditions, an RCD clamp reset circuit is designed based on (4.115) through (4.117). The largest duty ratio of the converter occurs with the minimum input voltage of $V_S = 24$ V

$$D = \frac{1}{n}\frac{V_O}{V_S} = \frac{1}{0.5}\frac{5}{24} = 0.416 \tag{4.118}$$

However, we set the maximum duty ratio at $D_{\max} = 0.46$ for a conservative design margin. Now, using the expression (4.117), the value of $R_c$ is determined as

$$R_c = \frac{2L_m}{(1 - D_{\max})^2 T_s} = \frac{2 \cdot 200 \times 10^{-6}}{(1 - 0.46)^2 10 \times 10^{-6}} = 137\ \Omega \tag{4.119}$$

The clamp capacitor is selected as $C_c = 5\ \mu$F to meet the condition $R_c C_c \gg T_s$ for an acceptable ripple component in the clamp voltage $v_C$.

**Table 4.28** Start-up Response of RCD Clamp Reset Forward Converter

PSIM model: RCD_Clamp_Reset Forward_Converter_01

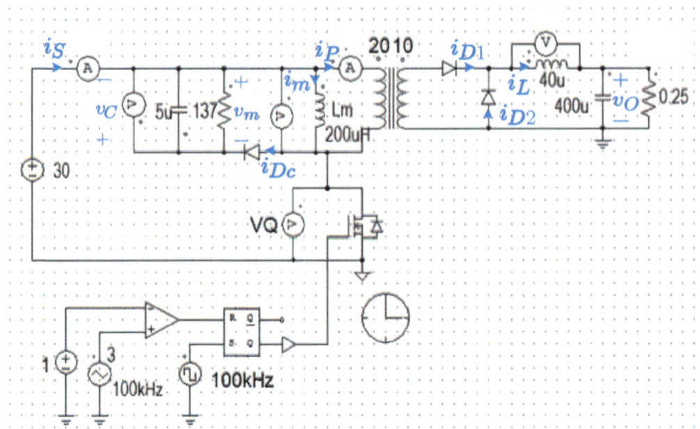


Simulation outcomes

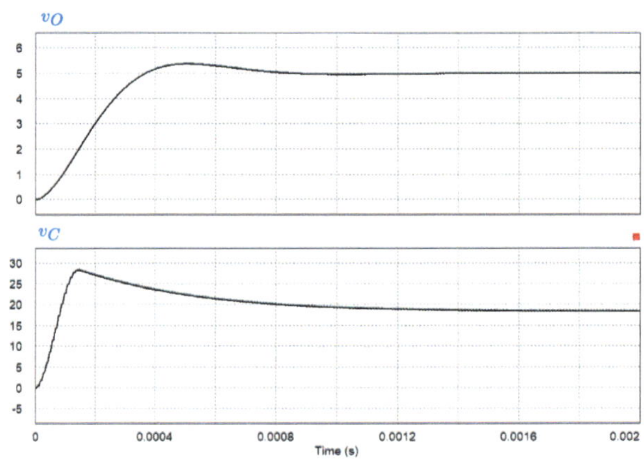


The duty ratio at the nominal operation with $V_S = 30$ V is determined as

$$D_{\text{normal}} = \frac{1}{n}\frac{V_O}{V_S} = \frac{1 \cdot 5}{0.5 \cdot 30} = \frac{1}{3} \tag{4.120}$$

The clamp voltage with $D_{\text{normal}}$ is evaluated from (4.115)

$$V_C = \sqrt{\frac{R_c T_s}{2L_m} D_{\text{normal}} V_S} = \sqrt{\frac{137 \cdot 10 \times 10^{-6}}{2 \cdot 200 \times 10^{-6}} \frac{1}{3} 30} = 18.5 \text{ V} \tag{4.121}$$

Table 4.28 shows the PSIM model and simulation results of the converter during the start-up period. The converter initiates its operation with $V_s = 30$V. The output voltage of converter terminates at $V_O = 5$ V. The clamp capacitor voltage settles into $V_C = 18.5$ V, as predicted in (4.121).

The steady-state responses of the converter with $V_s = 30$ V are shown in Table 4.29. The upper figure exhibits the response of the primary circuit waveforms, while the lower figure displays the circuit waveforms of the secondary circuitry of the converter.

### 4.8.6 Two-Switch Forward Converter

This section presents another forward converter, whose reset scheme is distinct from the previous forward converters. The converter is called the two-switch forward converter because it employs two active switches in the primary side of the isolation transformer. The circuit diagram of the converter is shown in Fig. 4.44.†

The converter employs two synchronized active switches, $Q_1$ and $Q_2$, and two diodes, $D_1$ and $D_2$, in the primary side of the switch network. The remaining part of the switch network contains a two-winding transformer and two rectification diodes. When the two synchronized active switches are turned-on, the converter presents the on-time subcircuit. The on-time subcircuit of this converter is functionally identical to that of the previous forward converters.

When the two synchronized switches are turned-off, the magnetizing current $i_m$ turns on the two diodes, $D_1$ and $D_2$, in the primary side of the circuit, and freewheels through the loop formed by $L_m$, $D_1$, $D_2$, and the voltage source $V_S$.

As $D_1$ and $D_2$ conduct, the input voltage $V_S$ is applied to the magnetizing inductance $L_m$ in the negative polarity. Thus, the magnetizing current ramps down with a slope of $-V_S/L_m$. This period is referred to as the reset period, and the diodes $D_1$ and $D_2$ are called the reset diodes. In the reset period, energy stored in the magnetizing inductance is recovered by the voltage source $V_S$, thereby offering a lossless reset.

As the condition $i_m = i_D = 0$ is reached, the reset diodes turn off and the recess period begins. Figure 4.44(b) shows the three subcircuits, each valid in the respective operational period of the converter. Figure 4.45 shows the major circuit waveforms of the converter, constructed using the subcircuits in Fig. 4.44(b). The shapes and peak values of the circuit waveforms are nearly identical to those of the previous tertiary-winding reset forward converter in Section 4.8.3. The only difference is the peak value of the switch voltage $v_Q$ during the reset and recess periods, which is half the previous case.

**PSIM® Two-switch forward converter — EXAMPLE 4.21**

This example illustrates the start-up behavior of a closed-loop controlled two-switch forward converter. The PSIM model of the converter is shown in Table 4.30. The

†The structure and operation of the two-switch forward converter closely resemble those of the non-dissipative solenoid drive circuit studied in Section 2.3.1.

**Table 4.29** Steady-State Response of RCD Clamp Reset Forward Converter

| PSIM model: RCD_Clamp_Reset Forward_Converter_01 |
|---|
| Simulation outcomes: Primary circuit waveforms |
| 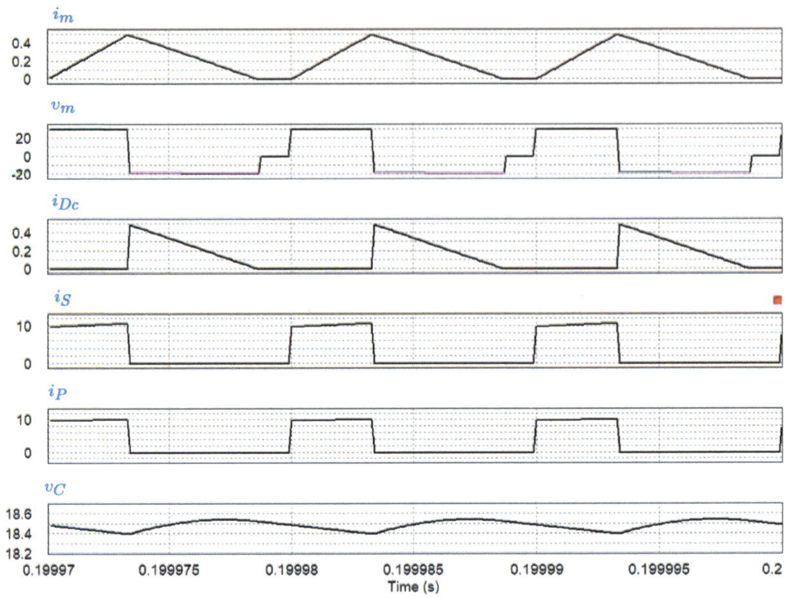 |
| Simulation outcomes: Secondary circuit waveforms |
| 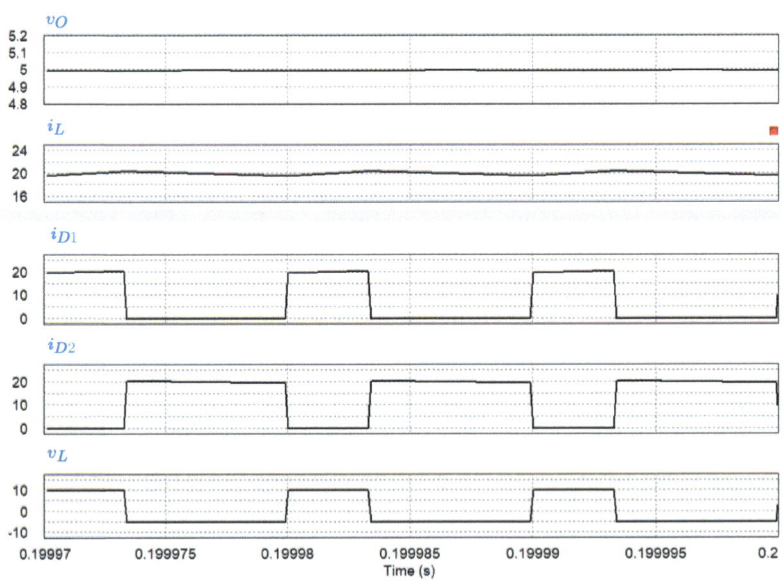 |

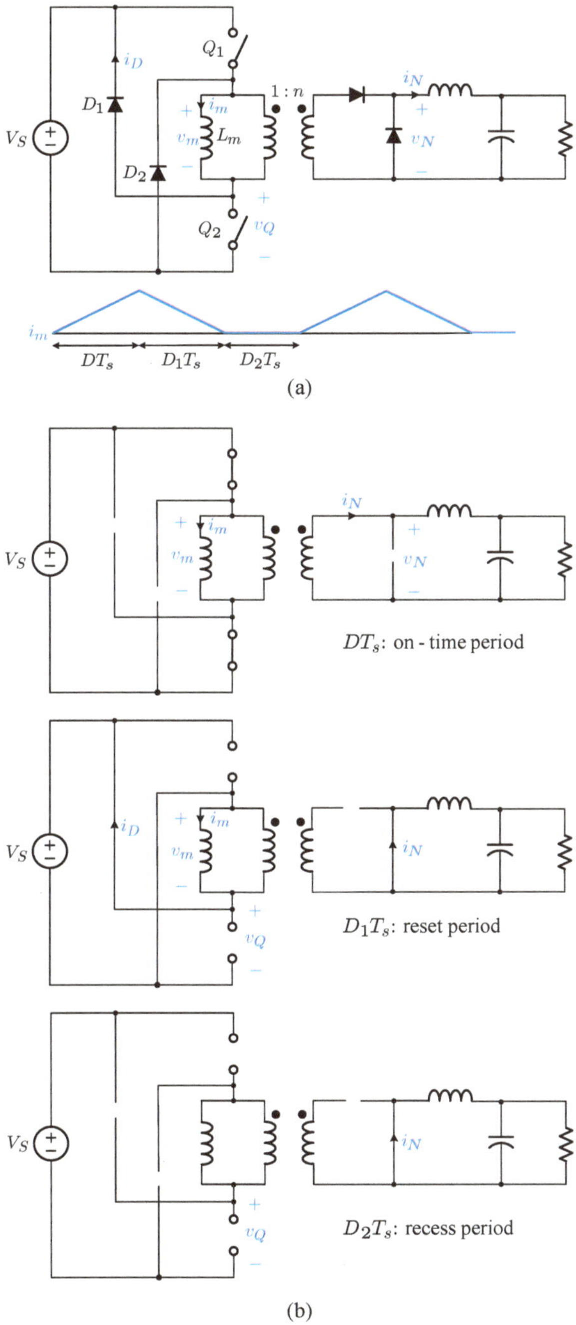


**Figure 4.44** Two-switch forward converter. (a) Circuit diagram. (b) Three subcircuits.

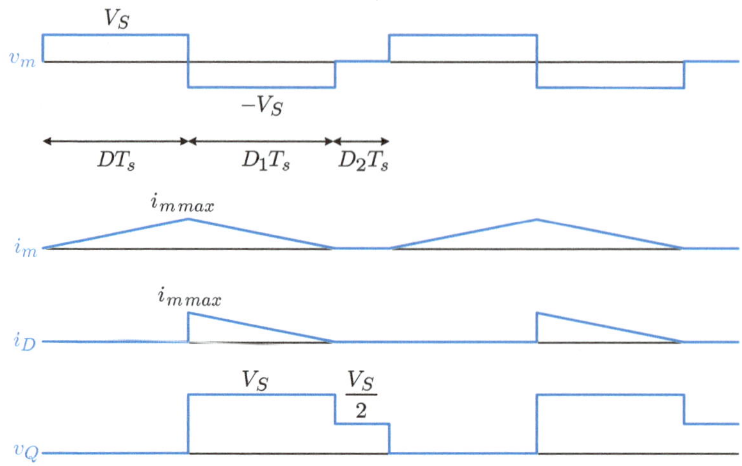


**Figure 4.45** Circuit waveforms of two-switch forward converter.

converter is regulated at

$$V_O = 2.5\left(1 + \frac{2200}{994}\right) = 8.0 \text{ V}$$

using voltage mode control, which is well suited for buck-derived forward converters. The feedback circuit employs an optocoupler isolated three-pole two-zero circuit. The converter operates with $V_S = 64$ V at $f_s = 50$ kHZ. The circuit parameters $L = 40\ \mu$H, $C = 400\ \mu$F, and $R = 2\ \Omega$. The transformer has the magnetizing current of 200 $\mu$H with a 20 : 10 turns ratio. The start-up response of the output voltage $v_O(t)$ and control voltage $v_{con}(t)$ are shown in Table 4.30. The output voltage reaches at the target value of $V_O = 8$ V with the steady-state duty ratio of $D = 0.25$. The height of the ramp signal in the PWM block is $V_{ramp}$ is 4 V. The control voltage settles in the average value of $\bar{v}_{con} = 1$ V, thus yielding the duty ratio of $D = 1/4 = 0.25$.

## EXAMPLE 4.22 PSIM® Steady-state responses

Table 4.31 shows the steady-sate response of the two-switch forward converter used in Example 4.21. Here, the DCM operation of the converter is considered. The load resistance is increased to $R = 15\ \Omega$ for DCM operation. Table 4.31 illustrates the converter operation in DCM. Readers are recommended to confirm the simulation results through the steady-state analysis.

**Table 4.30** Start-up Response of Two-Switch Forward Converter

PSIM model: Two-Switch Forward_Converter_01

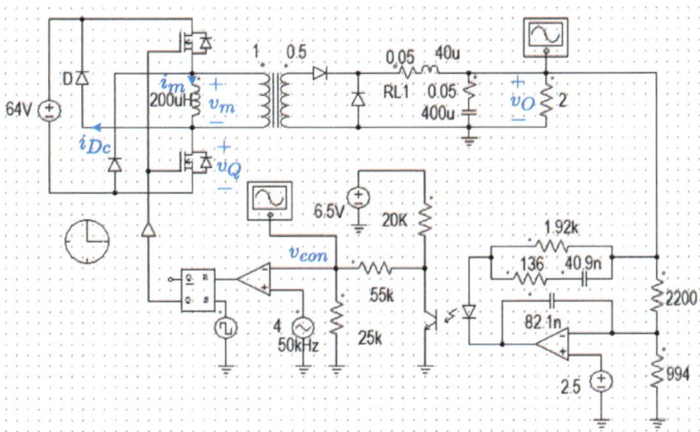

Simulation outcomes

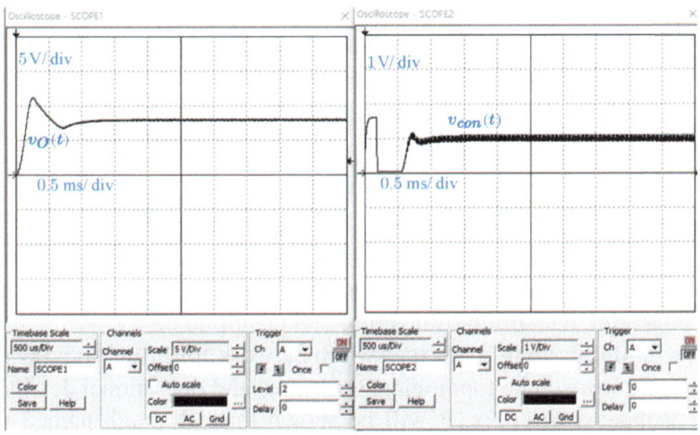

## PSIM® Transient response EXAMPLE 4.23

This example shows the transient response of the two-switch forward converter employing voltage mode control, which was used in the previous two examples. The PSIM model of the closed-loop controlled converter is shown in Table 4.32. As shown in the PSIM model, the load resistance $R$ is stepped up and down between $R = 2\,\Omega$ and $4\,\Omega$. The transient response of the output voltage $v_O$ and output filter current $i_L$ are shown in the table.

**Table 4.31** Steady-State Response of Two-Switch Forward Converter

PSIM model: Two-Switch_Forward_Converter_01

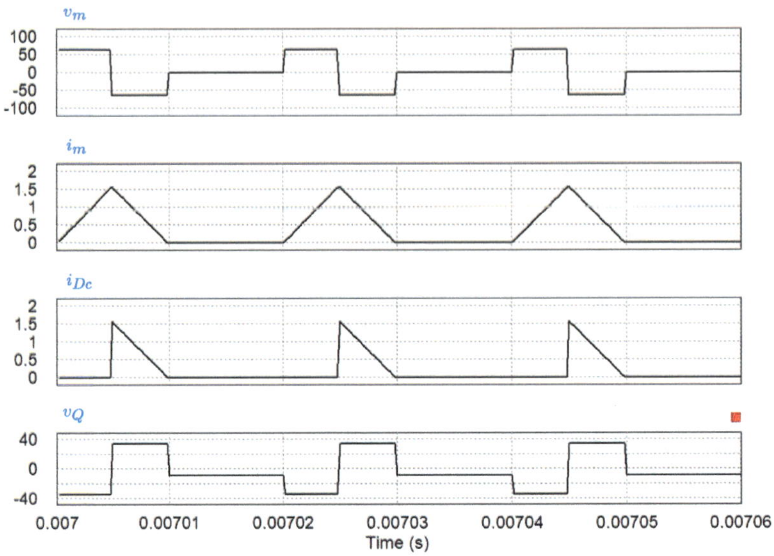


## 4.9 BRIDGE-TYPE ISOLATED CONVERTERS

Along with the forward converter, there is another class of buck-derived converters. The converters are collectively called the bridge-type converter due to their circuit structure. In this section, we study three major bridge-type converters, including the full-bridge converter, half-bridge converter, and push-pull converter.

We start with discussions about the switch network for bridge-type isolated converters. Then, we investigate topological structures and operational details about the bridge-type isolated converters. It will be shown that, although named differently due to the difference in the switch network, these three isolated converters are all functionally equivalent to the buck converter.

### 4.9.1 Switch Network for Bridge-Type Converters

This section introduces circuit configurations for the switch network. Figure 4.46 shows the generic structure of the switch network. The switch network is partitioned into two circuits using the isolation boundary as the borderline. The circuit located at the left-hand side of the isolation boundary is referred to as the primary circuit, while the circuit appearing at the other side is denoted as the rectification circuit.

There are various circuit configurations that can be employed as the primary circuit or rectification circuit. If a particular circuit configuration is selected, each for the

**Table 4.32** Transient Response of Two-Switch Forward Converter

PSIM model: Two-Switch_Forward_Converter_03

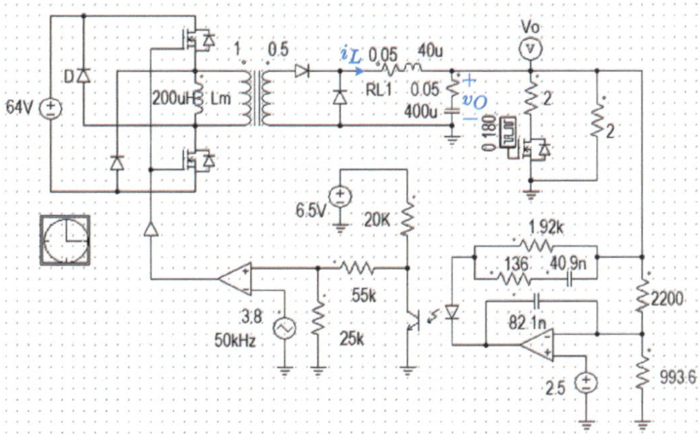


Simulation outcomes

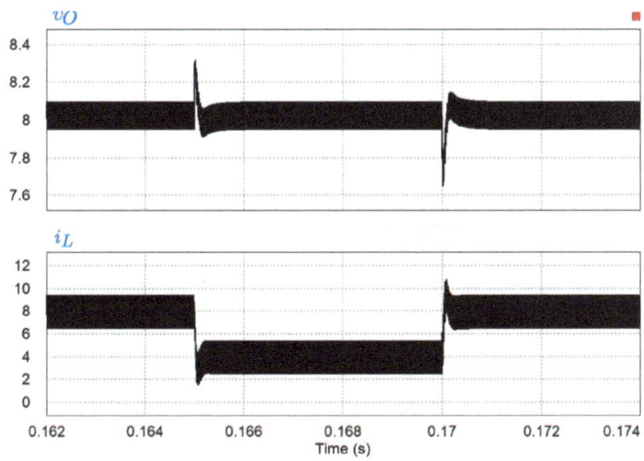


primary circuit and rectification circuit, a specific switch network will be created. Figure 4.46 illustrates the formation of the switch network, where three primary circuits and two rectification circuits are shown across the isolation boundary. An arbitrary pair of these circuits, one from the primary circuits and the other from the rectification circuits, could be merged to function as a switch network for a transformer-isolated buck-derived dc-to-dc converter.

The primary circuits are named as the full-bridge circuit, half-bridge circuit, and push-pull circuit. The term *full-bridge* stems from the structure of the circuit, which has a pair of vertical legs consisting of two active switches connected in series.

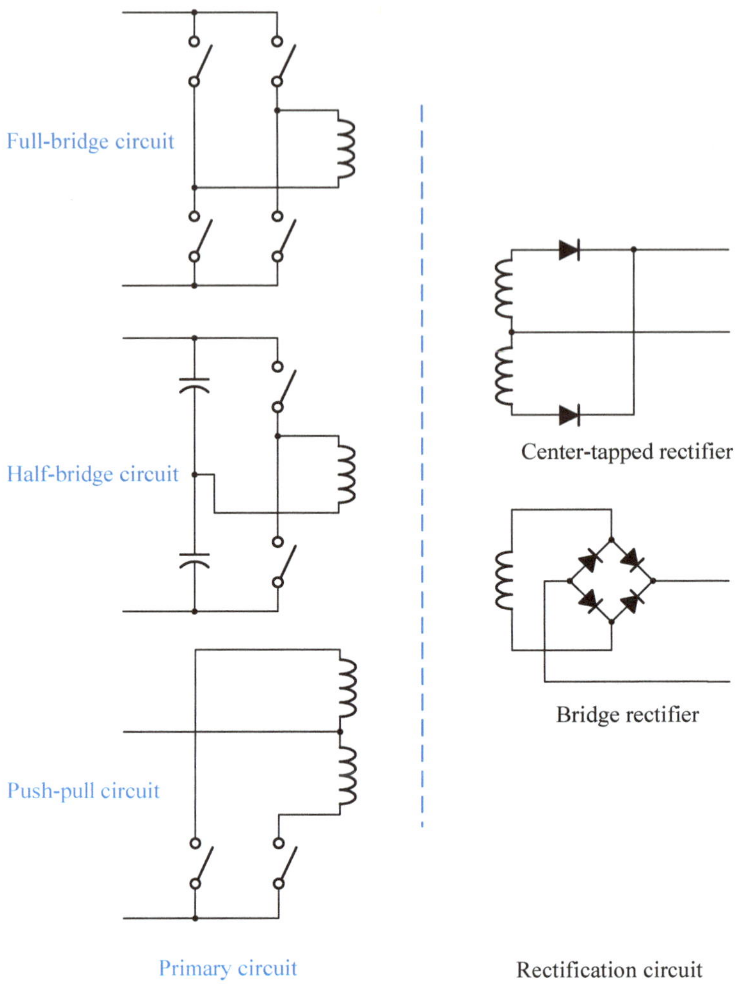


**Figure 4.46** Structure of switch network.

In the *half-bridge* circuit, only one active switch leg exists while the other leg is replaced with two capacitors connected in series. On the other hand, the *push-pull* circuit is named after its operational principle; one switch pushes the current into the transformer, while the other switch pulls the current out of the transformer.

The two rectifications circuits in Fig. 4.46 are called the center-tapped rectifier and bridge rectifier. It will be shown that these two rectifiers are equivalent in their function and thus are used interchangeably. A switch network could employ a two-, three-, or four-winding transformer, depending on the topologies of the primary and rectification circuits. The circuit models for multi-winding transformers were discussed in Section 4.7.2.

### 4.9.2 Full-Bridge Converter

As the first example of bridge-type isolated converters, Fig. 4.47 shows a circuit diagram of the full-bridge converter along with its major waveforms. The full-bridge circuit is employed as the primary circuit, as the name indicates. A three-winding transformer is used for the isolation transformer and the center-tapped rectifier is selected for the rectification circuit. The operation of the full-bridge converter is now discussed with aid of the circuit model of the three-winding transformer.

#### Operation with Ideal Transformer

The three-winding transformer in the middle of the switch network is assumed to be an ideal three-winding transformer with an $N_P : N_{S1} : N_{S2} = 1 : n : n$ turns ratio. This assumption simplifies the converter operation. The effects of non-ideal characteristics of a practical transformer will be discussed later. With the assumption of the $1 : n : n$ turns ratio, the ideal transformer equations are expressed as

$$v_P(t) : v_{S1}(t) : -v_{S2}(t) = 1 : n : n \tag{4.122}$$

$$i_P(t) = n\left(i_{D1}(t) - i_{D2}(t)\right) \tag{4.123}$$

according to the polarity/direction of the voltage/current waveforms shown in Fig. 4.47.

Four active switches in the primary circuit, $Q_1 - Q_4$, are grouped into two switch pairs, $\{Q_1\,Q_2\}$ and $\{Q_3\,Q_4\}$. Two individual switches in each switch pair are driven synchronously while the two switch pairs are operated complementary, as illustrated by the switch drive signals in Fig. 4.47(b). The time interval in which either the switches $\{Q_1\,Q_2\}$ or $\{Q_3\,Q_4\}$ are closed is referred to as the on-time period, $T_{on}$, while the time interval in which none of the switches is conducting is called the off-time period. *For the full-bride converter and other bridge-type converters, the duty ratio of the active switch is defined with respect to half the actual switching period of individual switches, referred to as the ripple period in literatures.* The definitions of the duty ratio $D$ and the ripple period $T_s$ for bridge-type PWM converters are illustrated with the switch drive signals in Fig. 4.47(b).

During an on-time period, the primary winding of the transformer is connected to the source, and the two secondary windings develop their terminal voltages according to the circuit equation of the ideal three-winding transformer. During an off-time period, the primary winding is isolated from the source, thereby making the terminal voltages of the three transformer windings identically zero.

The rectification circuit performs a full-wave rectification, thus providing the desired switch network voltage waveform $v_N$ for the low-pass filter circuit. The value of $v_N$ during on-time periods is given by the product of the input voltage and transformer turns ratio. The width of $v_N$ is determined by the duty ratio. Accordingly, the output voltage of the converter is given by

$$\boxed{V_O = \bar{v}_N(t) = nDV_S} \tag{4.124}$$

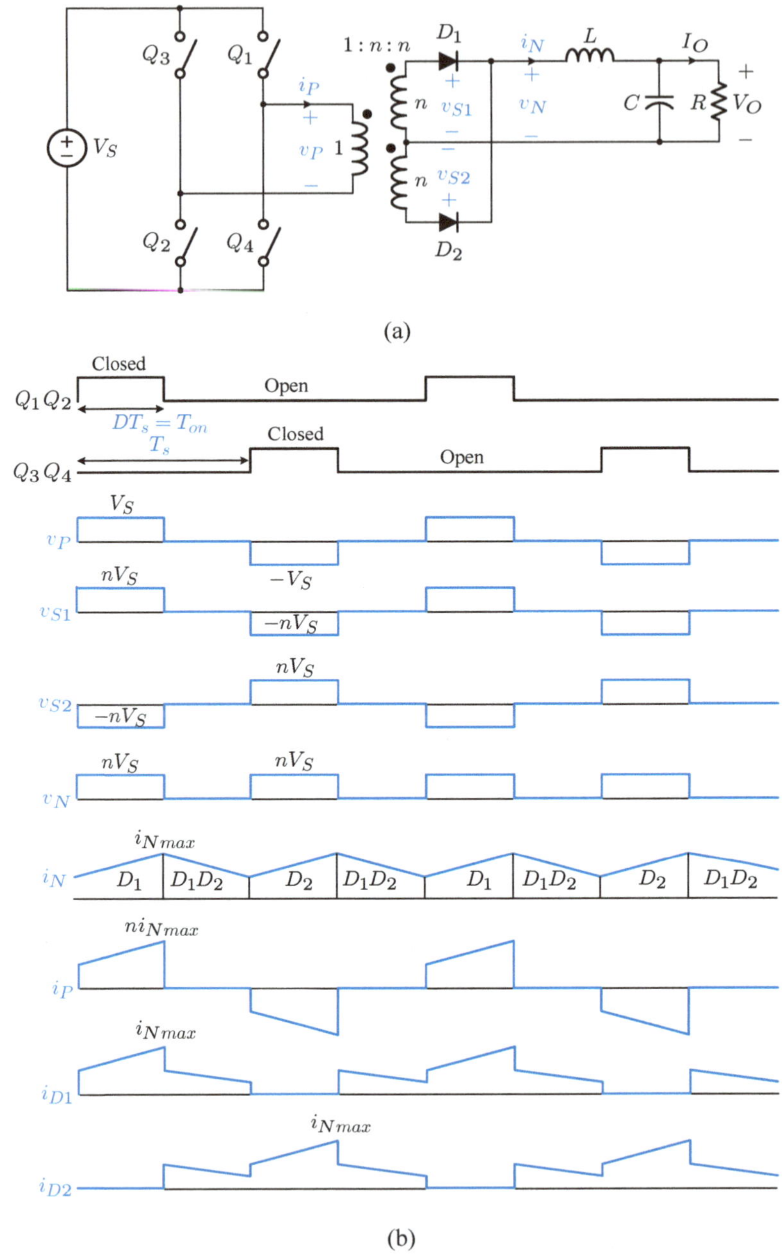


**Figure 4.47** Full-bridge converter. (a) Circuit diagram. (b) Major circuit waveforms.

During an on-time period, the output current of the switch network $i_N$ linearly increases with a slope $(nV_S - V_O)/L$. When $\{Q_1\,Q_2\}$ are closed in the on-time period, the primary current $i_P$ flows into the *dotted* end of the primary winding and the secondary current $i_N$ circulates through the upper diode $D_1$, while the lower diode $D_2$ is reverse-biased by the voltage of $2nV_S$. Conversely, when $\{Q_3\,Q_4\}$ are closed, $i_P$ reverses its direction and $i_N$ now circulates through $D_2$, while $D_1$ is reverse-biased. The specific diode(s), which is(are) conducting during the respective operational period, is(are) labeled on the waveform of $i_N$.

During an off-time period in which all four switches are open and the primary current $i_P$ is zero, the filter inductor current $i_N$ turns on $D_1$ and $D_2$ simultaneously to create a freewheeling path. The transformer winding voltages are zero during the off-time period. Accordingly, $i_N$ ramps down with a slope $-V_O/L$. The declining freewheeling current $i_N$ is equally divided into $i_{D1}$ and $i_{D2}$. Because $i_{D1}$ comes out of the *dotted* terminal and $i_{D2}$ enters the *dotted* terminal, while $i_P$ remains zero, the current equation for the ideal three-winding transformer, given by (4.123), is satisfied. The current waveforms are shown in Fig. 4.47(b).

The switch network produces the desired waveforms for $v_N$ and $i_N$, and the ideal three-winding transformer satisfies the circuit equations defined by (4.89) and (4.92). The average value of $i_N$ becomes the load current

$$I_O = \bar{i}_N(t) = \frac{V_O}{R} \tag{4.125}$$

Two underlying assumptions are employed in the previous analysis. First, CCM operation is assumed, which is true only when the condition $\Delta i_N/2 < I_O$ is satisfied. Yet, Fig. 4.47(b) could readily be modified to yield the circuit waveforms for DCM operation. In fact, the DCM operation of the buck converter, covered in Section 3.5.3, could be extended to the full-bridge converter with minor modifications. The second assumption is the ideal characteristics of the three-winding transformer. When a practical transformer is employed, the converter operation will be altered due to the presence of the magnetizing inductance. The operation with a magnetizing inductance is covered in the next section.

### Effects of Magnetizing Inductance

Figure 4.48(a) shows the circuit model of a practical three-winding transformer. The practical transformer is modeled as a parallel connection of the magnetizing inductance $L_m$ and ideal three-winding transformer. The primary winding current of the ideal transformer is renamed as $i'_P$. The current $i_P$ now denotes the input current to the practical transformer. It is easily deduced that the filter inductor current $i_N$ remains the same because the switch network voltage $v_N$ is the same as the previous case of ideal transformer. On the other hand, the other current waveforms will be altered due to the magnetizing current $i_m$ running through $L_m$.

The voltage waveform across the magnetizing inductance $L_m$ naturally meets the volt-sec balance condition and the magnetizing current $i_m$ thus is symmetrical.

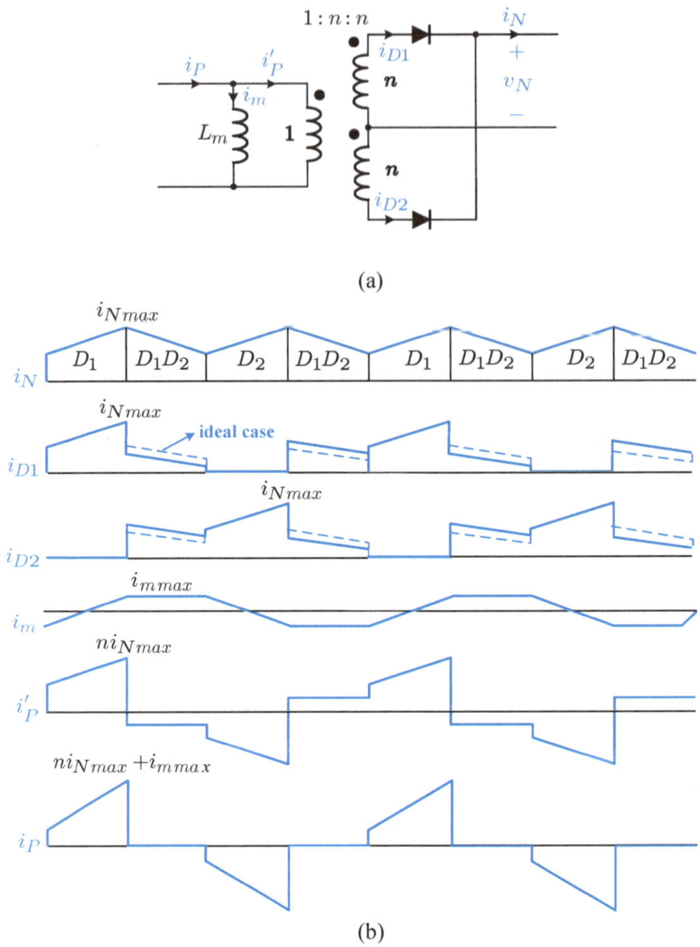


**Figure 4.48** Switch network with practical transformer. (a) Practical three-winding transformer. (b) Major waveforms.

During an on-time period, $i_m$ increases or decreases depending on the polarity of the voltage across $L_m$. The magnetizing current, commutating via the closed primary-side switches, runs through $L_m$ and does not enter the primary winding of the ideal transformer. Accordingly, during the on-time period, the winding currents of the ideal transformer, $i'_P$, $i_{D1}$, and $i_{D2}$, are identical to those of the previous case.

During an off-time period in which all the active switches are open, the magnetizing current $i_m$ now circulates through the primary winding of the ideal transformer. In this period, $i_m$ remains constant because the voltage across the magnetizing inductance is clamped at zero. With the presence of $i_m$ in the primary winding, the secondary currents, $i_{D1}$ and $i_{D2}$, are altered to meet the current equation of the ideal three-winding transformer

$$\boxed{i'_p(t) = -i_m(t) = n(i_{D1}(t) - i_{D2}(t))} \tag{4.126}$$

while maintaining their sum unchanged

$$\boxed{i_N(t) = i_{D1}(t) + i_{D2}(t)} \tag{4.127}$$

Figure 4.48(b) illustrates the current waveforms under this situation. The diode currents, $i_{D1}$ and $i_{D2}$, are *unevenly* distributed during the off-time period to satisfy the equations (4.126) and (4.127) at the presence of the magnetizing current. The dashed lines show the original $i_{D1}$ and $i_{D2}$ waveforms at the absence of the magnetizing current. The current waveform $i'_P$ is constructed using the fact $i'_P = ni_{D1}$ or $i'_P = ni_{D2}$ for the on-time period, and $i'_P = -i_m$ for the off-time period. The $i_P$ is obtained using the fact $i_P = i'_P + i_m$ for the on-time period and $i_P = 0$ for the off-time period.

When the magnetizing inductance $L_m$ approaches infinity, $i_m$ reduces to zero and the current waveforms return to those of the previous ideal transformer case. For most transformer fabrications, the magnetizing inductance is maximized within practical limits to enhance efficiency and performance of transformers. Accordingly, the magnetizing inductance is sufficiently large and the magnetizing current is negligibly small. For this reason, it is common practice to assume the ideal characteristics for the transformers employed in bridge-type isolated dc-to-dc converters.

Another justification for not considering the magnetizing inductance is that the input voltage of the switch network is symmetric and automatically meets the volt-sec balance condition on $L_m$. If this is not the case, the magnetizing inductance $L_m$, however large it might be, will be eventually saturated to cause a fatal failure of the converter. In theory, the transformer saturation will not happen to the bridge-type isolated converters. The assumption of the ideal transformer will be implicitly used in the upcoming discussions about other bridge-type isolated converters.

### Steady-state operation of full-bridge converter — EXAMPLE 4.24

This example shows the steady-state analysis and circuit waveforms of a full-bridge converter. The operational conditions and circuit parameters of the converter are $V_S = 64\,\text{V}$, $n = 0.5$, $L = 40\,\mu\text{H}$, $C = 400\,\mu\text{F}$, $R = 1\,\Omega$, $T_s = 20\,\mu\text{s}$, and $D = 0.25$. For the given operational conditions, the steady-state values for major circuit variables are calculated in Table 4.33, while the corresponding PSpice® simulations are shown in Fig. 4.49.

### PSIM® Transient response of full-bridge converter — EXAMPLE 4.25

Table 4.34 shows a PSIM model for a full-bridge converter. The model is customized for closed-loop simulations with an optocoupler-isolated voltage mode control. The PWM block is slightly modified to reduce the component count so that the model is suited for the license-free copy of PSIM® software.

Table 4.33 Steady-State Analysis of Full-Bridge Converter

| Circuit variable | Expression |
|---|---|
| $V_O$ | $nDV_S = 0.5 \cdot 0.25 \cdot 64 = 8\ \text{V}$ |
| $v_{N\,max}$ | $nV_S = 32\ \text{V}$ |
| $\Delta i_N$ | $\frac{nV_S - V_O}{L} D T_s = \frac{0.5 \cdot 64 - 8}{40 \times 10^{-6}} 0.25 \cdot 20 \times 10^{-6} = 3\ \text{A}$ |
| $i_{N\,max}$ | $\frac{V_O}{R} + \frac{1}{2}\Delta i_N = \frac{8}{1} + \frac{1}{2}3.0 = 9.5\ \text{A}$ |
| $i_{N\,min}$ | $\frac{V_O}{R} - \frac{1}{2}\Delta i_N = \frac{8}{1} - \frac{1}{2}3.0 = 6.5\ \text{A}$ |
| $i_{D1\,max}$ | $i_{Nmax} = 9.5\ \text{A}$ |
| $i_{D2\,max}$ | $i_{Nmax} = 9.5\ \text{A}$ |

The feedback controller employs voltage mode control, which is well suited for the buck and buck-derived converters. The optocoupler-isolated three-pole two-zero compensation is used for the stability and performance of the converter. Details of feedback circuit will be covered in a later volume of the book series.

The PSIM model is programmed for the step-load response of the converter. The transient response of the output voltage $v_O$ and inductor current $i_L$ in response to 50% changes in the load current are shown in Table 4.34.

### 4.9.3 Half-Bridge Converter

The half-bridge converter can be configured by combining the half-bridge circuit with either the bridge rectifier or center-tapped rectifier. The bridge rectifier is used for the half-bridge converter in Fig. 4.50(a). The two capacitors, $C_1$ and $C_2$ in Fig. 4.50(a), function as a voltage divider and each capacitor develops a dc voltage of $V_S/2$ at its terminals.

The switch drive signals and voltage waveforms across the primary and secondary windings of the two-winding transformer are shown in Fig. 4.50(b). When $Q_1$ is turned-on, $v_P$ and $v_S$ both become positive. Conversely, when $Q_2$ is closed, $v_P$ and $v_S$ are negative. During an off-time period in which both $Q_1$ and $Q_2$ are open, the

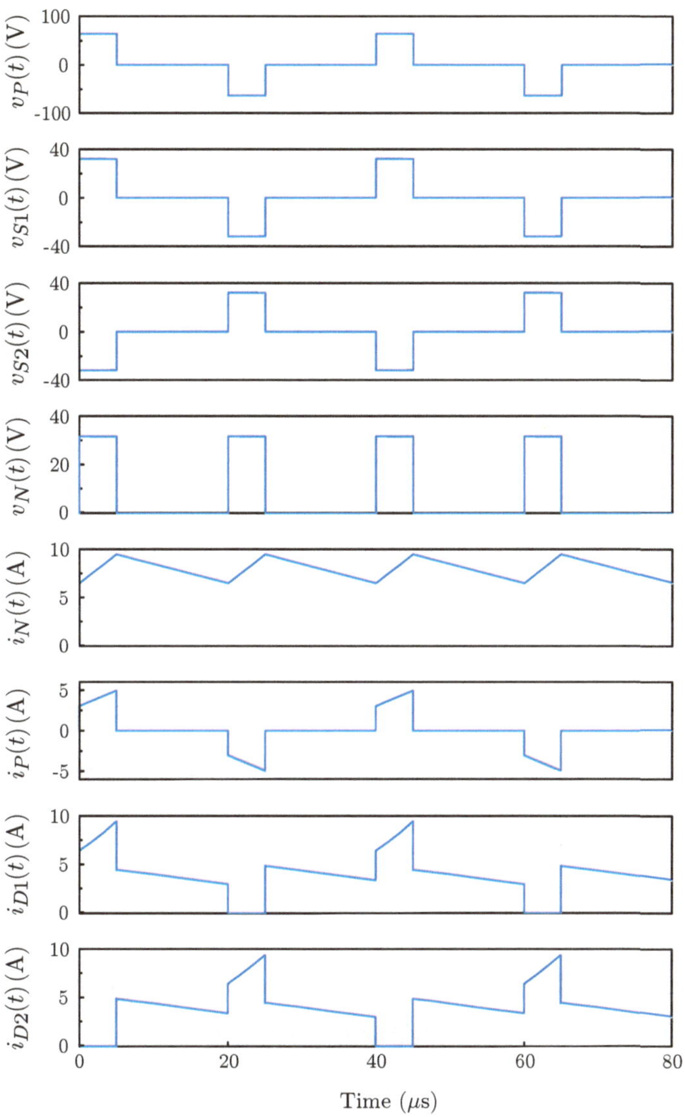


Figure 4.49 Circuit waveforms of full-bridge converter.

transformer is functionally isolated from the switch network, thereby yielding the conditions $v_P = v_S = 0$ and $i_P = i_S = 0$.

**Table 4.34** Transient Response of Full-Bridge Converter

PSIM model: Full_Bridge_Converter _01

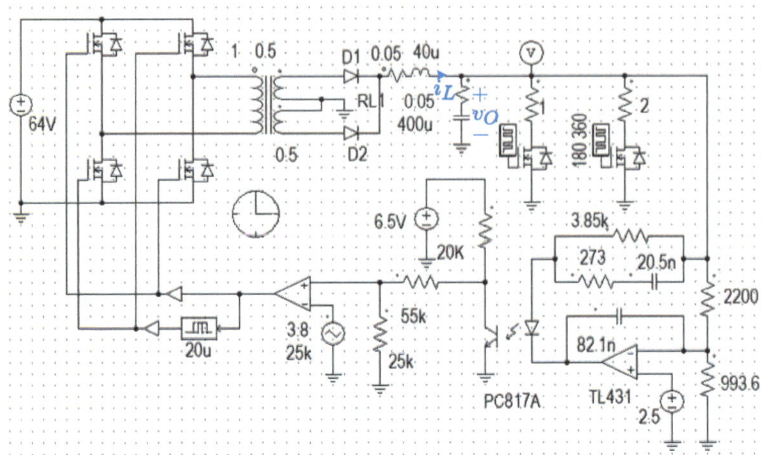


Simulation outcomes

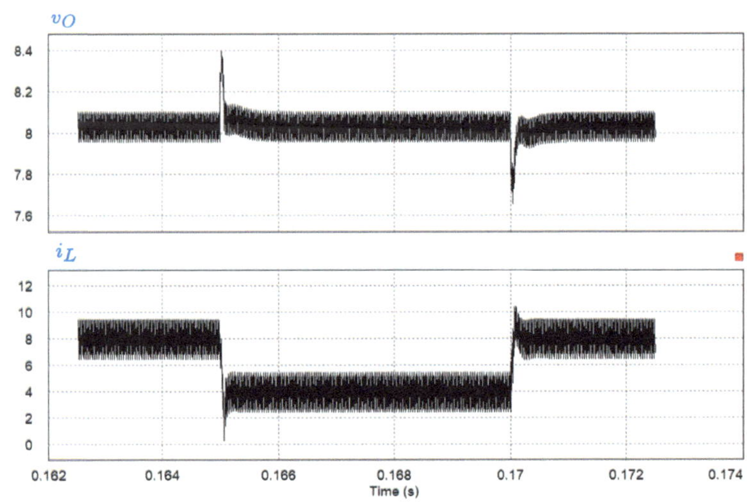


As shown in Fig. 4.50(b), the bridge rectifier circuit generates $v_N$ that alternates between $nV_S/2$ and zero. The output voltage of the converter then becomes

$$V_O = \bar{v}_N(t) = \frac{nDV_S}{2} \tag{4.128}$$

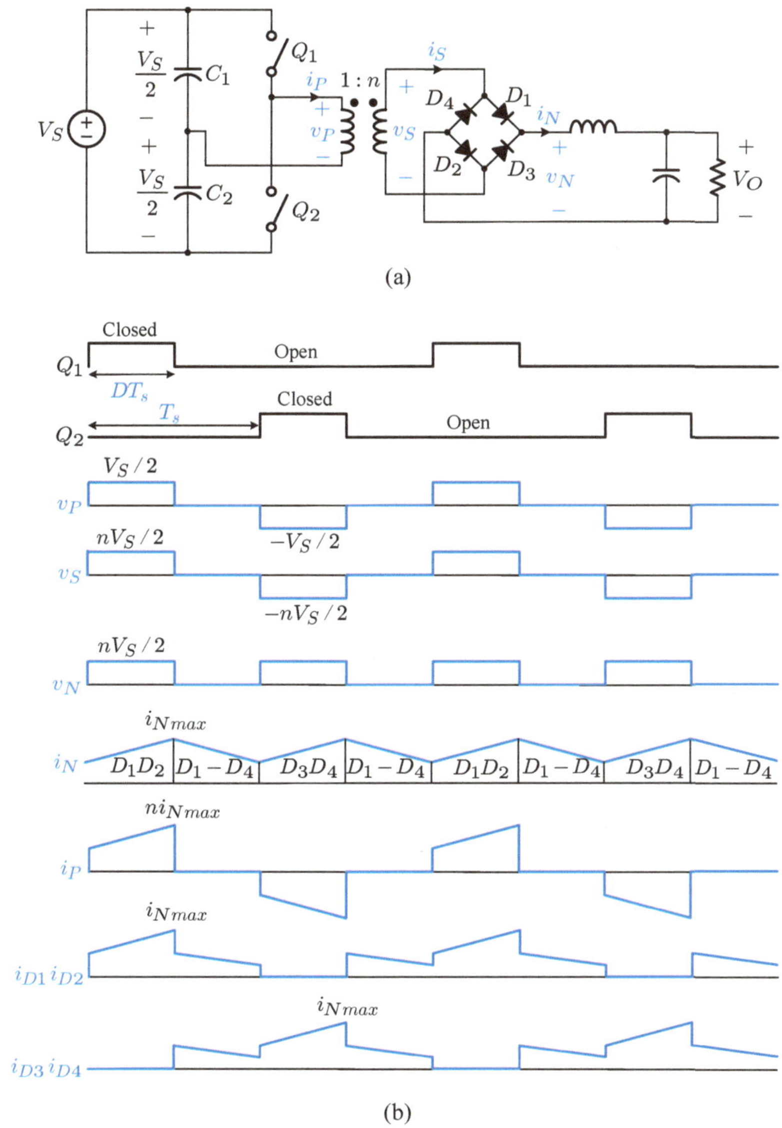


**Figure 4.50** Half-bridge converter. (a) Circuit diagram. (b) Major circuit waveforms.

The shape of $i_N$ is the same as the previous full-bridge converter case, yet the commutation of the diodes is different. When $Q_1$ is closed, $D_1$ and $D_2$ carry $i_N$. Conversely, when $Q_2$ is closed, $D_3$ and $D_4$ conduct. During an off-time period in which the transformer is functionally separated from the other part of the circuit, $i_N$ turns on all the four diodes simultaneously and freewheels through them. It is apparent that the converter could employ a center-tapped rectifier instead of the bridge

Table 4.35 PSIM Model for Half-Bridge Converter

PSIM model: Half_Bridge_Converter_01

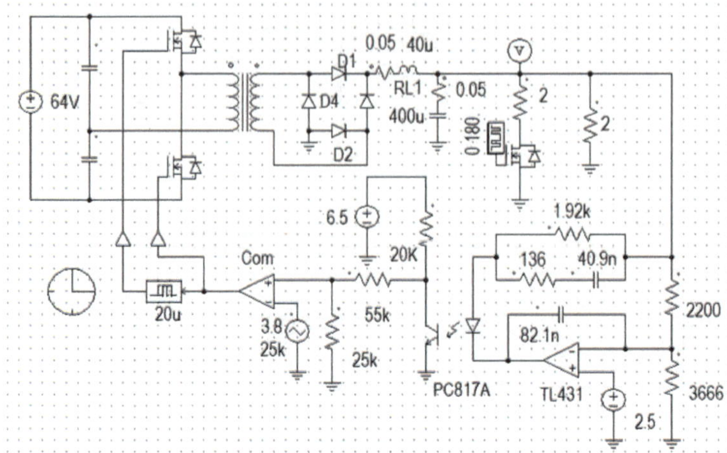


rectifier. When a center tapped rectifier is used, the current waveforms are identical to those of Fig. 4.47.

**EXAMPLE 4.26 PSIM® Model for half-bridge converter**

The PSIM model for a half-bridge converter is shown in Table 4.35. The converter employs voltage mode control with an optocoupler-isolated feedback circuit. Main features of the model are the same as the previous case of the full-bridge converter.

### 4.9.4 Push-Pull Converter

Another variation of the bridge-type converter is the push-pull converter shown in Fig. 4.51. The converter uses two active switches and one four-winding transformer. While a center-tapped rectifier is used in Fig. 4.51(a), the bridge rectifier could be employed as an alternative, provided that the four-winding transformer is replaced with a three-winding transformer.

When the switch $Q_1$ is closed, $v_{P1}$ and $v_{S1}$ become positive while $v_{P2}$ and $v_{S2}$ are negative. Conversely, when $Q_2$ is turned-on, $v_{P2}$ and $v_{S2}$ now become positive while the other two windings develop a negative voltage. The terminal voltages of the four windings are shown in Fig. 4.51(b). The rectification circuit performs a full-wave rectification, thus yielding the desired $v_N$ waveform. The current waveforms associated with the four-winding transformer are also shown in Fig. 4.51(b). The

**Table 4.36** PSIM Model for Push-Pull Converter

PSIM model: Push_Pull_Converter_01

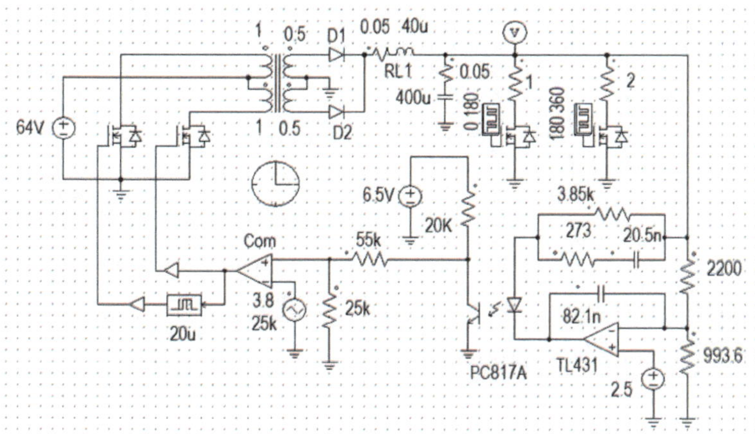


voltage and current waveforms in Fig. 4.51(b) satisfy the circuit equations for the four-winding transformer.

The push-pull converter is functionally the same as the full-bridge converter. In fact, the push-pull converter replicates the operation of the full-bridge converter using two active switches and a transformer with two primary windings.

**PSIM® Model for push-pull converter** **EXAMPLE 4.27**

Table 4.36 shows a PSIM model for a push pull converter. This model can be used to simulate the steady-state and transient responses of the converter with an optocoupler-isolated voltage mode control.

## 4.10 CHAPTER SUMMARY

In this chapter, we studied important families of dc-to-dc converters, focusing on their topological origin, operational principles, and circuit waveforms. We discussed the evolution of the buck converter into two other basic converters. The boost converter is viewed as a *reverse* buck converter, while the buck/boost converter is derived from the cascaded connection of the buck converter and boost converter. As another composite PWM converter, Ćuk converter is created by combining the upstream boost converter with downstream buck converter. The steady-state and transient responses of these non-isolated converters are theoretically analyzed and computationally simulated using PSpice® and PSIM® programs. This chapter also covered the recently-emerging

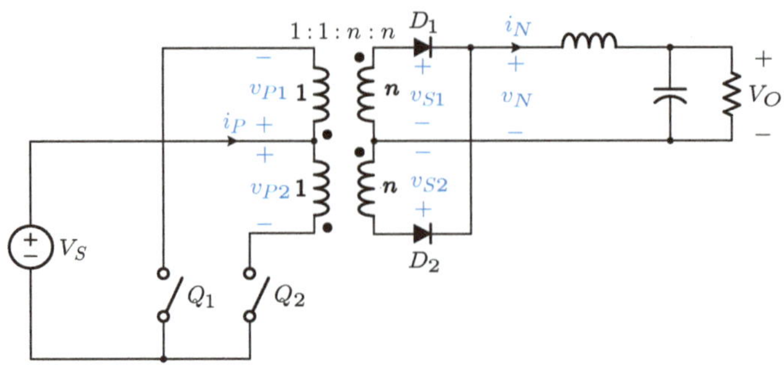


(a)

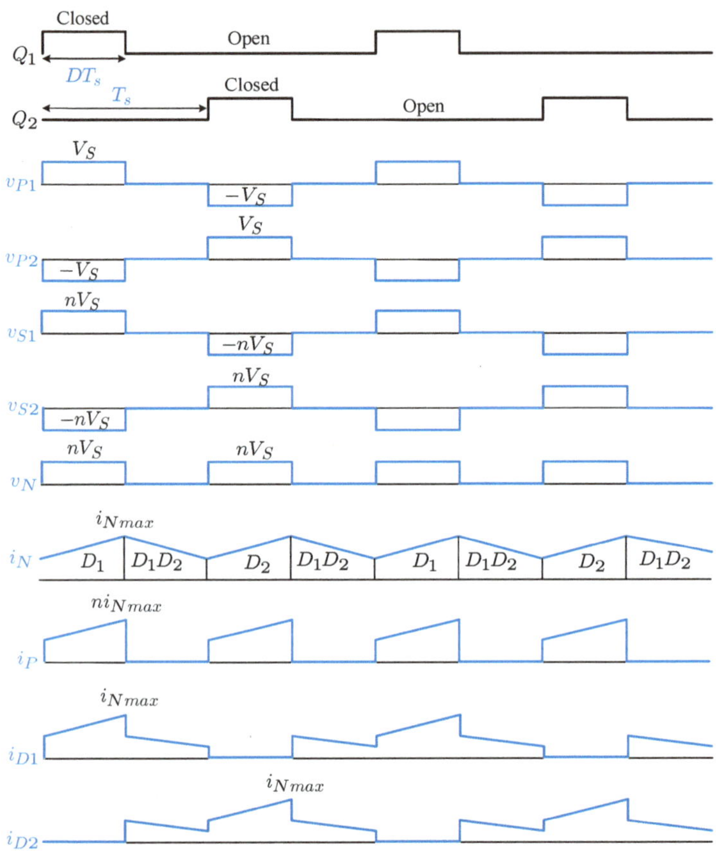


(b)

**Figure 4.51** Push-pull converter. (a) Circuit diagram. (b) Major circuit waveforms.

bidirectional power conversion. Operational details of the bidirectional boost converter are treated using the photovoltaic application example.

Addition of a transformer to the three basic converters creates a number of new isolated dc-to-dc converters. The simplest among them is the flyback converter whose operation closely mimics the buck/boost converter while providing an input-output isolation.

Many isolated dc-to-dc converters are derived from the buck converter by replacing the single-pole double-throw (SPDT) switch with a switch network. We first studied the forward converters, which employs a variety of auxiliary circuits to reset the transformer in the switch network. We discussed the topology and operation of the tertiary-winding reset forward converter, Zener diode reset forward converter, RCD clamp reset forward converter, and two-switch forward converter.

This chapter investigated another class of buck-derived isolated converters, called bridge-type converters. Depending on the structure of the switch network, these converters are classified as the full-bridge converter, half-bridge converter, and push-pull converter. The operation of these bridge-type buck-derived converters has shown to be equivalent to that of the buck converter. The operation of the bridge-type converters are summarized in a simple and systematic manner.

## REFERENCES

1. R. W. Erickson and D. Maksimović, *Fundamentals of Power Electronics*, 3rd Ed., Springer Nature, 2020.
2. R. Severns and G. Bloom, *Modern Dc-to-Dc Switchmode Power Converter Circuits*, Van Nostrand Reinhold, 1985.
3. M. K. Kazimierczuk, *Pulse-width Modulated DC-DC Power Converters*, John Wiley & Sons, Hoboken, NJ, 2008.

## PROBLEMS

**4.1** Consider the two converters shown in Fig. P4.1 and answer the questions.

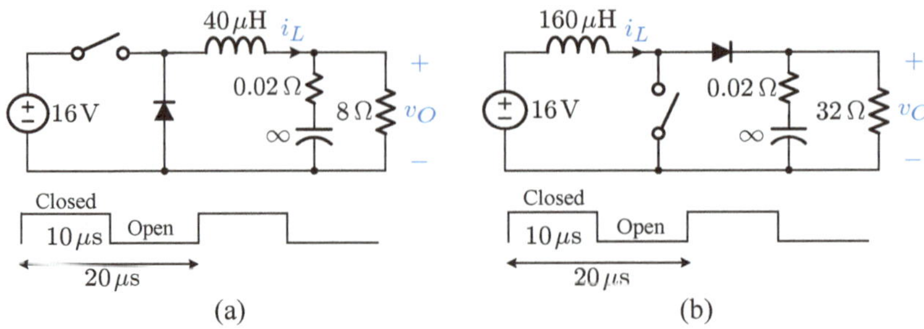


Fig. P4.1

a) For Converter (a), sketch the steady-state waveforms of the inductor current $i_L$ and output voltage ripple, $\tilde{v}_O = v_O - V_O$, for the two switching periods. Show the maximum and minimum values of the waveforms.

b) Repeat a) for Converter (b).

**4.2** Consider the two circuits shown in Fig. P4.2 and answer the questions.

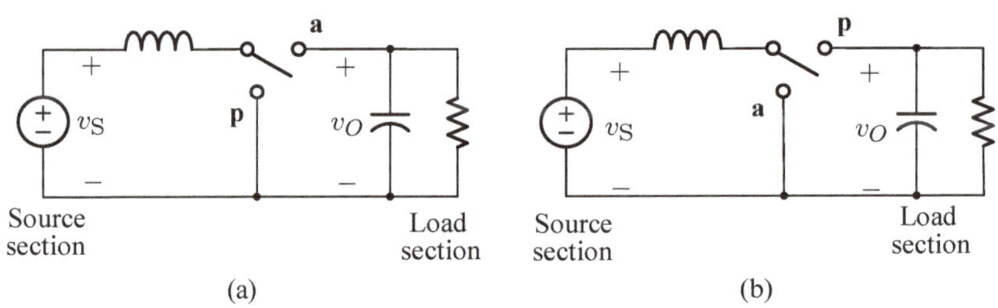


Fig. P4.2

The circuits operate with a constant-frequency trailing-edge modulation scheme by changing the on-time period of the MOSFET in the SPDT switch.

a) Circuit (a): Is the energy transfer from the source section to the load section is possible? If not, describe what will happen to the circuit. If so, justify the circuit operation.

b) Circuit (b): Is the energy transfer from the source section to the load section is possible under certain condition? *If so, what condition is needed for the energy transfer?* If not, describe what will happen to the circuit.

**4.3*** In the two converter circuits shown in Fig. P4.3, $Q_1$ and $Q_2$ are ideal *bidirectional* switches which deliver the current in the forward and backward directions. The switch drive signals shown in Fig. P4.3 are commonly applied to Converter (a) and Converter (b).

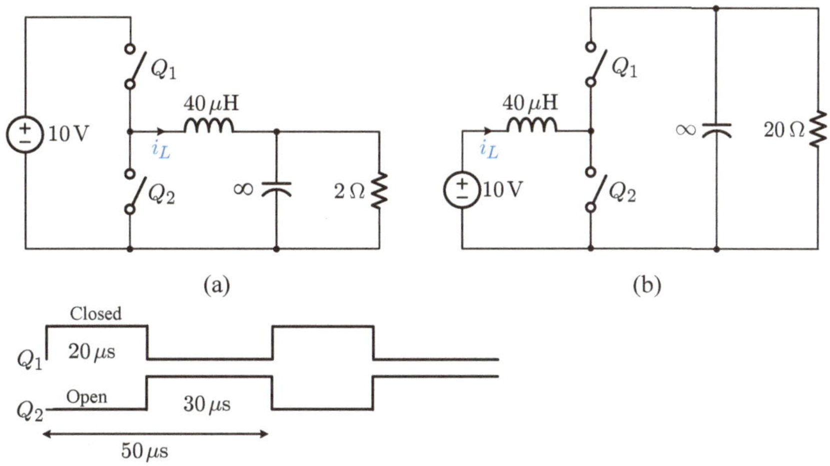


Fig. P4.3

a) For Converter (a), sketch $i_L$ for the two operational periods. Label the maximum and minimum values of the waveform.

b) Repeat a) for Converter (b).

**4.4** The maximum voltage that a switching device should block during the operation is called the voltage stress. Similarly, the maximum current that a switching device should carry is called the current stress. Referring to the boost converter shown in Fig. P4.4, answer the questions.

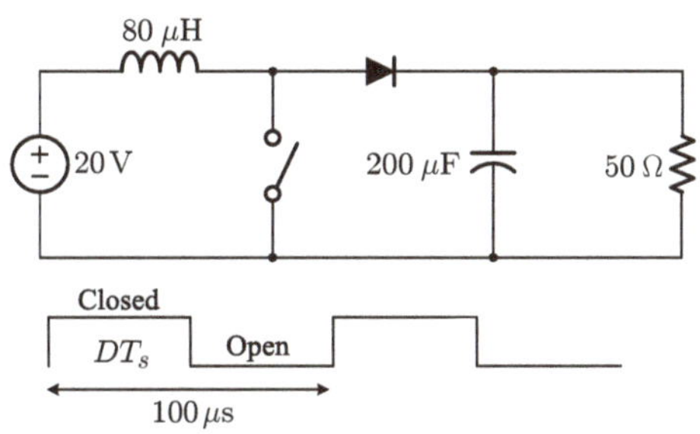


Fig. P4.4

a) Find the voltage stress and current stress of the MOSFET and diode when $D = 0.6$.

b) Repeat a) for $D = 0.9$.

**4.5** Referring to Fig. P4.5, answer the following questions.

a) For Converter (a), determine the following items:
   i) operational mode, CCM or DCM   ii) average inductor current $I_L$

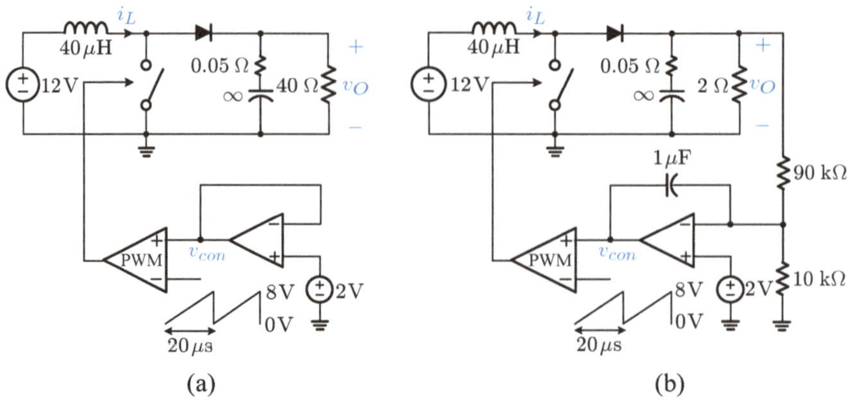


Fig. P4.5

iii) inductor current ripple $\Delta i_L$ iv) output voltage ripple $\Delta v_O$

v) average control voltage $\bar{v}_{con}$

b) Repeat a) for Converter (b).

**4.6** In the closed-loop controlled boost converter in Fig. P4.6, assume that the voltage feedback circuit is well designed for stability and good transient performance.

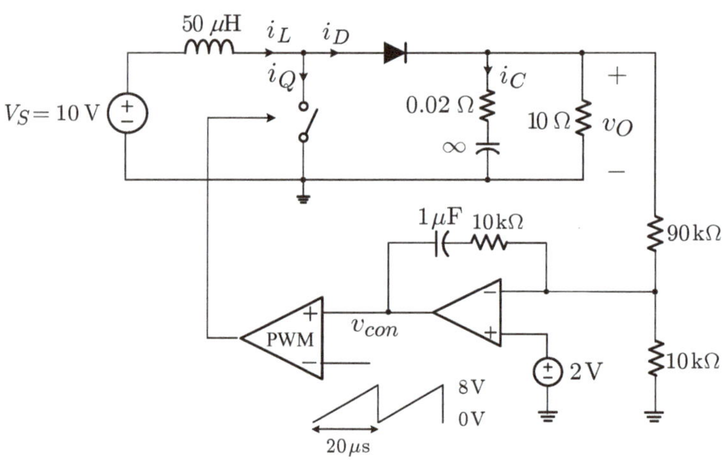


Fig. P4.6

a) Assume $v_S = 10$ V and sketch the steady-state waveform of $i_L, i_Q, i_D, i_C$, and $\bar{v}_{con}$ for the two operational periods. Show the maximum, minimum, and average value of the waveforms.

b) Assume $v_S = 10$ V and sketch the steady-state waveform of $\tilde{v}_O = v_O - \bar{v}_O$ for the two operational periods. Show the average value and shape/magnitude of the ripple component of the waveform.

c) Now assume the following step changes are occurred in the input voltage of the converter. Sketch the transient waveform of $v_O$, $i_L$ and $v_{con}$. Show all the characteristic features of the transient responses.

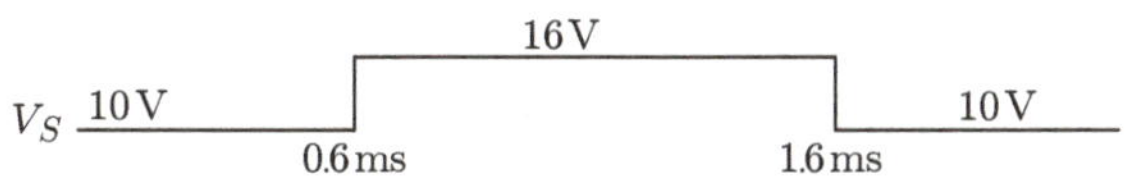


**4.7** Consider the two closed-loop controlled converters shown in Fig. P4.7 and answer the questions.

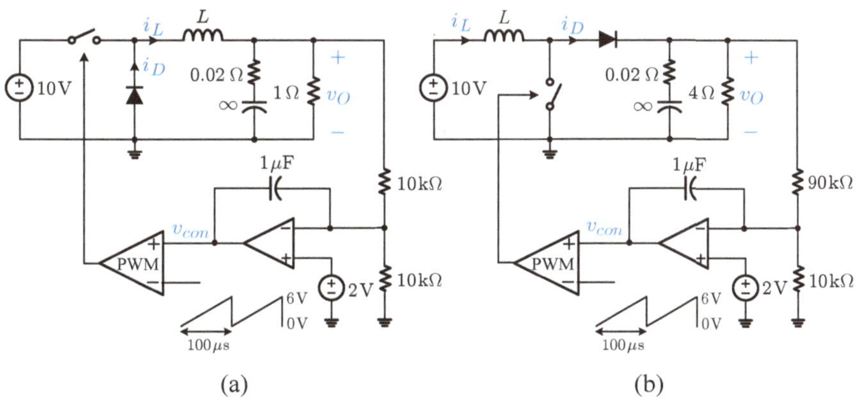


Fig. P4.7

a) For Converter (a), determine the value of the inductance $L$ so that the converter operates at the CCM/DCM borderline. You are not allowed to use the formula for the $L_{crit}$ expression.

b) Assume that Converter (a) operates with the inductance $L$ found in a). Under this condition, sketch the inductor current $i_L$, diode current $i_D$, control voltage $v_{con}$, and the ripple component of the output voltage, $\tilde{v}_O = v_O - V_O$, for the two operational periods. Show the maximum and minimum values.

c) Repeat a) and b) for Converter (b).

**4.8** Figure P4.8 shows a closed-loop controlled boost converter. Answer the questions.

a) Determine the value of the inductance $L$ so that the converter operates at CCM/DCM borderline with $V_S = 12$ V.

b) Sketch the voltage gain $V_O/V_S$ curve with respect to the duty ratio $D$, when $V_S$ varies between $6 \text{ V} < V_S < 30$ V. Use the inductance $L$ found in a).

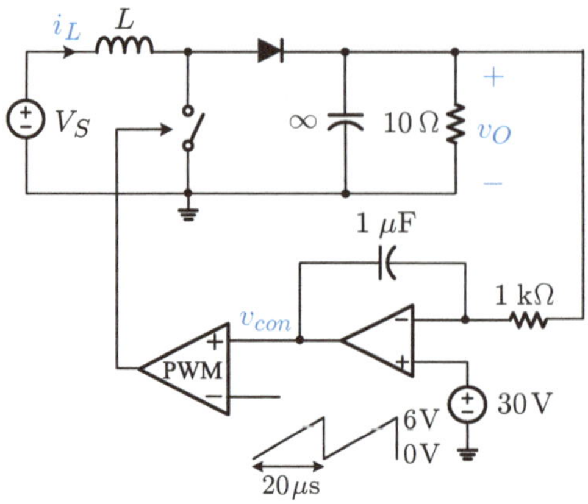


Fig. P4.8

c) Now assume that $V_S = 6$ V. Determine the operational mode of the converter. Sketch the steady-state waveforms of $i_L$ and $v_{con}$. Use the inductance $L$ found in a).

**4.9** Assume all the circuit components are ideal in the converter circuits shown in Fig. P4.9. For each converter circuit, calculate the average value for the control voltage $\bar{v}_{con}$ and output voltage $V_O$ in CCM operation.

**4.10*** Consider the four converter circuits shown in Fig. P4.10. For each converter circuit, calculate the reference voltage $V_{ref}$ that is required to produce the output voltage $V_O = 20$ V in CCM operation.

**4.11*** Figure P4.11 shows three converter circuits. Assume the same switch drive signal for the three converters.

a) Evaluate the average value of the three diode currents: $\bar{i}_{D1}$, $\bar{i}_{D2}$, and $\bar{i}_{D3}$.

b) Evaluate the average values of the three switch currents: $\bar{i}_{Q1}$, $\bar{i}_{Q2}$, and $\bar{i}_{Q3}$.

c) Find the voltage stress of the three switches: $v_{Q1\,max}$, $v_{Q2\,max}$, and $v_{Q3\,max}$.

**4.12*** Consider the four converter circuits shown in Fig. P4.12. For each converter circuit, find the average values of the control voltage $\bar{v}_{con}$ and output voltage $\bar{v}_O$.

**4.13** Consider the two closed-loop controlled converters shown in Fig. P4.13 and answer the questions.

a) For Converter (a), perform the following analyses.

i) Find the voltage stress of switch $v_{Q_{max}}$, average diode current $\bar{i}_D$, and average control voltage, $\bar{v}_{con}$.

ii) Sketch the steady-state waveform of output voltage ripple, $\tilde{v}_O = v_O - \bar{v}_O$, for two operational periods.

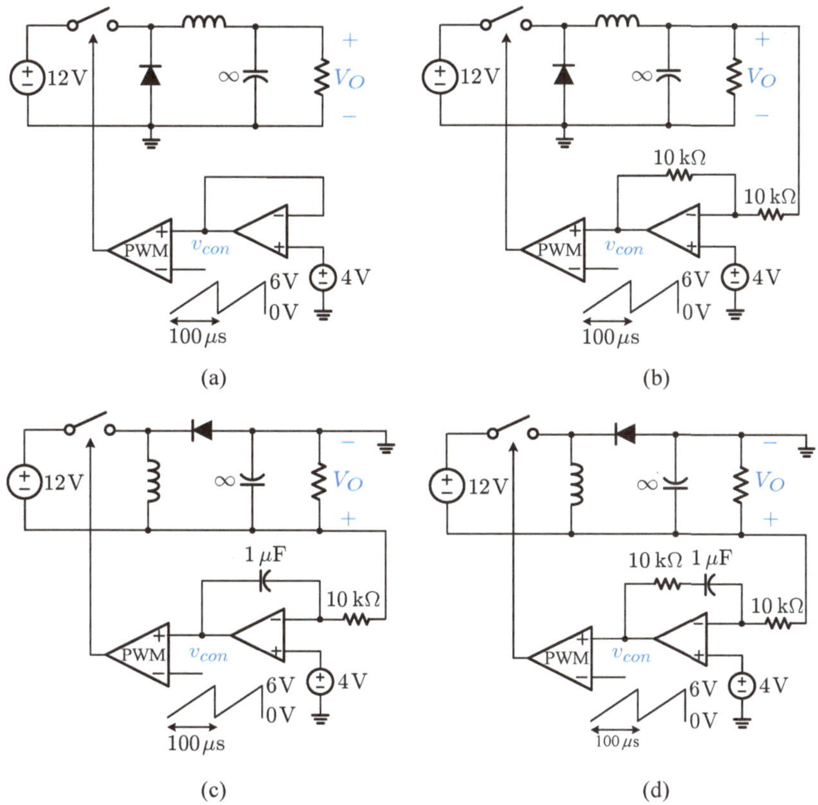


Fig. P4.9

b) Repeat a) for Converter (b).

**4.14*** Consider the non-inverting buck/boost converter shown in Fig. P4.14 and answer the questions. The two synchronized switches, $Q_1$ and $Q_2$, are derived with the common switch derive signal generated from the PWM block shown in the circuit.

a) Evaluate the average value of $v_O$ and $i_O$.

b) Evaluate the average value of $i_{D1}$, $i_L$, and $i_{Q2}$.

c) Sketch the waveforms of $i_L$, $i_{D1}$, and $i_{Q2}$ for the two operational periods. Show the maximum, minimum, and average values of the waveforms.

**4.15** Referring to the two converter circuits in Fig. P4.15, answer the following questions. The converters operate with $D = 0.6$ at the switching frequency $f_s = 50$ kHz.

a) Assume that all the circuit components are ideal. For each converter, evaluate the voltage stress of the active switch, $v_{Q_{max}}$, and the average value of the inductor current, $\bar{i}_L$.

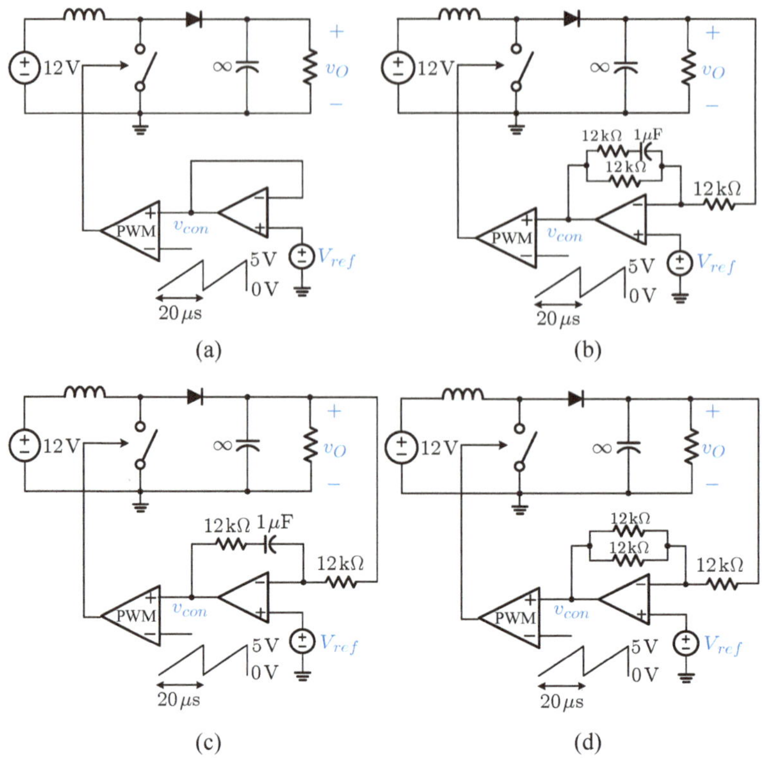


Fig. P4.10

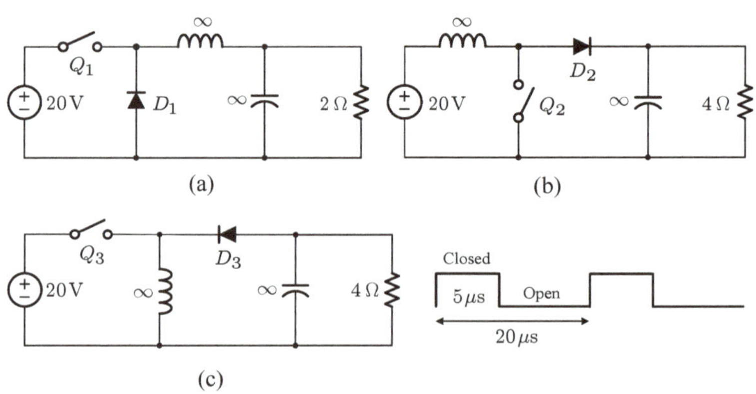


Fig. P4.11

b) For Converter (a), assume the following practical devices for the semiconductor switches:
$Q$: MOSFET switch with the channel resistance of $R_{DS}(on) = 0.5\ \Omega$
$D$: Schottky diode with the turn-on voltage of $V_D(on) = 0.5$ V

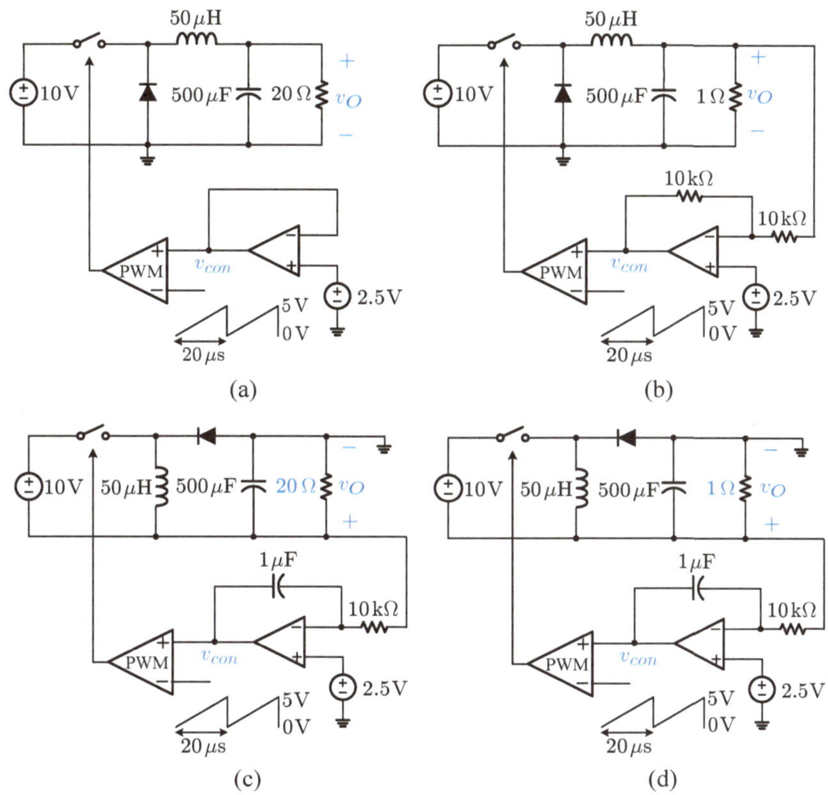


Fig. P4.12

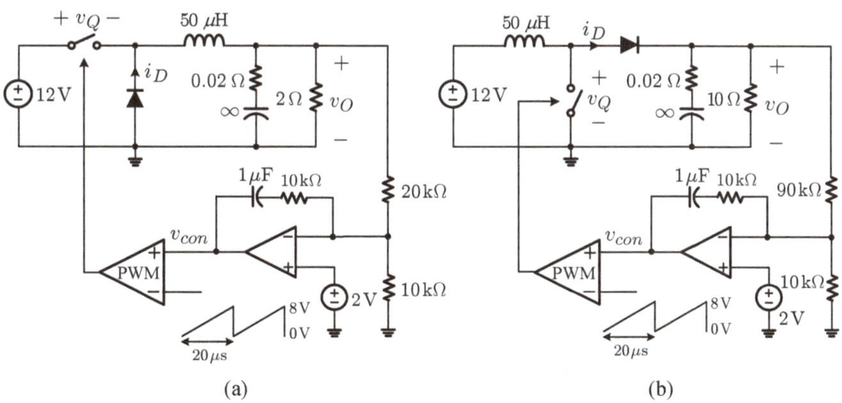


Fig. P4.13

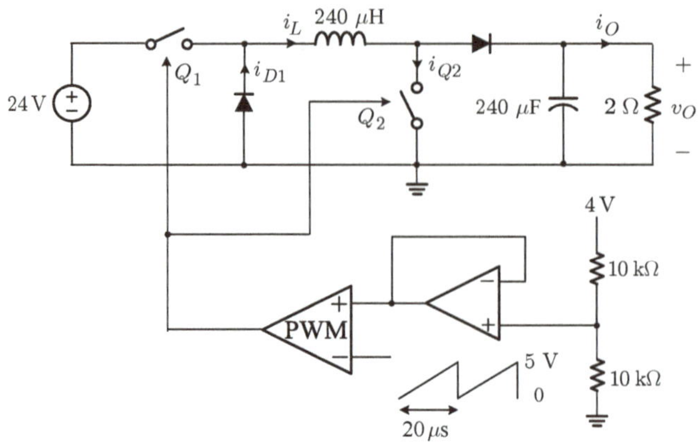


Fig. P4.14

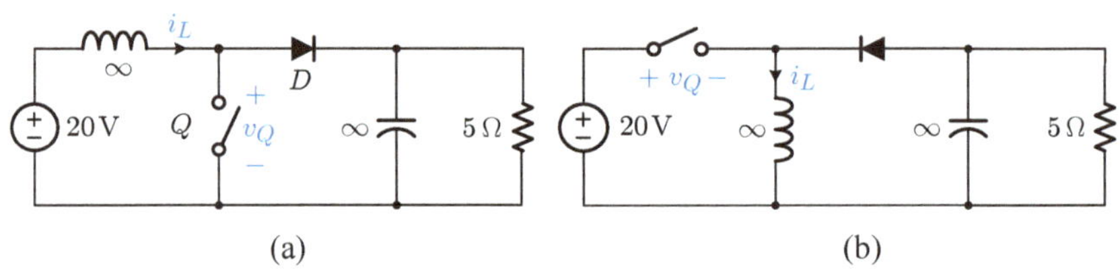


Fig. P4.15

The other circuit components are still assumed ideal. Evaluate the efficiency of Converter (a).

**4.16** The three converters in Fig. P4.16 generate the same inductor current waveform shown below. For each converter, evaluate the output voltage $V_O$, load resistance $R$, and inductance.

**4.17** Consider the two converter circuits shown in Fig. P4.17. The two converters generate the same inductor current waveform. For each converter, evaluate the output voltage $V_O$, load current $I_O$, inductance $L$, and load resistance $R$. You are not allowed to use the DCM gain formula in this problem.

**4.18** In the dc-to-dc conversion system shown in Fig. P4.18, assume that all the active switches are driven from the same switch drive signal. Also, assume CCM operation for all three converters. Evaluate the average current running through each semiconductor switch: $\bar{i}_{Q1}$, $\bar{i}_{D1}$, $\bar{i}_{Q2}$, $\bar{i}_{D2}$, $\bar{i}_{Q3}$, and $\bar{i}_{D3}$.

**4.19** A dc-to-dc power conversion system is shown in Fig. P4.19. All the converters operate in CCM with a duty ratio $D = 0.4$. Evaluate the average values of the inductor currents: $\bar{i}_{L1}$, $\bar{i}_{L2}$, and $\bar{i}_{L3}$.

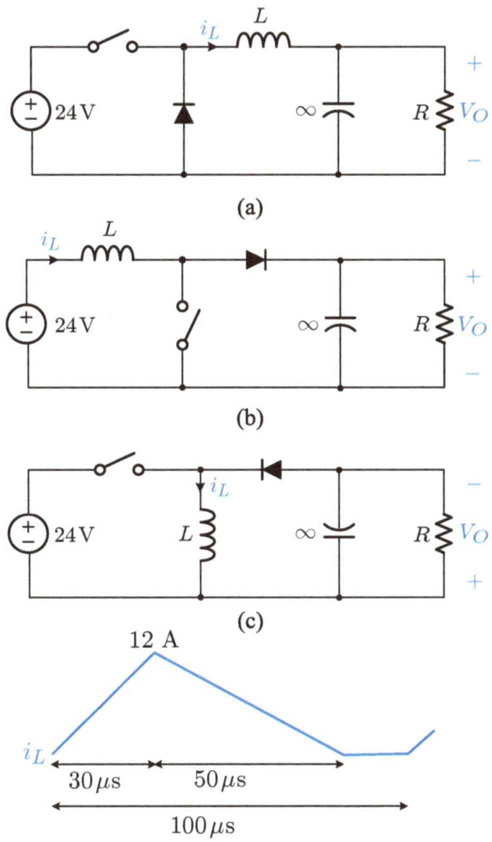


Fig. P4.16

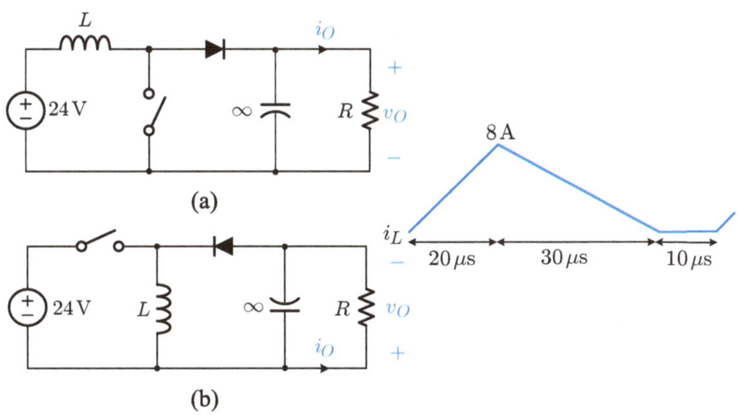


Fig. P4.17

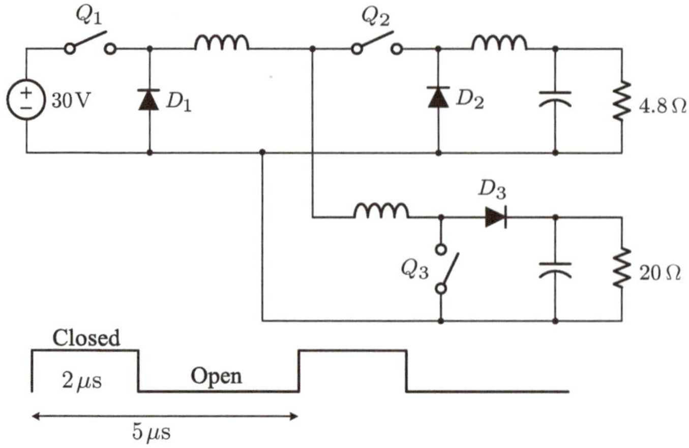


Fig. P4.18

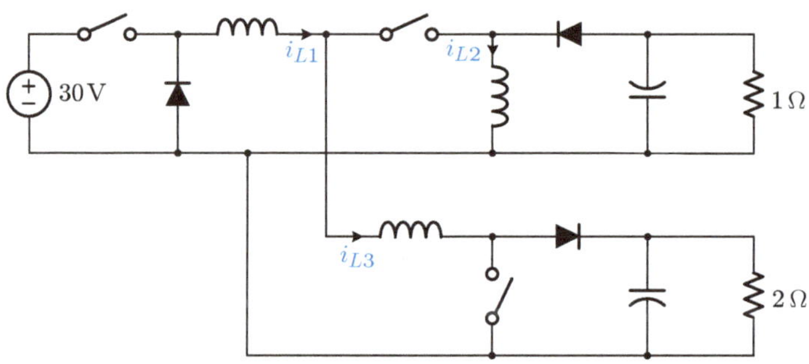


Fig. P4.19

**4.20** In the power conversion system shown in Fig. 4.20, the three converters all operate with $D = 0.4$ at the switching frequency $f_s = 50$ kHz. The three converters operate in CCM with the given circuit components.

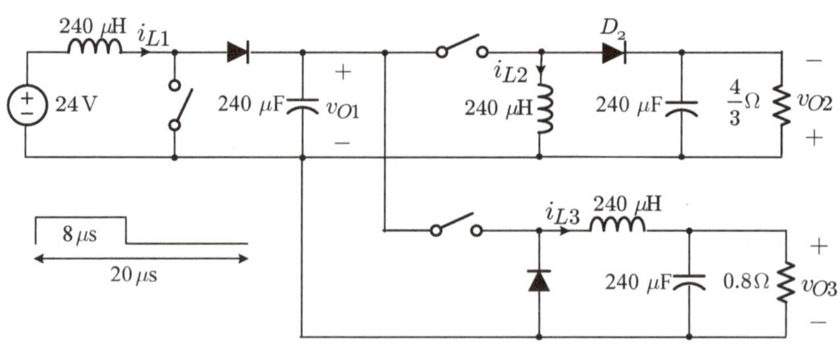


Fig. P4.20

a) Evaluate the average value of $v_{O1}$, $v_{O2}$, and $v_{O3}$.
b) Find the average value of $i_{L1}$, $i_{L2}$, and $i_{L3}$.
c) Sketch the waveforms of $i_{L1}$, $i_{L2}$, and $i_{L3}$ for the two operational periods. Show the maximum, minimum, and average values of each waveform.
d) Assume that the turn-on voltage of the diode $D_2$ is $V_{on} = 0.5$ V. Find the current stress, voltage stress, and power loss of $D_2$.

**4.21** Consider the power conversion system shown in Fig. P4.21. The three converters operate with $D = 0.5$ at the switching frequency $f_s = 50$ kHz. Assume that all the reactive components are so large that the converters are in CCM operation with negligible output voltage ripple.

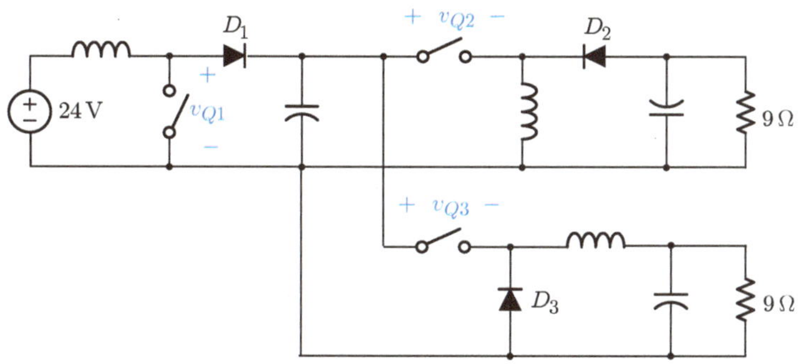


Fig. P4.21

a) Evaluate the voltage stress of the three active switches, $v_{Q1\ max}$, $v_{Q2\ max}$, and $v_{Q3\ max}$.
b) For the diodes in the circuit, assume the Schottky diode with the turn-on voltage of $V_D(on) = 0.5$ V. Evaluate the average power consumptions in diodes, $P_{D1}$, $P_{D2}$, and $P_{D3}$.

**4.22** For the three circuits shown in Fig. P4.22(a), assume that $Q_1$ and $Q_2$ are ideal *bidirectional* switches which deliver the current in both the forward and backward directions.

a) Assume that the average value of $\bar{v}_2$ is 14 V for all the three circuits. For the switch drive signals in Fig. P4.22(b), find the average value of $\bar{v}_1$ in the three circuits in Fig. P4.22(a).
b) Repeat a) for the switch drive signals in Fig. P4.22(c).

**4.23** Use the flux linkage balance condition and charge balance condition to derive the CCM voltage gain expression

$$\frac{V_O}{V_S} = \frac{D}{1-D}\left(\frac{1}{1+\dfrac{1}{(1-D)^2}\dfrac{R_l}{R}}\right)$$

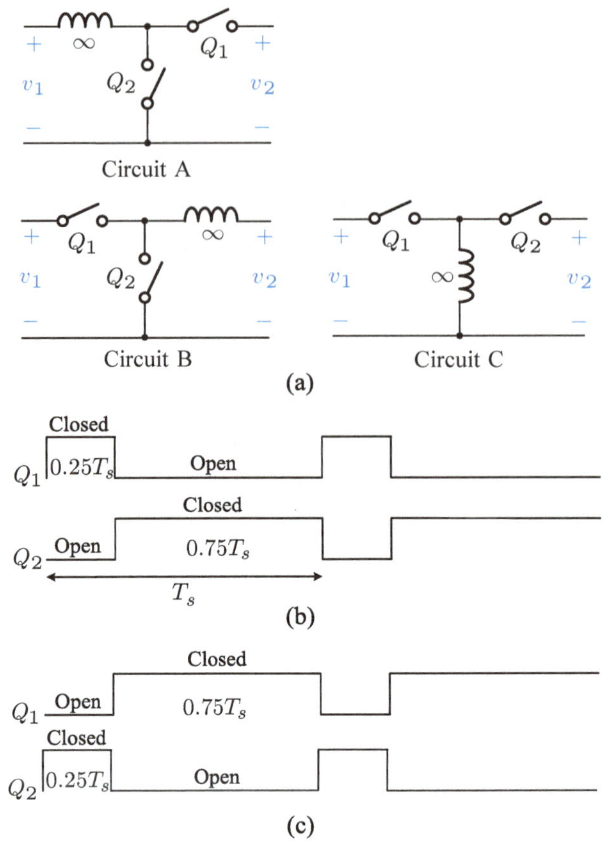


Fig. P4.22

for the buck/boost converter with an inductor winding resistance $R_l$.

**4.24*** Shown in Fig. P4.24 is a buck/boost converter with the non-ideal switches and practical passive components.

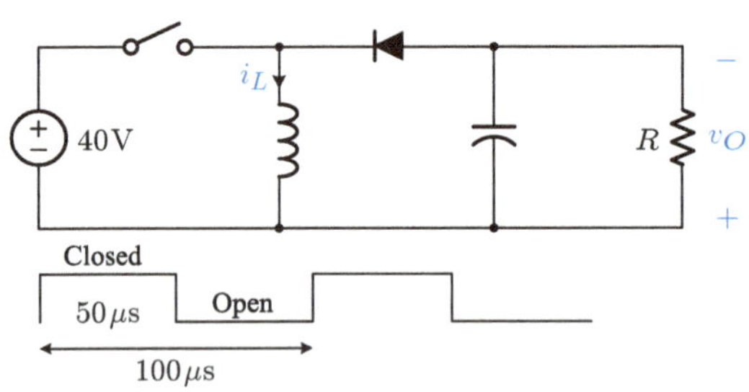


Fig. P4.24

a) During the normal operation, the load resistor is accidentally disconnected so that $R = \infty$. What do you think will happen to the circuit? If you claim a new steady-state operation, find the steady-state average values for the output voltage $v_O$ and inductor current $i_L$. If you claim a catastrophic failure of the circuit, identify the component that is responsible for the failure and state the origin/reason for the component breakdown.

b) Now assume that the load resistor is accidentally shorted, $R = 0$, during the normal operation of the converter. Repeat a) under this assumption.

**4.25** Figure P4.25 depicts a composite converter circuit and its switch drive signals. The composite converter is an integration of the buck converter and boost converter.

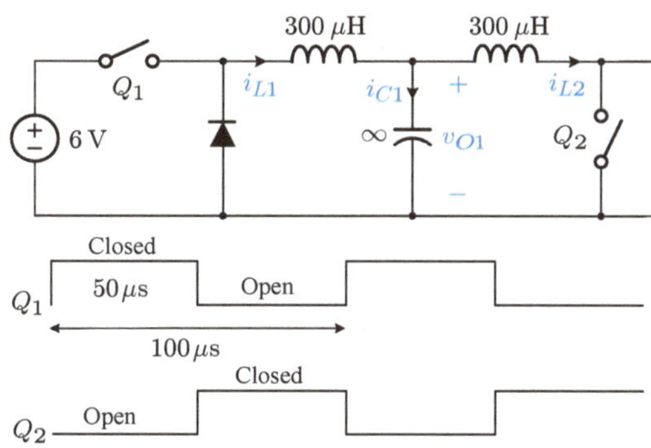


Fig. P4.25

a) Find the average values for $\bar{v}_{O1}$ and $\bar{v}_{O2}$.

b) Sketch the steady-state waveform of $i_{L1}$ and $i_{L2}$ for the two operational periods. Label the maximum and minimum values of each waveform.

c) Draw the steady-state waveform of $i_{C1}$ and $i_{C2}$ for the two operational periods. Label the maximum and minimum values of each waveform.

**4.26** A bidirectional boost converter is employed in a photovoltaic application. The system diagram and functional circuit diagram of the photovoltaic power system are shown in Fig. P4.26. Figure P4.26(a) is the functional block diagram of the system and Fig. P4.26(b) is its simplified circuit representation. Figure P4.26(c) shows the profile of the solar array current. The system operates with $V_B = 60$ V and $f_s = 5$ kHz. The circuit components of the converter are $L = 400$ μH and $C = 500$ μF. Answer the following questions regarding the operations of the bidirectional boost converter in the photovoltaic power system.

a) The payload requires a dc voltage of $V_{DC} = 150$ V for the operation and always draws a dc current of $I_{\text{Load}} = 50$ A. Over the operational period, the dc current supplied from the solar panel $I_{\text{Solar}}$ varies as follows. Perform the steady-state analysis of the bidirectional boost converter at the three

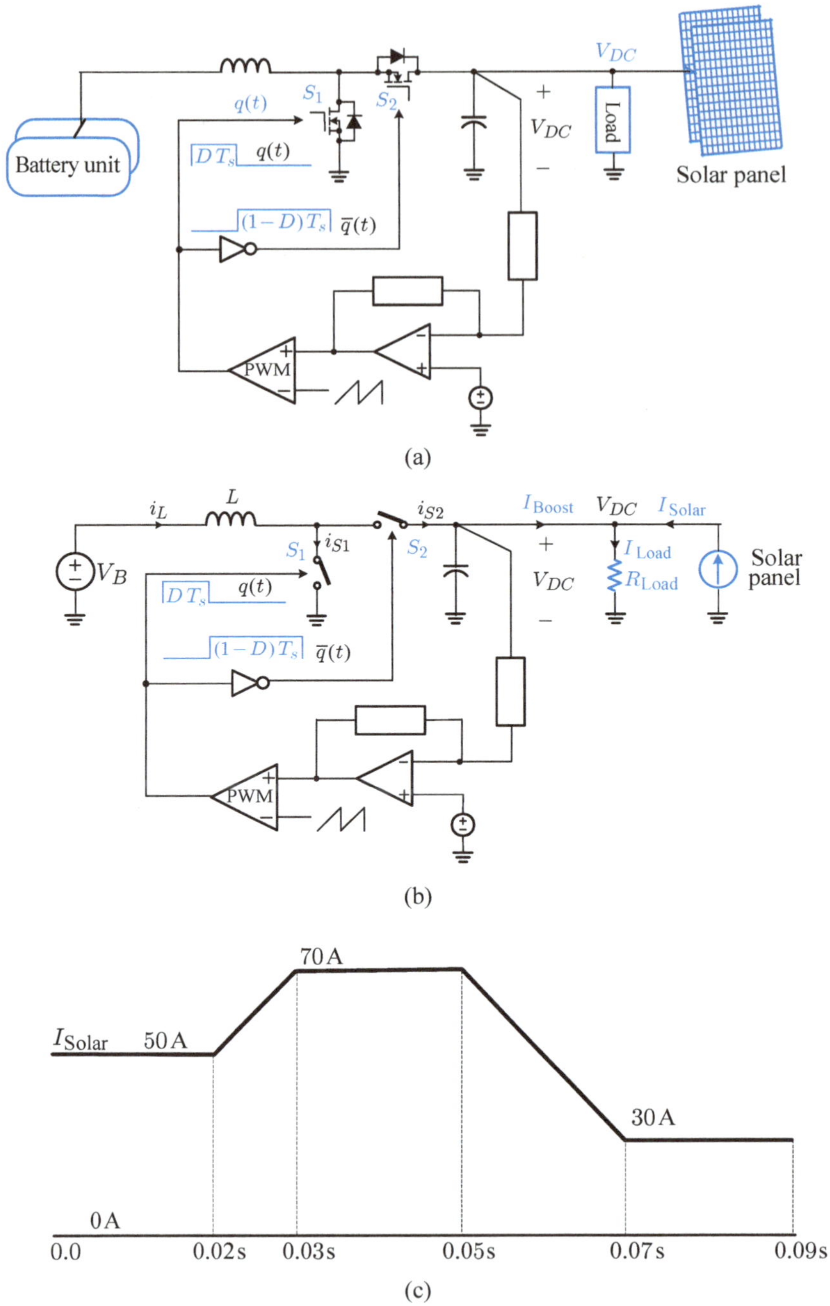

$V_{DC}$
$S_1$
$S_2$
$q(t)$
Battery unit
$DT_s$ $q(t)$
$(1-D)T_s$ $\overline{q}(t)$
$V_{DC}$
Load
Solar panel
PWM
(a)
$i_L$
$L$
$i_{S2}$
$I_{Boost}$
$V_{DC}$
$I_{Solar}$
$V_B$
$S_1$ $i_{S1}$
$S_2$
$I_{Load}$
$R_{Load}$
Solar panel
(b)
70 A
$I_{Solar}$ 50 A
30 A
0 A
0.0
0.02s
0.03s
0.05s
0.07s
0.09s
(c)


Fig. P4.26

different time instants of $t = 0.01$, $t = 0.04$, and $t = 0.08$ s. For each time instant, sketch the inductor current $i_L$ and the switch currents of the bidirectional switches, $i_{S1}$ and $i_{S2}$, for the two switching periods of the boost converter. Label the maximum, average, and minimum values of the waveforms.

b) Modify the PSIM®model **Bidirectional_Boost_Converter_03** in order to reflect the operational conditions at $t = 0.01$, $t = 0.04$, and $t = 0.08$ s. Perform the PSIM simulation using the modified PSIM model and plot the circuit waveforms of $i_L$, $i_{S1}$ and $i_{S2}$ for the two operational periods. Validate the accuracy of your analysis.

c) Modify **Bidirectional_Boost_Converter_02** so that the modified PSIM model produces the transitional responses of the system over the time period $0 < t < 0.09$s. Do not perform the simulation but show all the modifications you made.

d) Assume that you executed the PSIM®simulation using the model obtained in c). Sketch your predictions of the transient pattern of $I_{\text{Load}}$, $I_{\text{L}}$, $I_{\text{Boost}}$, and $I_{\text{Solar}}$ for $0 < t < 0.09$s, in the format of Fig. 4.21. Show all the characteristic features of the plots.

**4.27** Figure P4.27 depicts the circuit diagram of a composite converter, configured by cascading the boost converter and buck converter. Referring to the switch drive signals for $Q_1$ and $Q_2$, answer the questions.

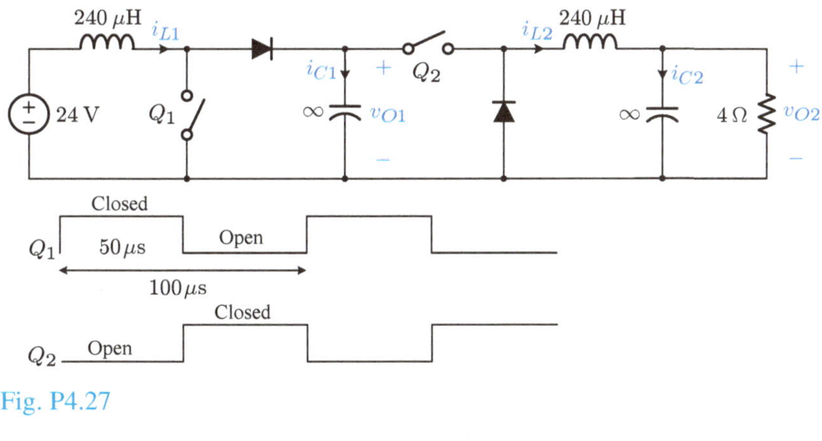


Fig. P4.27

a) Find the average values for $\bar{v}_{O1}$ and $\bar{v}_{O2}$.

b) Sketch the steady-state waveform of $i_{L1}$ and $i_{L2}$ for the two switching periods. Label the maximum, minimum, and average values of each waveform.

c) Draw the steady-state waveform of $i_{C1}$ and $i_{C2}$ for the two switching periods. Label the maximum, minimum, and average values of each waveform.

**4.28** Shown in Fig. P4.28 are 1) the circuit diagram of a bidirectional boost converter in a photovoltaic power system and 2) the profile of the output current of the boost converter, $I_{\text{Boost}}$, over the period of $0 < t < 0.09$ s.

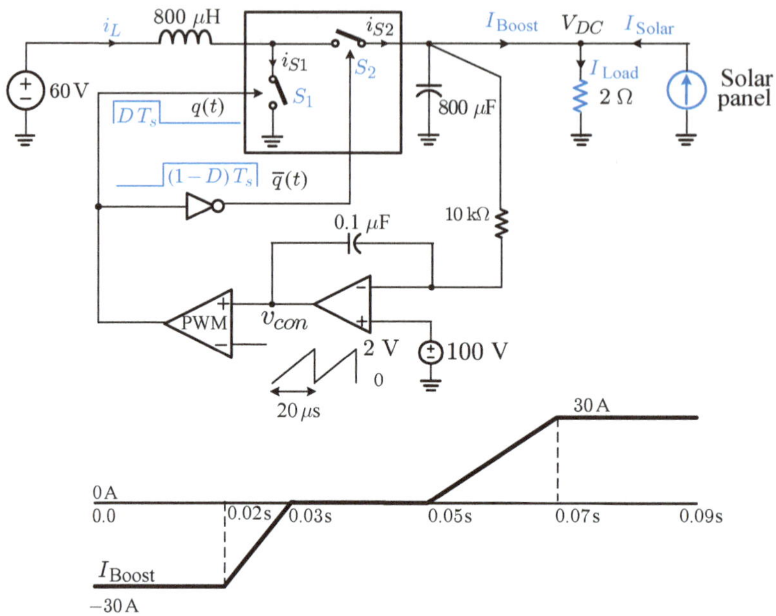


Fig. P4.28

a) Find the average value of $V_{\text{DC}}$ and control voltage $v_{con}$.

b) Sketch the profile of $I_{\text{Solar}}$, $I_{\text{Load}}$, $I_{\text{Boost}}$, and $I_L$ for $0.0 < t < 0.09$ s in the format of Fig. 4.21 in the text.

c) Consider the steady-state operation of the system at $t = 0.01$, $0.04$, and $0.08$ s. Sketch the inductor current $i_L$ and switch currents of the bidirectional switches, $i_{S1}$ and $i_{S2}$, at $t = 0.01$, $0.04$, and $0.08$ s for the two switching periods of the boost converter. Label the maximum, average, and minimum values of the waveforms.

d) Evaluate the voltage stress and current stress of the bidirectional switches $S_1$ and $S_2$.

e) Convert the functional diagram of the bidirectional switches $S_1$ and $S_2$, boxed in the circuit diagram, into the symbolic representation of semiconductor devices. Show the location and orientation of the devices.

**4.29** Shown in Fig. P4.29 are the circuit diagram of a closed-loop controlled buck/boost converter and flyback converter. Referring to the circuit diagrams, answer the questions. Assume that a 240 $\mu$F capacitor is used for the output filter of the two converters.

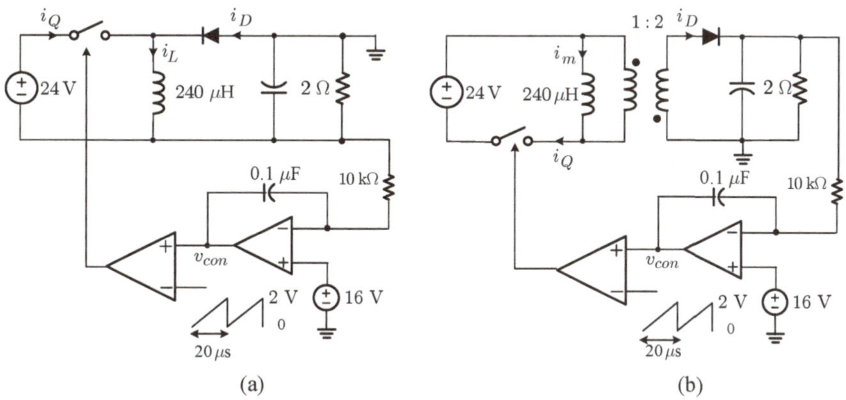


Fig. P4.29

a) For Converter (a), sketch the steady-state waveforms of $i_L$, $i_Q$, $i_D$, and $v_{con}$ for the two operational periods. Show the maximum, minimum, and average values of the waveforms.

b) For Converter (b), sketch the steady-state waveforms of $i_m$, $i_Q$, $i_D$, and $v_{con}$ for the two operational periods.

**4.30*** The forward converter shown in Fig. P4.30 is built using a practical transformer. The number of turns of the three-winding transformer is $N_P = N_S = N_T = 48$. The converter established steady-state operation with a duty ratio $D = 0.4$. Now assume that the following failure or change has occurred to the converter during its steady-state operation:

i) open-circuit failure of $D_C$ ii) open-circuit failure of $D_2$
iii) open-circuit failure of $C$ iv) decrease in the duty ratio to $D = 0.15$
v) increase in the duty ratio to $D = 0.65$

For each failure/change listed above, what do you think will happen to the converter? If you predict a fatal failure of the circuit, identify the component responsible to the failure and state the origin/reason of the component breakdown. If you claim a new steady-state operation, describe the major change in the circuit operation.

**4.31** The tertiary-winding reset circuit is employed in a forward converter, resulting in the tertiary-winding reset forward converter in Fig. P4.31. The converter operates with $V_S = 64$ V at the switching frequency $f_s = 50$ kHz. The circuit components of the converter are $L = 40\ \mu$H, $C = 400\ \mu$F, and $R = 1\ \Omega$. The turns ratio of the three-winding transformer is $N_P : N_S : N_T = 10 : 15 : 5$ and the magnetizing inductance of the transfer is $L_m = 200\mu$ H. The duty ratio of the converter can be selected at any value between $0.05 < D < 0.95$. Answer the following questions.

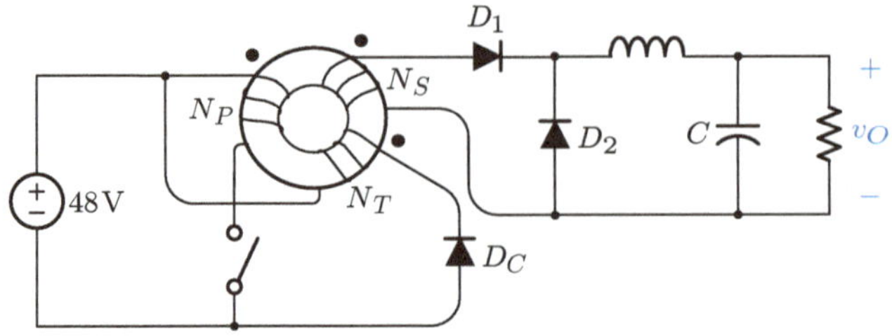


Fig. P4.30

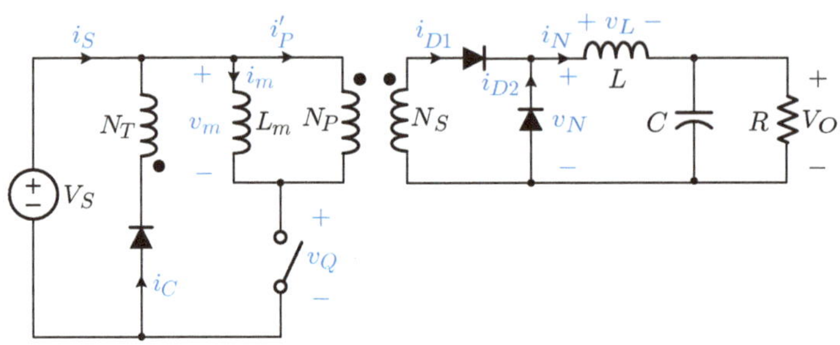


Fig. P4.31

a) For the given operational conditions, find the largest possible value for the output voltage $V_O$ of the converter. Also, specify the duty ratio that offers the largest output voltage.

b) Perform the steady-state analysis under the maximum output voltage condition found in a). Sketch the waveforms of $i_m$, $v_m$, $i_C$, $i_S$, $i'_P$, $v_Q$, $v_L$, $i_N$, $i_{D1}$ and $i_{D2}$ for the two operational periods. Show the maximum and minimum values for the waveforms.

c) Modify the PSIM®model **Forward_Converter_02** in the PSIM model library to reflect the operational conditions of the forward converter. Perform the PSIM simulation using the modified PSIM model and plot the circuit waveforms of $i_m$, $v_m$, $i_C$, $i_S$, $i'_P$, $v_Q$, $v_L$, $i_N$, $i_{D1}$ and $i_{D2}$ for the three operational periods. Validate the accuracy of your analysis.

**4.32** The Zener diode reset circuit is combined with the forward converter, resulting in the Zener-diode reset forward converter in Fig. P4.32. The converter operates with $V_S = 64$ V at the switching frequency $f_s = 50$ kHz. The circuit components of the converter are $L = 40\ \mu$H, $C = 400\ \mu$F, and $R = 1\ \Omega$. The turns ratio of the transformer windings is $N_P : N_S = 10 : 15$ and the magnetizing inductance of the transformer is $L_m = 200\ \mu$H. The Zener breakdown voltage of the Zener diode $D_Z$ is $V_Z = 32$ V. The duty ratio of the converter can be selected at any value between $0.05 < D < 0.95$. Answer the following questions.

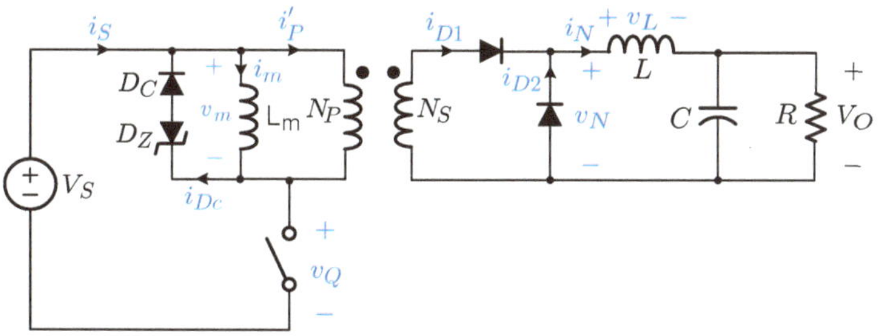


Fig. P4.32

a) For the given operational conditions, find the largest possible value for the output voltage $V_O$ of the converter. Also, specify the duty ratio that offers the largest output voltage $V_O$.

b) Perform the steady-state analysis under the maximum output voltage condition found in a). Sketch the waveforms of $i_m$, $v_m$, $i_{Dc}$, $i_S$, $i'_P$, $v_Q$, $v_L$, $i_N$, $i_{D1}$ and $i_{D2}$ for the two operational periods. Show the maximum and minimum values for the waveforms.

c) Modify the PSIM®model **Forward_Converter_01** in the PSIM model library to reflect the operational conditions of the forward converter. Perform the PSIM simulation using the modified PSIM model and plot the circuit waveforms of $i_m$, $v_m$, $i_{Dc}$, $i_S$, $i'_P$, $v_Q$, $v_L$, $i_N$, $i_{D1}$ and $i_{D2}$ for the three operational periods. Validate the accuracy of your analysis.

**4.33*** For the four isolated dc-to-dc converters shown in Fig. P4.33, assume that the transformer has a finite magnetizing inductance.

a) Answer the following questions.
   i) Which converter would result in the worst power conversion efficiency?
   ii) Which converter would offer the largest power handling capacity?
   iii) Which converter would minimize the manufacturing cost?

b) For each dc-to-dc converter, find the range of the duty ratio that guarantees proper operation of the converter. Express your answer in the form of $(\quad) < D < (\quad)$.

c) Now assume that all the converters are closed-loop controlled to produce an output voltage of 6 V. Evaluate the voltage stress of the active switch employed in each circuit.

**4.34*** The three forward converter circuits in Fig. P4.34 employ three different reset circuits. Referring to the converter schematics, answer the questions. The values of the reactive components of the converters are all identical and sufficiently large enough to put the converters in CCM operation. The transformers of the

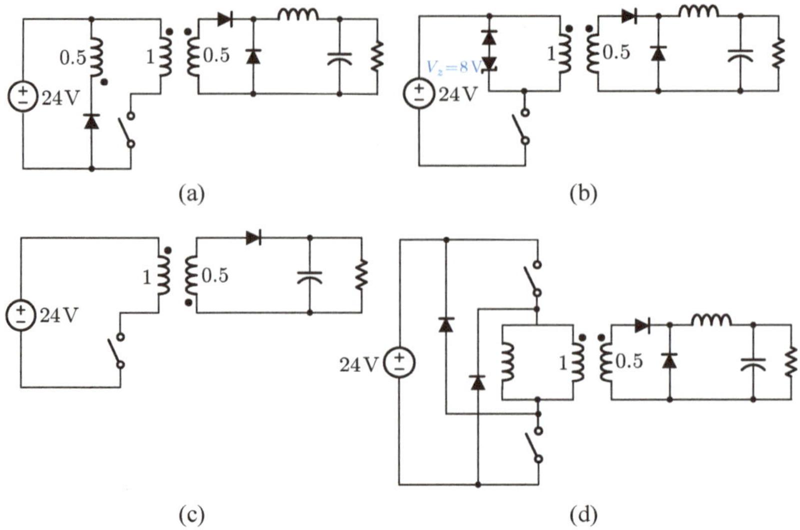


Fig. P4.33

three converters are built on the same magnetic core. The three transformers have identical winding turns of $N_P = 15$ and $N_S = 30$, yet the three-winding transformer in Converter (a) has a tertiary winding turns $N_T$. The three converters produce the same magnetizing current $i_m$ shown in the circuit diagram.

a) Find the magnetizing inductance $L_m$ and turns of the tertiary winding $N_T$ of Converter (a).
b) Evaluate the average value of $i_N$ and $V_O$ in the three converters.
c) Evaluate the breakdown voltage of the Zener diode $D_{Z1}$ in Converter (b) and breakdown voltages of the two identical Zener diodes, $D_{Z2}$ and $D_{Z3}$, in Converter (c).
d) Evaluate the power loss in the three converters.
e) Now, assume that input voltages of the three converter are all increased to $V_S = 60$. What will happen to the three converters? Justify your answer as precisely and convincingly as possible.

**4.35*** The following questions concern the topological structure, circuit components, and operational principles of various PWM converters. Answer the questions in a concise and precise manner.

a) The transformer employed in a flyback converter is typically built on the magnetic core to which an air gap is intentionally introduced. Explain the necessity of this practice.
b) While an appropriate reset circuit is required for the forward converter, the flyback converter and bridge-type converters do not require any reset circuit. Explain the reasons for this.

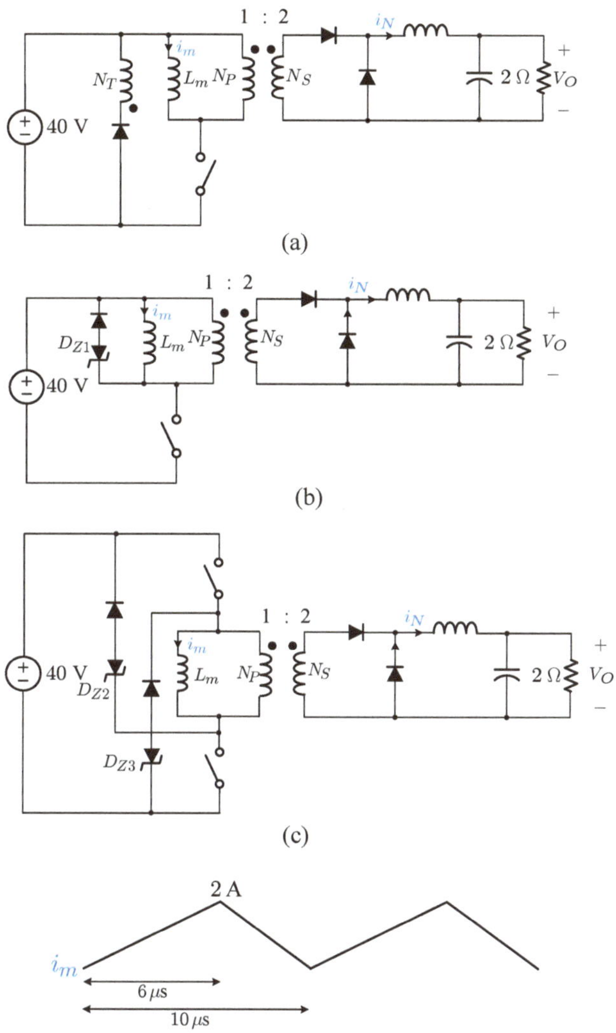


Fig. P4.34

c) Figure P4.35 shows six different converter topologies. For each converter, find the maximum possible value for the output voltage and the voltage stress of the active switch employed in the converter. The duty ratio of the converter will vary for this problem.

**4.36*** Shown in Fig. P4.36 are the five isolated PWM dc-to-dc converters. Assume all the components are ideal except for the practical transformer. For the five PWM converters, assume that
i) the magnetizing inductance of the practical transformer is $L_m = 200\ \mu$H
ii) the duty ratio is $D = 0.2$, with respective to the switching period $T_s =$

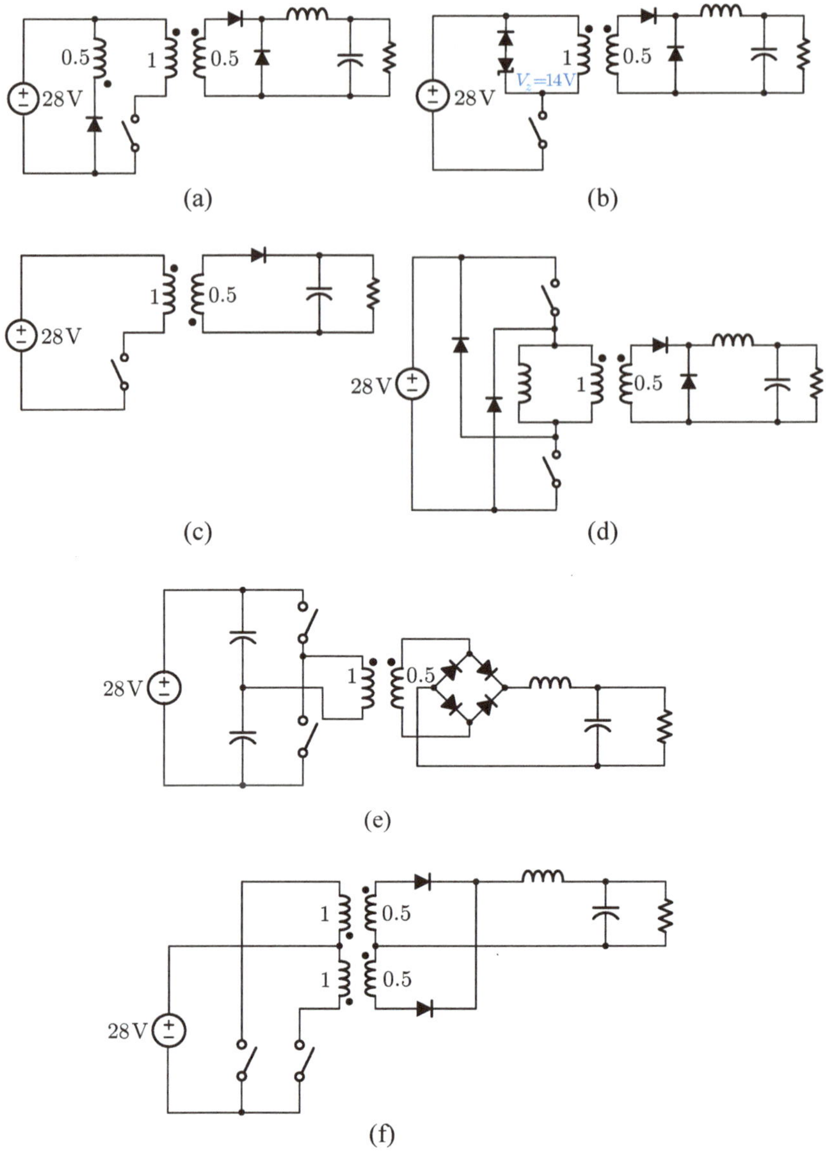


Fig. P4.35

50 $\mu$s for Converters (a) through (d), and with respective to the ripple period $T_s = 50\ \mu$s for Converter (e), thus yielding the turn-on time of $T_{on} = 10\ \mu$s for all five converters

For each converter, evaluate the output voltage $V_O$, the peak value of the switch

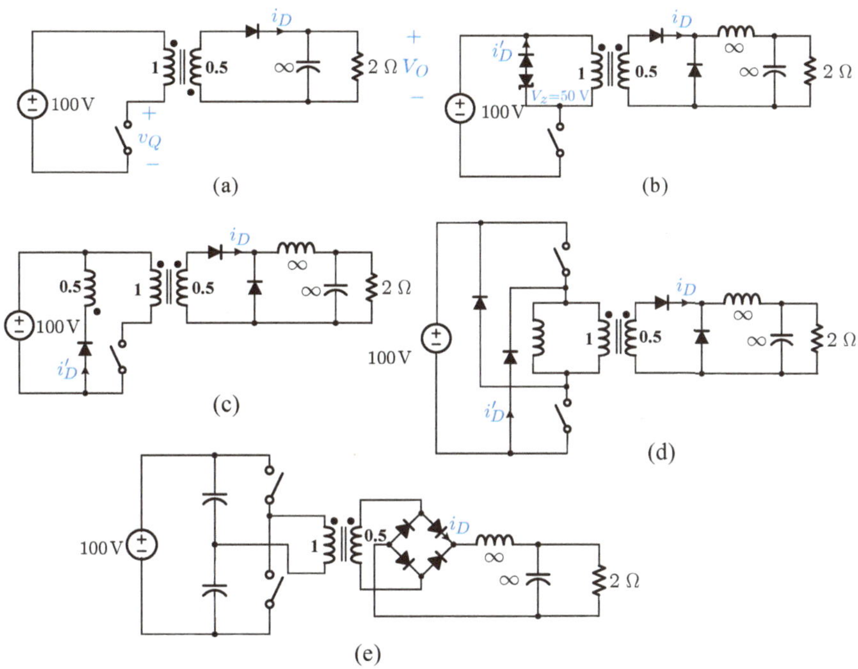


Fig. P4.36

voltage $v_{Q\ peak}$, and the average values of the diode current, $\bar{\imath}_D$ and $\bar{\imath}'_D$, as appropriate.

# APPENDIX A

# ANSWERS TO END-OF-CHAPTER PROBLEMS

Some design problems and graphical problems do not have a unique solution. For these problems, the answers in this list are a typical illustration and other solutions are equally acceptable.

## CHAPTER 2

2.1 **a)** $i_{L\,max} = 10$ A $i_{L\,min} = -10$ A **b)** $v_{C\,max} = 8$ V $v_{C\,min} = -4$ V

2.2 **a)** (a) $V_S = 12$ V (b) $V_S = 50$V (c) $V_S = 20$V
**b)** (a) $i_{L\,max} = 7.2$ A (b) $i_{L\,max} = 12$A (c) $i_{L\,max} = 12$ A

2.4 **a)** $L = 40$ mH **b)** $i_{L\,max} = 20$ A **c)** $t = 130$ ms **d)** $19.2\text{ V} \leq V_X$

2.5 **a)** i) $i_{L\,max} = 1.5$ A $i_{L\,min} = 0$ A ii) $i_{L\,max} = 0.5$ A $i_{L\,min} = 0$ A
**b)** i) $v_{C\,max} = 1.5$ V $v_{C\,min} = 0$ V ii) $v_{C\,max} = 0.5$ V $v_{C\,min} = 0$ V

2.6 **a)** (a) $v_2/v_1 = 1 - D$ (b) $v_2/v_1 = 1/D$ (c) $v_2/v_1 = -(1 - D)/D$
**b)** (d) $i_2/i_1 = (1 - D)/D$

2.7 **a)** $D = 0.25$ $T_s = 120\ \mu$s for (a) and (b) **b)** $D = 0.2$ $T_s = 150\ \mu$s for (c) and (d)

2.8 **a)** $\bar{v}_C = 50$ V **b)** $V_2 = 20$ V **c)** $\bar{\imath}_{L1} = 4/3$ A
**d)** $i_{L1\,max} = (4/3 + 1/8)$ A $i_{L1\,min} = (4/3 - 1/8)$ A $i_{L2\,max} = 2.5$ A $i_{L2\,min} = 1.5$ A $i_{C\,max} = (4/3 + 1/8)$ A $i_{C\,min} = -2.5$ A $v_{C\,max} = (50 + 1/3)$ V $v_{C\,min} = (50 - 1/3)$ V $\Delta v_C = 2/3$ V

**2.9** **a)** $\bar{v}_C = 60\ \text{V}$ $V_2 = 40\ \text{V}$
**b)** $i_{L1\,max} = 5\ \text{A}$ $i_{L1\,min} = 3\ \text{A}$ $i_{L2\,max} = 4\ \text{A}$ $i_{L2\,min} = 0\ \text{A}$
**c)** $i_{C\,max} = 5\ \text{A}$ $i_{C\,min} = -4\ \text{A}$ $v_{C\,max} = 62\ \text{V}$ $v_{C\,min} = 58\ \text{V}$

**2.10** **a)** $\bar{v}_C = 40\ \text{V}$ $V_2 = 20\ \text{V}$ **b)** $i_{L1\,max} = 3\ \text{A}$ $i_{L1\,min} = -3\ \text{A}$ $i_{L2\,max} = 3\ \text{A}$ $i_{L2\,min} = -3\ \text{A}$ $i_{C\,max} = 3\ \text{A}$ $i_{C\,min} = -3\ \text{A}$ $v_{C\,max} = 40.56\ \text{V}$ $v_{C\,min} = 39.44\ \text{V}$

**2.12** **a)** $i_P = 1.75\cos t$ $v_S = 2\sin t$
**b)** $i_P = 2.75\cos t$ $i_m = 1.5\cos t$ $v_m = 2\sin t$ $v_O = 4\sin t$

**2.13** **a)** $L_m = 0.5\ \text{H}$ $n = 0.25$
**b)** (a) $i_p = -40\cos t$ (b) $i_p = -42.5\cos t$ (c) $i_p = -40\cos t$
(d) $i_p = (32/3)\sin t$ (e) $i_p = 32\sin t$

**2.14** **a)** i) $L = 2.5\ \text{mH}$ ii) $n = 5$
**b)** i) $V_{S\,max} = 120\ \text{V}$ ii) $T_{s\,max} = 4\ \text{ms}$ iii) $i_{P\,max} = 3\ \text{A}$ $i_{P\,min} = -3\ \text{A}$
**c)** $i_{S\,max} = 1\ \text{A}$ $i_{S\,min} = -1\ \text{A}$ $i_{P\,max} = 3.5\ \text{A}$ $i_{P\,min} = -3.5\ \text{A}$

**2.15** **a)** $L_1 = 2\ \text{mH}$ $L_2 = 5\ \text{mH}$ $n = 7/5$ **b)** $i'_p = -0.5 \times 10^3 \cos t$

**2.16** **a)** $i_{P\,max} = 1.25\ \text{A}$ $i_{P\,min} = -1.25\ \text{A}$ $i_{S\,max} = 0.5\ \text{A}$ $i_{S\,min} = -0.5\ \text{A}$
**b)** $i_{P\,max} = 1.25\ \text{A}$ $i_{P\,min} = -1.25\ \text{A}$ $i_{S\,max} = 0.5\ \text{A}$ $i_{S\,min} = -0.5\ \text{A}$

**2.17** **a)** $i_{P\,max} = 12\ \text{A}$ $i_{P\,min} = 0\ \text{A}$ $i_{Q\,max} = 12\ \text{A}$ $i_{Q\,min} = 0\ \text{A}$ $i_{D\,max} = 12\ \text{A}$ $i_{D\,min} = 0\ \text{A}$ $v_{S\,max} = 0\ \text{V}$ $v_{S\,min} = -48\ \text{V}$
**b)** $i_{P\,max} = 6\ \text{A}$ $i_{P\,min} = 0\ \text{A}$ $i_{D\,max} = 6\ \text{A}$ $i_{D\,min} = 0\ \text{A}$ $v_{S\,max} = 48\ \text{V}$ $v_{S\,min} = -48\ \text{V}$
**c)** $i_{P\,max} = 6\ \text{A}$ $i_{P\,min} = 0\ \text{A}$ $i_{D\,max} = 6\ \text{A}$ $i_{D\,min} = 0\ \text{A}$ $v_{S\,max} = 96\ \text{V}$ $v_{S\,min} = -48\ \text{V}$

**2.18** $i_{S\,max} = 2\ \text{A}$ $i_{S\,min} = 0\ \text{A}$ $i_{P\,max} = 6\ \text{A}$ $i_{P\,min} = 0\ \text{A}$
$i_{Q\,max} = 6\ \text{A}$ $i_{Q\,min} = 0\ \text{A}$ $i_{D\,max} = 6\ \text{A}$ $i_{D\,min} = 0\ \text{A}$

**2.19** **a)** $L = 48\ \text{mH}$ $R = 5\ \Omega$ **b)** i) 0.6 J ii) 0.6 J iii) 10 W
**c)** i) $i_{L\,max} = 7.5\ \text{A}$ $i_{L\,min} = 0\ \text{A}$ $v_{L\,max} = 15\ \text{V}$ $v_{L\,min} = -10\ \text{V}$
ii) $i_{L\,max} = 10\ \text{A}$ $i_{L\,min} = 0\ \text{A}$ $v_{L\,max} = 15\ \text{V}$ $v_{L\,min} = -5\ \text{V}$
Not workable due to saturation of the solenoid inductance

**2.20** **a)** i) $i_L = 4e^{-(t-0.01)/0.01}$ ii) 16 W
**b)** i) $i_{L\,max} = 4\ \text{A}$ $i_{L\,min} = 0\ \text{A}$ ii) Transferred to the voltage source $V_{S2}$
**c)** i) $v_C = 126.5\ \text{V}$ ii) $i_{L\,max} = 4\ \text{A}$ $v_C\,(200\ \text{ms}) = 178.8\ \text{V}$
iii) $v_C$ continues to grow until the capacitor explodes.

**2.21** **a)** $i_{P\,max} = i_{Q\,max} = i_{D\,max} = 18\ \text{A}$
**b)** $i_{P\,max} = i_{Q\,max} = i_{D\,max} = 12\ \text{A}$
**c)** $i_{P\,max} = i_{Q\,max} = i_{D\,max} = 12\ \text{A}$

**2.22** $i_{P\,max} = 3.5\ \text{A}$ $i_{D1\,max} = 1.5\ \text{A}$ $i_{D2\,max} = 2\ \text{A}$ $i_{D3\,max} = 2\ \text{A}$
$i_{P\,min} = i_{D1\,min} = i_{D2\,min} = i_{D3\,min} = 0\ \text{A}$

**2.23** (a) $i_{T\,max} = 10$ A $i_{T\,min} = 0$ A $i_{S\,max} = 10$ A $i_{S\,min} = -10$ A
(b) $i_{T\,max} = 10$ A $i_{T\,min} = 5$ A $i_{S\,max} = 10$ A $i_{S\,min} = -10$ A

**2.24** **a)** $i_{S\,max} = 2$ A $i_{S\,min} = -2$ A $v_{L\,max} = 40$ V $v_{L\,min} = -80$ V
**b)** i) 0.4 mJ ii) 0.2 mJ iii) 0.2 mJ

**2.25** $i_{L\,max} = 2$ A $i_{L\,min} = 0$ A $i_{D1\,max} = 2$ A $i_{D1\,min} = 0$ A $i_{D2\,max} =$ 2 A $i_{D2\,min} = 0$ A $i_{Q1\,max} = 6$ A $i_{Q1\,min} = 0$ A $v_{X\,max} =$ 24 V $v_{X\,min} = 0$ V

**2.26** **a)** i) $L_m = 200\ \mu$H ii) $n = 0.5$ iii) $D = 0.4$ iv) $C = 4.5\ \mu$F
**b)** i) 0.114 J ii) $v_C = 253$ V iii) $t = 1.4$ ms
**c)** Turn-on the switch as soon as $i_D$ is reduced to zero, while keeping the on-time constant.

**2.27** **c)** $i_{L\,max} = 60$ A $i_{D\,max} = 60$ A $v_{C\,max} = 170$ V $i_{L\,min} = i_{D\,min} =$ $v_{C\,min} = 0$
**d)** $v_C(100 \times 10^{-3}) = 268$ V

**2.28** **a)** $i_{P\,max} = 6$ A $i_{P\,min} = 0$ A $v_{S\,max} = 0$ V $v_{S\,min} = -48$ V
**b)** $i_{P\,max} = 3$ A $i_{P\,min} = 0$ A $v_{S\,max} = 48$ V $v_{S\,min} = -96$ V
**c)** $i_{P\,max} = 6$ A $i_{P\,min} = 0$ A $v_{S\,max} = 48$ V $v_{S\,min} = -48$ V
**d)** $i_{P\,max} = 3$ A $i_{P\,min} = 0$ A $v_{S\,max} = 4.24$ V $v_{S\,min} = 0$ V

**2.29** **a)** i) 1.2 mJ ii) 0.8 mJ iii) 0 mJ iv) 0.4 mJ
**b)** i) 0.9 mJ ii) 0.45 mJ iii) 0.45 mJ iv) 0 mJ

**2.30** **a)** i) 6 mJ ii) 5 mJ iii) 1 mJ iv) $v_C = 100$ V
**b)** i) 6 mJ ii) 1 mJ iii) 5 mJ iv) $v_C = 44.7$ V

**2.31** **a)** $v_C(100 \times 10^{-3}) = 190$ V $v_C(200 \times 10^{-3}) = 190$ V
**b)** $v_C(100 \times 10^{-3}) = 190$ V $v_C(200 \times 10^{-3}) = 268$ V
**c)** I: The same operation as the original circuit
II: A negative voltage charging
III: Inoperative (Transformer saturation)

**2.32** **a)** $L = 2$ mH **b)** $V_{Z\,min} = 4$ V **c)** $i_{D1\,max} = 12$ A $i_{D1\,min} = 0$ A $i_{D2\,max} = 6$ A $i_{D2\,min} = 0$ A $v_{C\,max} = 58.8$ V $v_{C\,min} = 0$ V

**2.33** **a)** i) 7.2 mJ ii) 2.16 mJ iii) 1.44 mJ iv) 3.6 mJ v) $v_C = 65.7$ V
**b)** $i_{D1\,max} = 12$A $i_{D1\,min} = 0$A $i_{D2\,max} = 6$A $i_{D2\,min} = 0$A $i_{D3\,max} =$ 12 A $i_{D3\,min} = 0$ A **c)** $\bar{i}_{D1} = 1.5$ A $\bar{i}_{D3} = 0.6$ A

**2.34** **a)** $i_{L1\,max} = 6$ A $i_{L1\,min} = 0$ A $i_{D1\,max} = 6$ A $i_{D1\,min} = 0$ A $i_{T\,max} = 6$ A $i_{T\,min} = 0$ A $i_{D2\,max} = 3$ A $i_{D2\,min} = 0$ A
$v_{C\,max} = 26.8$ V $v_{C\,min} = 0$ V
**b)** i) $L = 2$ mH ii) $V_{z\,min} = 4.8$ V iii) $i_{L2\,max} = 4.8$ A $i_{L2\,min} = 0$ A $i_{D3\,max} = 4.8$ A $i_{D3\,min} = 0$ A

**2.35** **a)** i) 83.2 mJ ii) 27.6 mJ iii) 83.2 mJ iv) 35.3 V v) 6.9 W vi) $V_Z \geq 4.3$ V

**b)** $i_{D1\,max} = i_{D2\,max} = 45/7$ A $i_{D3\,max} = 4.3$ A $i_{D1\,min} = i_{D2\,min} = i_{D3\,min} = 0$ A
**c)** $\bar{i}_{D1} = 1.6$ A $\bar{i}_{D3} = 0.37$ A

## CHAPTER 3

**3.2** **a)** $\Delta v_O = 7.64\ \mu$V **b)** $L_1 = 3.18$ mH $C_1 = 3.18$ mF **c)** The two-stage filter requires only 11% of the total inductance and capacitance of the single-stage filter for the same attenuation.

**3.3** **a)** $V_O = 4$V **b)** $v_{L\,max} = 6$V $v_{L\,min} = -4$V $i_{D\,max} = 5$A $i_{D\,min} = 0$A
**c)** $L = 60\ \mu$H $R = 1\,\Omega$

**3.4** **a)** $\bar{i}_{Q1} = 1.6$ A $\bar{i}_{D1} = 2.4$ A $\bar{i}_{Q2} = 2.4$ A $\bar{i}_{D2} = 1.6$ A
**b)** $P_{Q1} = 3.2$ W $P_{D1} = 1.2$ W $P_{Q2} = 4.8$ W $P_{D2} = 0.8$ W **c)** $\eta = 0.89$

**3.5** **a)** $\Delta v_O = 0.064$ V **b)** $\Delta v_O = 0.062$ V

**3.6** **a)** $V_O = 4.8$ V **b)** $V_O = 9.6$ V **c)** Transformer saturation
**d)** $V_O = 9.6$ V **e)** Transfer saturation **f)** $V_O = 9.6$ V **g)** Merit: Galvanic isolation
Demerit: Power loss in Zener diode

**3.7** **a)** $L = 200\ \mu$H $C = 25\ \mu$F **b)** $L = 2$ mH

**3.8** **a)** $\Delta v_O = 57$ mV
**b)** First circuit: $\Delta v_O = 5.09$ V Second circuit: $\Delta v_O = 0.81$ V

**3.9** **a)** Converter A: 0.6 V Converter B: 0.06 V Converter C: 0.005 V
**b)** Converter A: 0.43 V Converter B: 0.043 V Converter C: 0.0043 V

**3.10** **b)** $\Delta v_O(t) \approx 1.53 \times 10^{-5}$ V
**c)** $L = 1.67$ mH $C = 1.67$ mF
**d)** $L = 1/3$ H

**3.12** **c)** $i_{L\,max} = 1.5$ A $i_{L\,min} = -0.5$ A

**3.13** **a)** $L = 12.5\ \mu$H **b)** $R = 2.5\,\Omega$ **c)** $V_O$ is proportional to $R$ **d)** $R = 5\,\Omega$
**e)** $R = \infty$ **f)** $v_{O\,min} = 9.9$ V $v_{O\,max} = 10.1$ V

**3.14** $V_O = 16$ V $I_O = 5$ A $R = 3.2\,\Omega$ $L = 40\ \mu$F

**3.15** **a)** $v_{L\,max} = 3$ V $v_{L\,min} = -9$ V **b)** $L = 40\ \mu$H $R = 5\,\Omega$
**c)** $i_{C\,max} = 2.7$ A $i_{C\,min} = -1.8$ A **d)** $\Delta v_O = 0.54$ V
**e)** $\Delta v_O = 25.9$ mV

**3.16** **a)** $i_{L\,max} = 10$ A $i_{L\,min} = 0$ A $\bar{i}_L = 5$ A **b)** $i_{L\,max} = 7$ A $i_{L\,min} = 0$ A $\bar{i}_L = 2.97$ A **c)** $i_{L\,max} = 3.6$ A $i_{L\,min} = 0$ A $\bar{i}_L = 1.8$ A

**3.17** **a)** $L = 40\ \mu$H **b)** $L > 100\ \mu$H **c)** $D_{crit} = 0.4$

**3.18** **a)** i) $i_{L\,max} = (4 + 5.3/2)$ A $i_{L\,min} = (4 - 5.3/2)$ A $v_{con} = 2$ V ii) $i_{L\,max} = 4$ A $i_{L\,min} = 0$ A $v_{con} = 1$ V
**b)** i) $i_{L\,max} = 4$ A $i_{L\,min} = 0$ A $v_{con} = 1$ V ii) $i_{L\,max} = (4 + 5.3/2)$ A $i_{L\,min} = (4 - 5.3/2)$ A $v_{con} = 2$ V

**3.19** Converter A: $\bar{v}_{con} = 3$ V $V_O = 6$ V Converter B: $\bar{v}_{con} = 1.5$ V $V_O = 3$ V
Converter C: $\bar{v}_{con} = 1.5$ V $V_O = 3$ V Converter D: $\bar{v}_{con} = 2$ V $V_O = 4$ V

**3.20** **a)** $R = 5\,\Omega : \bar{i}_L = 1$ A slope: $1.25 \times 10^5$ $i_{L\,peak} = 1.625$A
$R = 8\,\Omega : \bar{i}_L = 0.625$ A slope: $1.25 \times 10^5$
$R = 12\,\Omega : \bar{i}_L = 0.47$ A slope: $1.08 \times 10^5$
**b)** $R = 5\,\Omega : \bar{i}_L = 0.4$ A slope: $2 \times 10^5$ $i_{L\,peak} = 0.8$A
$R = 8\,\Omega : \bar{i}_L = 0.25$ A slope: $2 \times 10^5$
$R = 12\,\Omega : \bar{i}_L = 0.17$ A slope: $2 \times 10^5$

**3.21** **a)** $i_{L\,max} = 6.5$ A $i_{L\,min} = 3.5$ A $v_{con} = 1.25$ V **b)** $1\text{ V} < v_{con} < 5/3$ V
**c)** $|Z_2(j\,0)/Z_1(j\,0)| = 1 \neq \infty$ **d)** $V_{con} = 2$ V $V_O = 8$ V

**3.22** **a)** $0.16 < D < 0.9$ $0.64\text{ V} < v_{con} < 3.6$ V **b)** $L = 200\,\mu$H
**c)** $V_S = 20$ V : $i_{L\,max} = 1.35$ A $i_{L\,min} = 0.45$ A
$V_S = 22.5$ V : $i_{L\,max} = 1.8$ A $i_{L\,min} = 0$ A
$V_S = 60$ V : $i_{L\,max} = 3.36$ A $i_{L\,min} = 0$ A

**3.23** **a)** i) $8\text{ V} < V_S < 10$ V $1.6\text{ V} < v_{con} < 2$ V
ii) $10\text{ V} < V_S < 20$ V $0.692\text{ V} < v_{con} < 1.6$ V
**b)** i) $V_O = 4$ V $V_{con} = 1.6$ V ii) No change because $|Z_2(j\,0)/Z_1(j\,0)| = \infty$ iii) $V_{con} = 2.29$ V $V_O = 5.71$ V

**3.24** **a)** $i_{L\,max} = 6$ A $i_{L\,min} = 0$ A $\tilde{v}_{O\,max} = 0.6$ V $\tilde{v}_{O\,min} = -0.6$ V
**b)** i) $0.25 < D < 0.5$ ii) $1.25\text{ V} < v_{con} < 2.5$ V iii) $3\text{ A} < \Delta i_L < 6$ A
iv) $0.06\text{ V} < \Delta v_O < 0.12$ V

**3.25** **a)** $i_{L\,max} = 4.6$ A $i_{L\,min} = 3.4$ A $\bar{i}_L = 4$ A $i_{Q\,max} = 4.6$ A $i_{Q\,min} = 3.4$ A $\bar{i}_Q = 1.6$ A $i_{D\,max} = 4.6$ A $i_{D\,min} = 3.4$ A $\bar{i}_D = 2.4$ A $i_{C\,max} = 0.6$ A $i_{C\,min} = -0.6$ A $\bar{i}_C = 0$ $\bar{v}_{con} = 2$ V $\tilde{v}_{O\,max} = 0.03$ V $\tilde{v}_{O\,min} = -0.03$ V $\bar{\tilde{v}}_O = 0$

**3.26** A: energy deficit with a smaller input voltage
B: reduced input voltage and increased duty ratio
C: energy surplus with a larger input voltage
D: larger duty ratio with a reduced input voltage
F: smaller duty ratio in DCM operation

**3.27** **a)** $L = 100\mu$ H **b)** $v_{O\,max} = 12.3$ V $v_{O\,min} = 11.7$ V **c)** $v_{O\,max} = 6.15$V $v_{O\,min} = 5.85$V **e)** $v_{O\,max} = 12.3$V $v_{O\,min} = 11.7$V $v_{con} = 2$V

## CHAPTER 4

4.1 **a)** $i_{L\,max} = 2$ A $i_{L\,min} = 0$ A $\tilde{v}_{O\,max} = 0.02$ V $\tilde{v}_{O\,min} = -0.02$ V
**b)** $i_{L\,max} = 2.5$ A $i_{L\,min} = 1.5$ A $\tilde{v}_{O\,max} = 0.03$ V $\tilde{v}_{O\,min} = -0.02$ V

4.2 **a)** No, the inductor will be saturated.
**b)** Yes, the energy transfer is possible with the condition $V_O > V_S$.

4.3 **a)** $i_{L\,max} = 3.5$ A $i_{L\,min} = 0.5$ A **b)** $i_{L\,max} = 6.875$ A $i_{L\,min} = -0.625$ A

4.4 **a)** For both MOSFET and diode: voltage stress: 77.8 V current stress: 15 A
**b)** For both MOSFET and diode: voltage stress: 200 V current stress: 51.3 A

4.5 **a)** i) DCM operation ii) $I_L = 0.57$ A iii) $\Delta i_L = 1.5$ A iv) $\Delta v_O = 0.075$ V v) $\bar{v}_{con} = 2$ V
**b)** i) CCM operation ii) $I_L = 16.7$ A iii) $\Delta i_L = 2.4$ A iv) $\Delta v_O = 0.895$ V v) $\bar{v}_{con} = 3.2$ V

4.6 **a)** $i_{L\,max} = 5$ A $\bar{\imath}_L = 4$ A $i_{Q\,max} = 5$ A $\bar{\imath}_Q = 2$ A $i_{D\,max} = 5$ A $\bar{\imath}_D = 2$ A $i_{C\,max} = 3$ A $i_{C\,min} = -2$ A $v_{con} = 4$ V
**b)** $\tilde{v}_{O\,max} = 0.06$ V $\tilde{v}_{O\,min} = -0.04$ V $\bar{v}_O = 0$

4.7 **a)** $L = 30\,\mu$H **b)** $i_{L\,max} = 8$ A $i_{L\,min} = 0$ A $i_{D\,max} = 8$ A $i_{D\,min} = 0$ A $\bar{v}_{con} = 2.4$ V $\tilde{v}_{O\,max} = 0.08$ V $\tilde{v}_{O\,min} = -0.08$ V
**c)** $L = 25\,\mu$H $i_{L\,max} = 20$ A $i_{L\,min} = 0$ A $i_{D\,max} = 20$ A $i_{D\,min} = 0$ A $\bar{v}_{con} = 3$ V $\tilde{v}_{O\,max} = 0.3$ V $\tilde{v}_{O\,min} = -0.1$ V

4.8 **a)** $L = 9.6\,\mu$H **b)** $D_{crit} = 0.6$ $D_{max} = 0.8\,(V_S = 6\text{ V})$ $D_{min} = 0\,(V_S = 30\text{ V})$ **c)** CCM $(D = 0.8)$ $i_{L\,max} = 20$ A $i_{L\,min} = 10$ A $\bar{v}_{con} = 4.8$ V

4.9 **a)** $\bar{v}_{con} = 3$ V $V_O = 8$ V **b)** $\bar{v}_{con} = 8/3$ V $V_O = 16/3$ V **c)** $\bar{v}_{con} = 1.0$ V $V_O = 4$ V **d)** $\bar{v}_{con} = 1.5$ V $V_O = 4$ V

4.10 (a) $V_{ref} = 2$ V (b) $V_{ref} = 11$ V (c) $V_{ref} = 20$ V (d) $V_{ref} = 8$ V

4.11 **a)** $\bar{\imath}_{D1} = 1.88$ A $\bar{\imath}_{D2} = 6.67$ A $\bar{\imath}_{D3} = 1.67$ A
**b)** $\bar{\imath}_{Q1} = 0.63$ A $\bar{\imath}_{Q2} = 2.22$ A $\bar{\imath}_{Q3} = 0.56$ A
**c)** $v_{Q1\,max} = 20$ V $v_{Q2\,max} = 26.7$ V $v_{Q3\,max} = 26.7$ V

4.12 (a) $\bar{v}_{con} = 2.5$ V $\bar{v}_O = 6.2$ V (b) $\bar{v}_{con} = 5/3$ V $\bar{v}_O = 10/3$ V
(c) $\bar{v}_{con} = 0.625$ V $\bar{v}_O = 2.5$ V (d) $\bar{v}_{con} = 1.0$ V $\bar{v}_O = 2.5$ V

4.13 **a)** i) $v_{Q\,max} = 12$V $\bar{\imath}_D = 1.5$A $\bar{v}_{con} = 4$V ii) $\tilde{v}_{O\,max} = 0.012$V $\tilde{v}_{O\,min} = -0.012$ V
**b)** i) $v_{Q\,max} = 20$V $\bar{\imath}_D = 3$A $\bar{v}_{con} = 3.2$V ii) $\tilde{v}_{O\,max} = 0.026$V $\tilde{v}_{O\,min} = -0.06$ V

4.14 **a)** $\bar{v}_O = 16$ V $\bar{\imath}_O = 8$ A **b)** $\bar{\imath}_{D1} = 8$ A $\bar{\imath}_L = 13.3$ A $\bar{\imath}_{Q2} = 5.32$ A
**c)** $i_{L\,max} = 13.7$ A $i_{L\,min} = 12.9$ A $\bar{\imath}_L = 13.3$ A $i_{D1\,max} =$

13.7 A $i_{D1\,min} = 0$ A $\bar{\imath}_{D1} = 8$ A $i_{Q2\,max} = 13.7$ A $i_{Q2\,min} =$ 0 A $\bar{\imath}_{Q2} = 5.32$ A

**4.15** **a)** Converter A: $v_{Q\,max} = 50$ V $\bar{\imath}_L = 25$ A Converter B: $v_{Q\,max} =$ 50 V $\bar{\imath}_L = 15$ A **b)** $\eta = 0.95$

**4.16** (a) $V_O = 9$ V $R = 1.875\,\Omega$ $L = 37.5\,\mu$H (b) $V_O = 38.4$ V $R = 12.8\,\Omega$ $L = 60\,\mu$H (c) $V_O = 14.4$ V $R = 4.8\,\Omega$ $L = 60\,\mu$H

**4.17** Converter A: $V_O = 40$ V $I_O = 2$ A $L = 60\,\mu$H $R = 20\,\Omega$
Converter B: $V_O = 16$ V $I_O = 2$ A $L = 60\,\mu$H $R = 8\,\Omega$

**4.18** $\bar{\imath}_{Q1} = 0.83$ A $\bar{\imath}_{D1} = 1.24$ A $\bar{\imath}_{Q2} = 0.4$ A $\bar{\imath}_{D2} = 0.6$ A $\bar{\imath}_{Q3} = 0.67$ A $\bar{\imath}_{D3} = 1$ A

**4.19** $\bar{\imath}_{L1} = 22$ A $\bar{\imath}_{L2} = 13.3$ A $\bar{\imath}_{L3} = 16.7$ A

**4.20** **a)** $\bar{v}_{O1} = 40$ V $\bar{v}_{O2} = 26.7$ V $\bar{v}_{O3} = 16$ V
**b)** $\bar{\imath}_{L1} = 35.6$ A $\bar{\imath}_{L2} = 33.3$ A $\bar{\imath}_{L3} = 20$ A
**c)** $i_{L1\,max} = 36$ A $\bar{\imath}_{L1} = 35.6$ A $i_{L1\,min} = 35.2$ A $i_{L2\,max} = 34$ A $\bar{\imath}_{L2} = 33.3$ A $i_{L2\,min} = 32.6$ A $i_{L3\,max} = 20.4$ A $\bar{\imath}_{L3} = 20$ A $i_{L3\,min} = 19.6$ A
**d)** $\imath_{D2\,peak} = 34$ A $v_{D2\,peak} = 26.7$ V $P_{loss} = 10$ W

**4.21** **a)** $v_{Q1\,max} = 60$ V $v_{Q2\,max} = 150$ V $v_{Q3\,max} = 60$ V
**b)** $P_{D1} = 8.7$ W $P_{D2} = 5$ W $P_{D3} = 0.8$ W

**4.22** **a)** Circuit A: boost mode: $\bar{v}_1 = 3.5$ V Circuit B: buck mode: $\bar{v}_1 = 56$ V Circuit C: buck/boost mode: $\bar{v}_1 = 42$ V
**b)** Circuit A: boost mode: $\bar{v}_1 = 10.5$ V Circuit B: buck mode: $\bar{v}_1 = 18.7$ V Circuit C: buck/boost mode: $\bar{v}_1 = 4.67$ V

**4.24** **a)** $v_O$ continues to grow until the capacitor explodes with an overvoltage condition.
**b)** $i_L$ continues to grow until the inductor saturates

**4.25** **a)** $\bar{v}_{O1} = 3$ V $\bar{v}_{O2} = 6$ V
**b)** $i_{L1\,max} = 4.25$ A $i_{L1\,min} = 3.75$ A $i_{L2\,max} = 4.25$ A $i_{L2\,min} = 3.75$ A
**c)** $i_{C1\,max} = 0.5$ A $i_{C1\,min} = -0.5$ A $i_{C2\,max} = 2.25$ A $i_{C2\,min} = -2$ A

**4.27** **a)** $\bar{v}_{O1} = 48$ V $\bar{v}_{O2} = 24$ V
**b)** $i_{L1\,max} = 8.5$ A $i_{L1\,min} = 3.5$ A $i_{L2\,max} = 8.5$ A $i_{L2\,min} = 3.5$ A
**c)** $i_{C1\,max} = 5$ A $i_{C1\,min} = -5$ A $i_{C2\,max} = 2.5$ A $i_{C2\,min} = -2.5$ A

**4.28** **a)** $V_{DC} = 100$ C $\bar{v}_{con} = 0.8$ V
**d)** $i_{S1\,max} = i_{S2\,max} = 50.3$ A $v_{S1\,max} = v_{S2\,max} = 100$ V

**4.29** **a)** $i_{L\,max} = 13.7$ A $\bar{\imath}_L = 13.3$ A $i_{D\,max} = 13.7$ A $\bar{\imath}_D = 8$ A $i_{Q\,max} =$ 13.7 A $\bar{\imath}_Q = 5.32$ A $v_{con} = 0.8$ V
**b)** $i_{m\,max} = 21.55$ A $\bar{\imath}_m = 21.3$ A $i_{Q\,max} = 21.55$ A $\bar{\imath}_Q = 5.3$ A $i_{D\,max} =$ 10.78 A $\bar{\imath}_D = 8$ A $v_{con} = 0.5$ V

**4.30** i) $D_c$ open: no reset for the magnetizing inductance
ii) $D_2$ open: no freewheeling path for the inductor current
iii) $C$ open: new steady-state with an increased current ripple
iv) $D = 0.15$: new steady-state with a reduced output voltage
v) $D = 0.65$: saturation of the magnetizing inductance

**4.33** **a)** i) (b) ii) (d) iii) (c)
**b)** (a) $0 < D < 48/72$ (b) $0 < D < 8/32$ (c) $0 < D < 1$ (d) $0 < D < 1/2$
**c)** (a) 72 V (b) 32 V (c) 28.8 V (d) 24 V

**4.34** **a)** $L_m = 120\ \mu\text{H}$ $N_T = 10$
**b)** $\bar{i}_N = 24$ A $V_O = 48$ V
**c)** $V_{Z1} = 60$ V $V_{Z2} = V_{Z3} = 10$ V
**d)** (a) $P_{loss} = 0$ (b) $P_{loss} = 24$ W (c) $P_{loss} = 8$ W
**e)** (a) $i_{m\,max} = 3$ A (b) Transformer saturation
(c) Transformer saturation

**4.35** **a)** To obtain a controllable magnetizing inductance
**b)** Flyback converter: use of magnetizing inductance as a functional inductor
Bridge-type converter: automatic magnetizing inductance reset
**c)** (a) $v_{O\,max} = 9.3$ V $v_{Q\,max} = 84$ V (b) $v_{O\,max} = 4.7$ V $v_{Q\,max} = 42$ V
(c) $v_{O\,max} = \infty$ $v_{Q\,max} = \infty$ (d) $v_{O\,max} = 7$ V $v_{Q\,max} = 28$ V
(e) $v_{O\,max} = 7$ V $v_{Q\,max} = 28$ V (f) $v_{O\,max} = 14$ V $v_{Q\,max} = 56$ V

**4.36** **a)** $V_O = 12.5$ V $v_{Q\,max} = 125$ V $\bar{i}_D = 6.25$ A
**b)** $V_O = 10$ V $v_{Q\,max} = 150$ V $\bar{i}_D = 2.5$ A $\bar{i}'_D = 1$ A
**c)** $V_O = 10$ V $v_{Q\,max} = 300$ V $\bar{i}_D = 2.5$ A $\bar{i}'_D = 0.25$ A
**d)** $V_O = 10$ V $v_{Q\,max} = 100$ V $\bar{i}_D = 2.5$ A $\bar{i}'_D = 0.5$ A
**e)** $V_O = 5$ V $v_{Q\,max} = 100$ V $\bar{i}_D = 1.25$ A

# Index